# US AIR FORCE
# SECRET SPACE PROGRAM

SHIFTING EXTRATERRESTRIAL ALLIANCES & SPACE FORCE

# Also By DR. MICHAEL SALLA

**Antarctica's Hidden History:**
**Corporate Foundations of Secret Space Programs**
— Book Three of the Secret Space Programs Series —

**The U.S. Navy's Secret Space Program**
**& Nordic Extraterrestrial Alliance**
— Book Two of the Secret Space Programs Series —

**Insiders Reveal Secret Space Programs**
**& Extraterrestrial Alliances**
— Book One of the Secret Space Programs Series —

**Kennedy's Last Stand:**
Eisenhower, UFOs, MJ-12 & JFK's Assassination

**Galactic Diplomacy**
Getting to Yes with ET

**Exposing U.S. Government Policies**
**on Extraterrestrial Life**

**Exopolitics:**
Political Implications of Extraterrestrial Life

# US AIR FORCE
# SECRET SPACE PROGRAM
SHIFTING EXTRATERRESTRIAL ALLIANCES & SPACE FORCE

Michael E. Salla, Ph.D.

Hawaii, USA

# US AIR FORCE SECRET SPACE PROGRAM
## SHIFTING EXTRATERRESTRIAL ALLIANCES & SPACE FORCE

Exopolitics Consultants
PO Box 478
Holualoa, HI 96725 USA

Printed in the United States of America

Managing Editor: Angelika Whitecliff
Cover Design: Rene McCann

ISBN 978-0-9986038-4-1

Library of Congress Control Number: 2019940219

Author's website: www.exopolitics.org

# TABLE OF CONTENTS

TABLE OF FIGURES..........................................................................IX

PREFACE .......................................................................................XIII

CHAPTER 1 ........................................................................................1

THE WRIGHT BROTHERS, US ARMY AIR CORPS & SECRET
STUDIES OF CRASHED EXTRATERRESTRIAL SPACECRAFT...........1
MASTERING THE SKIES!..............................................................1
ITALIAN AND AMERICAN CRASH-RETRIEVALS ...........................5

CHAPTER 2 ......................................................................................13

THE INTERPLANETARY PHENOMENON UNIT .............................13

CHAPTER 3 ......................................................................................31

PROJECT RAND AND NON-TERRESTRIAL TECHNOLOGIES .........31
PROJECT RAND .......................................................................38
WILLIAM TOMPKINS AND PROJECT RAND'S UNDISCLOSED MISSION........40
STRUGGLE FOR US AIR FORCE SUPREMACY IN SPACE ..................44

CHAPTER 4 ......................................................................................55

THE ROSWELL CRASH AND THE CREATION OF MJ-12 .................55
THE TRUTH & COVER-UP OF THE ROSWELL CRASH.....................57
MAJESTIC DOCUMENTS ON THE ROSWELL CRASH......................65
ROSWELL EVENT LEADS TO MAJIC .............................................71

CHAPTER 5 ......................................................................................81

DIVERGING PATHS: .......................................................................81

RAND'S ROOTS WITHIN THE USAF AND US NAVY'S SECRET
SPACE PROGRAMS ........................................................................81

CHAPTER 6 ......................................................................................91

THE GERMAN CONNECTION TO UFO'S OVER THE UNITED
STATES .........................................................................................91
FLYING SAUCERS FROM ANTARCTICA ........................................91

AIR FORCE INVESTIGATION ENCOUNTERS THE GERMAN CONNECTION.......97

**CHAPTER 7** ..........................................................................109

**HIDDEN LOSSES: THE SHOOT DOWN POLICY AGAINST ALIEN & ANTARCTIC GERMAN SPACECRAFT**...........................109
    UFO INTERFERENCE WITH ROCKET PROGRAM LEADS TO DEADLY
    RETALIATION.................................................................109
    THE ESCALATION OF FLYING SAUCER ENGAGEMENTS...............116

**CHAPTER 8** ..........................................................................123

**DID EXTRATERRESTRIAL OR GERMAN SAUCERS FLY OVER WASHINGTON DC?** .........................................................123
    GEORGE VAN TASSEL'S WARNING ABOUT UFO OVERFLIGHTS ......127
    ANTARCTIC GERMAN COLONY AND THE WASHINGTON DC FLYOVER.......134

**CHAPTER 9** ..........................................................................143

**PRESIDENT EISENHOWER MEETS WITH THE NORDICS: NUCLEAR NEGOTIATIONS GO AWRY** ...........................143
    ATOMS FOR PEACE – EISENHOWER'S OVERTURE TO THE GALACTIC
    FEDERATION ...............................................................143
    EISENHOWER'S SECRET MEETING WITH THE NORDICS..............147
    WHISTLEBLOWERS AND THE UNTOLD STORY........................151

**CHAPTER 10** ........................................................................167

**EXTRATERRESTRIAL & GERMAN DELEGATIONS AT HOLLOMAN AFB** ..................................................................167
    EYEWITNESS ACCOUNTS AT HOLLOMAN .............................169
    AGREEMENT WITH GRAY EXTRATERRESTRIALS .......................175
    AGREEMENT REACHED WITH THE ANTARCTIC GERMANS.............181

**CHAPTER 11** ........................................................................187

**USAF BEGINS TECHNOLOGY ASSISTANCE PROGRAMS WITH GRAY EXTRATERRESTRIALS** ...........................187
    SHORT GRAYS AND USAF TECHNOLOGY ..............................195
    CHARLES HALL ON EXTRATERRESTRIAL ASSISTANCE FOR USAF ATOMIC
    POWERED SPACECRAFT .................................................199

**CHAPTER 12** .................................................................................207

**NASA CREATED AS A FRONT**.....................................................207
The Antarctic German Space Program and NASA...........................209
CIA Funds Antarctic German Space Program Through Apollo
Program........................................................................................219

**CHAPTER 13** .................................................................................225

**MANNED ORBITING LABORATORY AND SECRET SPACE
PROGRAMS**...................................................................................225
Polar Orbit Missions Designed to Spy on Antarctic Germans, not
Russians ........................................................................................229
MOL, NASA, and the Secret Astronaut Corps .............................231
Advanced Phase of MOL Links Multiple Modules Together........233
How the Public was Misled: Cancellation of MOL ......................236
Von Braun and the Real Purpose Behind the USAF/NRO/NASA
Space Station..................................................................................241

**CHAPTER 14** .................................................................................245

**USAF ANTIGRAVITY ALIEN REPRODUCTION VEHICLES** ...........245
Reverse Engineering a Better Delivery System .............................249
Three Kinds of Alien Reproduction Vehicles ...............................250
Sources Corroborating Sorensen's 1988 ARV Sighting.................255
Comparison of ARVs with Recovered Nazi Flying Saucers...........257
USAF Flying Saucers and Superluminal Travel .............................263

**CHAPTER 15** .................................................................................267

**USAF FLYING TRIANGLES AND THE TR-3B**...............................267

**CHAPTER 16** .................................................................................277

**STARGATE SG-1 AND USAF SANCTIONED SOFT-DISCLOSURE OF
TRAVERSABLE WORMHOLES** .......................................................277
Carl Sagan and Scientific Validation for Traversable Wormholes
via Singularity Points.....................................................................288
Dr. Eric Davis and Traversable Wormholes via Exotic Matter.291

**CHAPTER 17** .................................................................................299

**THE SHOCK OF THE NAVY'S PARALLEL SECRET SPACE PROGRAM** .......................................................................................... **299**

SUPPORT FOR TOM DELONGE'S "SEKRET MACHINES INITIATIVE" PULLED .................................................................................. 306

USAF SSP TAKES STEPS TOWARD FULL DISCLOSURE ............................ 310

**CHAPTER 18** ................................................................ **313**

**THE HISTORY OF HUMAN-LOOKING EXTRATERRESTRIALS ON EARTH** ................................................................. **313**

MAJESTIC DOCUMENT REVEALS US DIPLOMATIC RELATIONS WITH HUMAN-LOOKING EXTRATERRESTRIALS ................................................................. 314

AUTHENTICATION OF THE DIA DOCUMENT ...................................... 326

BEHIND THE LEAK ..................................................................... 330

**CHAPTER 19** ................................................................ **335**

**USAF SSP PHOTO DISCLOSURE OF MACDILL'S ANTIGRAVITY VEHICLES** ................................................................. **335**

JP'S PHOTOS OF CRAFT NEAR MACDILL AIR FORCE BASE ...................... 337

FLYING TRIANGLE SIGHTINGS CONTINUE AS HURRICANE IRMA APPROACHES .......................................................................... 342

DID MASER SATELLITES STEER HURRICANE IRMA IN WEATHER WAR AGAINST U.S.? ......................................................................... 348

Q & A SESSION WITH COREY GOODE ON HURRICANE IRMA, MASER WEAPONS & FLYING TRIANGLES ................................................. 352

CYLINDRICAL UFO NEAR MACDILL AFB AFFIRMS USAF-NORDIC ALLIANCE .......................................................................... 358

COVERT DISCLOSURE OF ANTIGRAVITY CRAFT NEAR MACDILL AFB ....... 361

CONTACT WITH A NORDIC EXTRATERRESTRIAL WORKING WITH USAF SSP .......................................................................... 368

CONCLUSION .......................................................................... 374

**CHAPTER 20** ................................................................ **377**

**USAF SSP SHOOTS DOWN DEEP STATE "FALSE FLAG" MISSILE ATTACK ON HAWAII** ................................................................. **377**

STATE OF HAWAII ISSUES BALLISTIC MISSILE ALERT ON JANUARY 13, 2018 .......................................................................... 377

ALTERNATIVE MEDIA REPORTS & WITNESSES POINT TO GENUINE MISSILE

ATTACK .................................................................................... 385
"Q" EXPOSES FALSE FLAG ATTACK MEANT TO START WWIII .................. 389
"Q" PULLS THE CURTAIN ON THE PUPPET MASTERS CONTROLLING THE
FALSE FLAG ATTACKS ............................................................... 393
USAF SPACE-BASED WEAPONS PLATFORM ...................................... 396
DID THE USAF RECEIVE HELP FROM NORDIC EXTRATERRESTRIALS IN
SHOOTING DOWN THE MISSILE? ................................................. 399
CONCLUSION ........................................................................... 401

**CHAPTER 21** ........................................................................ **403**

**SPACE FORCE OPENS THE DOOR FOR USAF SSP DISCLOSURE. 403**
PRESIDENT TRUMP PROPOSES SPACE FORCE AS NEW BRANCH OF US
MILITARY ................................................................................ 403
DEPARTMENT OF DEFENSE REPORT FOR THE CREATION OF A SPACE FORCE
.............................................................................................. 406
SPACE FORCE, Q ANON, AND SECRET SPACE PROGRAM DISCLOSURE ....... 412
DISCLOSURE OF USAF SECRET SPACE PROGRAM AND NORDIC ALLIANCE
.............................................................................................. 414
OFFICIAL DISCLOSURE THROUGH SPACE FORCE ............................. 417

**ACKNOWLEDGMENTS** .......................................................... **423**
**ABOUT THE AUTHOR** ........................................................... **425**
**ENDNOTES** ......................................................................... **427**
**INDEX** ................................................................................ **467**

# TABLE OF FIGURES

FIGURE 1. SOLDIERS WATCHING TEST OF MILITARY FLYER IN 1909 ............................... 3
FIGURE 2. THE AERODYNE VEHICLE ................................................................................ 9
FIGURE 3. "THE GREAT LOS ANGELES AIR RAID" .......................................................... 14
FIGURE 4. FEB 26, 1942 MEMORANDUM FROM MARSHALL TO ROOSEVELT ................. 18
FIGURE 5. RE-TYPED ROOSEVELT MEMORANDUM TO MARSHALL ................................. 22
FIGURE 6. ALLEGED MEMO FROM GENERAL MARSHAL TO PRESIDENT ROOSEVELT ...... 25
FIGURE 7. ORDER ISSUED TO RICO BOTTA TO TRAVEL TO WRIGHT FIELD ON THE DAY OF THE LA
    AIR RAID ........................................................................................................ 28
FIGURE 8. "DOUBLE TOP SECRET" DOCUMENT SIGNED BY PRESIDENT ROOSEVELT ................. 34
FIGURE 9. LEAKED "WHITE HOT REPORT" DATED SEPTEMBER 19, 1947. SOURCE: THE
    MAJESTIC DOCUMENTS ................................................................................ 43
FIGURE 10. DOCUMENT REVEALING SCHUMANN WAS PART OF OPERATION PAPERCLIP .......... 49
FIGURE 11. RAND PRELIMINARY DESIGN STUDY ......................................................... 52
FIGURE 12. MAJOR JESSE MARCEL CROUCHES OVER WEATHER BALLOON REMAINS ....... 61
FIGURE 13. EISENHOWER BRIEFING DOCUMENT, P. 2. THE MAJESTIC DOCUMENTS ............. 76
FIGURE 14. INITIAL MEMBERS OF MAJESTIC 12 GROUP .............................................. 78
FIGURE 15. CUTLER-TWINING MEMO. SOURCE: THE MAJESTIC DOCUMENTS ............. 80
FIGURE 16. DR. KLEMPERER LETTER ON" UNCONVENTIONAL PROPULSION" ................. 85
FIGURE 17. NAVAL HISTORICAL CENTER DRAWING ...................................................... 86
FIGURE 18. LEE VAN ATTA ARTICLE ON OPERATION HIGHJUMP ................................. 93
FIGURE 19. COMPARISON: UFO DRAWING BY KENNETH /PHOTO OF HORTON FLYING WING
    ..................................................................................................................... 96
FIGURE 20. CIA DOCUMENT SHOWING ROLE OF WRIGHT PATTERSON AFB AS ONLY USAF
    FACILITY INVESTIGATING FLYING SAUCER PHENOMENON ............................. 100
FIGURE 21. CIA DOCUMENT ON USSR CONNECTION TO FLYING SAUCERS ............................. 102
FIGURE 22. CIA DOCUMENT ON SAUCER FORMATIONS NEAR ANTARCTICA ............................ 105
FIGURE 23. LUBBOCK LIGHTS - POSSIBLE PHOTOGRAPH OF GERMAN FLYING SAUCER
    FORMATION FROM ANTARCTICA ................................................................... 106
FIGURE 24. NEWS REPORT OF V-2 ROCKET AFFECTED BY UFO ............................... 111
FIGURE 25. IN THE 72 HOUR PERIOD FOLLOWING THE V-2 JUAREZ INCIDENT, 29
    AIRPLANES CRASHED AROUND THE WORLD KILLING 198 PEOPLE. GRAPHIC ©
    1994 BY J. ANDREW KISSNER. ...................................................................... 115
FIGURE 26. SUMMER OF 1952 NEWSPAPER HEADLINES ........................................... 124

FIGURE 27. PHOTO OF 1952 WASHINGTON DC FLYOVER ........................................ 140

FIGURE 28. EISENHOWER'S COVER STORY .................................................................. 149

FIGURE 29. 1982 BRITISH NEWSPAPER STORY ON EISENHOWER–EXTRATERRESTRIAL MEETING. ................................................................................................................ 156

FIGURE 30. BILL KIRKLIN'S SERVICE RECORD SHOWING HE WAS STATIONED AT HOLLOMAN AFB FROM MARCH 1, 1954 TO AUGUST 5, 1955 ........................... 169

FIGURE 31. COL PHILIP CORSO'S SERVICE RECORD SHOWS HE SERVED ON THE OPERATIONS COORDINATING BOARD .................................................................................... 180

FIGURE 32. "SPECIAL OPERATIONS MANUAL" ....................................................... 190

FIGURE 33. ALLEGED PHOTO OF GRAY ALIEN. CREDIT: ROBERT DEAN .................. 194

FIGURE 34. ARTISTS IMPRESSION OF THE "TALL WHITES". CREDIT: TERESA BARBATELLI .... 200

FIGURE 35. HEINRICH HIMMLER, VISITS PEENEMÜNDE IN APRIL 1943. WERNHER VON BRAUN IS STANDING BEHIND HIMMLER AND WEARING A BLACK NAZI SS UNIFORM. ................. 214

FIGURE 36. DOCUMENTARY EVIDENCE FROM W. M. TOMPKINS ......................... 218

FIGURE 37. PROPOSED DESIGN OF "MANNED ORBITING LABORATORY" ....................... 228

FIGURE 38. VON BRAUN SPACE STATION ................................................................. 235

FIGURE 39. GENERAL ELECTRIC PROPOSAL FOR SPACEBORNE COMMAND POSTS ....... 239

FIGURE 40. GENERAL ELECTRIC'S ILLUSTRATION OF A THREE SECTION STATION BUILT WITH MOL MODULES THAT COULD ACCOMMODATE 40 ASTRONAUTS ................. 240

FIGURE 41. SPACE STATION COMPRISED OF NINE CYLINDRICAL SECTIONS. CREDIT: SPHERE BEING ALLIANCE ................................................................................ 244

FIGURE 42. DECLASSIFIED NRO DOCUMENT SHOWING GENERAL ELECTRIC PROPOSAL FOR ASSEMBLING A "COMMAND POST" SPACE STATION FROM MOL MODULES. 246

FIGURE 43 DECLASSIFIED NRO DOCUMENT SHOWING GENERAL ELECTRIC PROPOSAL FOR ASSEMBLING SPACE STATION OVER 1970'S ................................................. 248

FIGURE 44. CUTAWAY OF ARV SHOWING 48 LARGE CAPACITOR STACKS OF EIGHT PLATES EACH AND OXYGEN CYCLINDERS. COPYRIGHT (MARCH, 1989): MARK MCCANDLISH. ALL RIGHTS RESERVED. ......................................................................................... 252

FIGURE 45. PROVO, UTAH, UFO – JULY 1966 .......................................................... 257

FIGURE 46. PRODUCTION STATISTICS FOR NAZI SS FLYING SAUCERS ....................... 260

FIGURE 47. SLIDE BEN RICH SHOWED WHEN DISCUSSING TAKING ET HOME ............. 264

FIGURE 48. DOCUMENT CONFIRMING EDGAR FOUCHE WORKED AT NELLIS AFB ADJACENT TO AREA 51 ................................................................................... 270

FIGURE 49. ILLUSTRATION OF TR-3B. ADAPTED EDGAR FOUCHE ORIGINAL VERSION ................................................................................................................ 272

FIGURE 50. CREDITS SHOWING USAF & US SPACE COMMAND COOPERATION WITH STARGATE SG-1 PRODUCTION ..................................................................... 279

FIGURE 51. INTRA-UNIVERSE WORMHOLE AS A HYPERSPACE SHORTCUT THROUGH CONVENTIONAL SPACE ................................................................................. 292

FIGURE 52. ARTISTIC ILLUSTRATION OF A STARGATE IN NYC DISPLAYING A DESTINATION WORLD .......................................................................................................297

FIGURE 53. COURTESY: COSMIC DISCLOSURE/GAIA TV ...............................302

FIGURE 54. COVER PAGE OF LEAKED "ASSESSMENT OF THE SITUATION/STATEMENT OF POSITION ON UNIDENTIFIED FLYING OBJECTS" DOCUMENT...................315

FIGURE 55. PAGE FROM DIA DOCUMENT IDENTIFIES MJ-1 AS OFFICER IN CHARGE...............317

FIGURE 56. FIRST 3 PHOTOS IN SERIES SHOW START OF PULSING (ZOOM INSERTS ADDED). TAKEN BY JP ON AUG 31, 2017 ......................................................340

FIGURE 57. LAST 3 PHOTOS IN SERIES SHOW CIRCULAR PULSE EMANATING FROM CRAFT (ZOOM INSERTS ADDED). TAKEN BY JP ON AUG 31, 2017 ...........................340

FIGURE 58. PHOTOS OF FLYING TRIANGLE (ZOOM INSERTS ADDED) TAKEN BY JP ON SEPT 4, 2017 ......................................................................................................341

FIGURE 59. PHOTO OF TRIANGULAR CRAFT (ZOOM INSERT ADDED) TAKEN BY JP ON SEPT 5, 2017 ......................................................................................................343

FIGURE 60. THREE PHOTOS OF TRIANGULAR CRAFT (ZOOM INSERTS ADDED) TAKEN BY JP ON SEPT 6, 2017 ......................................................................................................344

FIGURE 61. THREE OF THE PHOTOS BY JP OF FLYING TRIANGLE AND HELICOPTER (ZOOM INSERTS ADDED). TAKEN ON SEPT 8, 2017 ...........................................345

FIGURE 62. EVACUATION ORDER FOR MACDILL AFB ISSUED SEPTEMBER 8, 2017 ..............347

FIGURE 63. THREE PHOTOS OF CYLINDER-SHAPED CRAFT (ZOOM INSERTS ADDED) TAKEN BY JP ON SEPT 14, 2017 ................................................................................359

FIGURE 64. PHOTO OF RECTANGULAR-SHAPED CRAFT (ZOOM INSERT ADDED) TAKEN BY JP ON OCT 19, 2017......................................................................................363

FIGURE 65. PHOTO OF FLYING RECTANGLE AND FLYING TRIANGLE (ZOOM INSERT ADDED) TAKEN BY JP ON OCT 19, 2017 ......................................................................364

FIGURE 66. JP'S SKETCH OF CORRIDOR INSIDE THE RECTANGULAR-SHAPED PLATFORM UFO .......................................................................................................366

FIGURE 67. PATCH USED BY AIR FORCE SPECIAL OPERATIONS....................367

FIGURE 68. PHOTO COURTESY OF "JP" (ZOOM INSERT ADDED)......................375

FIGURE 69. EMERGENCY ALERT MESSAGES ISSUED BY THE CIVIL AUTHORITY TO HAWAII RESIDENTS .................................................................................................378

FIGURE 70. COMPARISON OF AEGIS AND OTHER DEFENSE SYSTEMS. SOURCE: MISSILE DEFENSE AGENCY .......................................................................................................397

FIGURE 71. SCREENSHOT OF UFO'S SEEN DURING LIVE CNN BROADCAST FROM HAWAII. COURTESY OF SOLARIS MODALIS ......................................................................400

# PREFACE

Perhaps some of us have to go through dark and devious ways before we can find the river of peace or the highroad to the soul's destination.

— Joseph Campbell

The United States Air Force (USAF) is experiencing a profound shift with its major space assets soon being moved into a new military branch called "Space Force", which is designed to aggregate the assets from all the military services under one authority for activities in space. The gradual emergence of Space Force offers an opportunity for the Air Force to publicly reveal the most well-guarded secrets in its arsenal of exotic spacecraft, unconventional weapons and futuristic technologies, assembled over the more than seventy years since its official emergence as a military service back on September 18, 1947. What is driving this profound change is the realization by Air Force leaders that they had been deceived by their major allies in the research and development of advanced aerospace technologies. Countries such as the United Kingdom, Canada or Australia are not the culprits, but rather the corporations and "Deep State" affiliated groups behind the development and building of advanced space technologies.

In addition, extraterrestrial allies participated in the deception over space technologies that were built and deployed. These groups encouraged USAF leaders to believe that their arsenal of space weapons and spacecraft under secret development would make the Air Force a space power with which to be reckoned. The discovery of this deception made USAF leaders realize that, rather than being the "tip of the spear" of the military's projection into deep space, they had essentially become something analogous to a space-based "coast guard" – lagging far behind their primary competitors and other military forces with deep space capabilities.

In this book, you will learn about the untold trials and triumphs, power-hungry maneuvers and public frauds comprising the history of the Air Force's secret space program (SSP), which ultimately lead to its redemptive future and re-aligned national mission. Spanning from its humble roots in the US Army's Signal Corps; to its unexpected and paradigm shifting recovery of extraterrestrial craft in the 1940's; to secret agreements made with a breakaway German colony in Antarctica (the Fourth Reich) and different extraterrestrial groups in the 1950's; to its complicity in the German infiltration of NASA and the Military-Industrial Complex in the 1960's; its development of stealth space stations from the 1970's; deployment of squadrons of saucer, triangular and rectangular-shaped craft beginning from the 1980's to the 1990's; and finally, its great awakening made possible by its 2016 discovery of the grand deception that had been perpetrated upon it.

The story that emerges is one of the development of a secret space program that was co-opted into serving the interests of the Deep State and their shadowy Fourth Reich allies, rather than becoming a genuine space power serving US national interests. After discovering they had been falsely and disgracefully betrayed, Air Force leaders woke up and courageously made the momentous decision to shift alliances. In the process, they began taking globally significant steps to realign the Air Force's covert space program with a more ethical group of extraterrestrials who had previously helped its sister military service, the US Navy, to become a genuine power in deep space. This stalwart change led to the Air Force taking bold steps to reveal the existence of its arsenal of spacecraft to the US public, and to taking action to prevent a false flag operation by the Deep State/CIA designed to start World War III.

In the three previous books of this Secret Space Program

series, readers have been given detailed information about the evolution of space programs belonging to the US Navy (Solar Warden), an Antarctica-based German colony (Dark Fleet), and a transnational corporate entity (Interplanetary Corporate Conglomerate). This was made possible by the groundbreaking testimonies of William Tompkins, Corey Goode, and other insiders who revealed their participation in, or direct knowledge of, "20 and back" programs in which highly advanced technologies were used on personnel serving in the SSPs for 20-year periods. At the end of their covert space service, these personnel were allegedly "mind-wiped" and "age-regressed" before being returned to the time when they were initially recruited, most often just after they had started their regular military service. The role of the US Air Force was only briefly mentioned in previous volumes since the practice of a "20 and back" was not used on personnel in its classified space operations, which were limited to near-Earth operations rather than deep space. In the book before you, the full historical details are laid out to show exactly how and why the Air Force developed a very different space program to that of the Navy, and why the Air Force made the decision early on to throw in its lot with the German Antarctic colony and extraterrestrial groups that were allied with it. Dramatic events led the Air Force to break free of this cunning yoke that usurped its authority and clouded its visionary goals for decades, with the colossal move of strategic separation finally taking place from 2016 to 2018.

It's no accident that in the midst of a profound strategic shift in extraterrestrial alliances by USAF leaders, the idea of a Space Force has emerged and is being aggressively pushed by President Donald Trump and his administration. Aggregating US space assets into one unified military department will make it easier for the USAF and the Trump administration to combat the Deep State and its Fourth Reich allies who were caught off guard by the Air Force's strategic realignment. Handing control over space assets in Earth-orbit to Space Force should not present a major problem for most of the military services. The big question, however, is when will the

Navy hand over control of its "deep space" assets, the battle groups belonging to Solar Warden, as Space Force is brought to life through executive orders and Congressional legislation. On February 19, 2019, Space Policy Directive-4 was issued by President Trump stating that Space Force would initially be located under the Department of the Air Force, and he privately affirmed that a Department of the Space Force would eventually be created. This indicates that Solar Warden, unknown to Air Force leaders until 2016, will not be integrated into Space Force while it is under the administrative control of the Department of the Air Force. Only after the creation of a fully independent Department of the Space Force will the Navy hand over control of its space battle groups that have been secretly deployed since the early 1980's.

Space Force will pave the way for the Air Force, and eventually the Navy, to enter into a more transparent and accountable era where their true space capabilities are gradually revealed to the general public, and funding provided by the US Congress is done in a manner consistent with the US Constitution. This will do away with the corrupting influence of black budget funds provided by the Deep State/CIA through illicit activities that deeply compromise the ethos and values of all branches of the US military which have had to rely on these funds to secretly build their respective space programs. It can be anticipated that as Space Force emerges as a separate military branch first under the Department of the Air Force, and eventually becoming independent under the Department of the Space Force, advanced spacecraft and aerospace technologies already deployed in Earth-orbit and deep space will gradually be revealed.

The general public will be exposed for the first time to highly-advanced technologies that have been exclusively developed for secret space operations by select military and corporate entities dating back to the 1940's. This will transform the civilian aviation and space industries; provide free energy sources that liberate humanity from its dependence on fossil fuels; and revolutionize the medical industry with the release of holographic and electromagnetic healing technologies covertly used for decades within the secret space programs. The release of such technologies will be vital in helping humanity satisfactorily deal with predicted catastrophic Earth changes arising from a number of possible events including: a "solar flash" (aka "micronova"), magnetic and geophysical pole shifts, and the infusion of cosmic rays from the galactic region into which our solar system is currently entering. Space Force will provide an opportunity for US authorities to disclose the truth about visiting extraterrestrial life and kickstart diplomatic relations that will be urgently needed to help humanity deal with the many challenges that lie ahead, leading us to become a worthy member of a richly diverse galactic community.

Michael E. Salla, Ph.D.
March 9, 2019

# CHAPTER 1

## The Wright Brothers, US Army Air Corps & Secret Studies of Crashed Extraterrestrial Spacecraft

> Man must rise above the Earth – to the top of the atmosphere and beyond – for only thus will he fully understand the world in which he lives.
>
> – Socrates

### Mastering the Skies!

On December 17, 1903, at Kitty Hawk, North Carolina, Wilbur and Orville Wright started the aviation revolution with the first successful flight of a controlled power driven heavier-than-air "aeroplane". Their "Wright Flyer", piloted by Wilbur, flew for a brief 12 seconds over a distance of 120 feet, yet it was a momentous flight, nonetheless. Longer flights were performed over the next two years while reporters and the scientific community largely ignored their pioneering efforts of having conducted the world's first 'heavier-than-air' flights. Famously, *Scientific American* wrote two skeptical stories in 1904 and 1905 about the *alleged* accomplishment. It was only on December 15, 1906, that *Scientific American* finally confirmed that the impossible had happened: "In all the history of invention, there is probably no parallel to the unostentatious manner in which the

Wright brothers of Dayton, Ohio ushered into the world their epoch-making invention of the first successful aeroplane flying-machine."[1]

At a time when the horse and buggy was still the dominant form of transportation for most Americans, and Ford Model T automobiles would only start rolling off the production line in 1908, the development of the first airplane was remarkable. Like the 1969 Apollo moon landing, the Wright brothers' accomplishment at Kitty Hawk would become a defining moment in shaping human consciousness. Like Icarus, precocious young men soon began dreaming of the day they could fly into the heavens – modern style – in the newly invented aeroplane (Greek for "air wanderer"). Included among those envisaging new opportunities made possible by the innovative invention were officers of the US Army who, after *Scientific American's* startling confirmation, came to watch what the Wright brothers were doing.

On June 28, 1909, the Wright brothers demonstrated their "Military Flyer" for the Aeronautical Division of the US Army Signal Corps. Immediately recognizing the military potential of the airplane, the Army offered a contract for a two-seater model that could fly 40 mph (64 km/hour) for a distance of 125 miles (201 km). After passing rigorous testing, the Signal Corps accepted the Military Flyer Model A on August 2, 1909, and paid the brothers $30,000 ($830,789.01 USD in 2018).[2]

The brothers next established their Wright Company on November 22, 1909, with its first factory built in their home town of Dayton, Ohio. Soon after, the "Wright Flying School" was launched in 1910 at Huffman Prairie Flying Field, and it would train over a hundred pilots. Up to its merger with the Glenn L. Martin Company in 1916, the Wright Company built approximately 120 airplanes using patents for different models designed for the nascent aviation industry, which included the militaries of other nations that by 1914 had become belligerents in the First World War.

Among the leading military aviators trained by the brothers was Henry Harley "Hap" Arnold, who graduated as a pilot in 1911, and subsequently became one of their test pilots. Decades later, Arnold would rise to the apex of the US Army Air Force and become the only five-star general recognized by two military services – the US Army and the US Air Force (newly created in September 1947). After World War II, Arnold became a key figure in launching a new military aviation revolution – aerospace operations through his sponsorship of Project RAND (an acronym for "research and development").

Figure 1. Soldiers watching test of Military Flyer in 1909

The U.S. formally entered hostilities of WWI on April 6, 1917, and three major Army Signals Corps installations were quickly built in the vicinity of Dayton, Ohio. Since Wilbur Wright passed away in 1912, Orville Wright was widely considered to be the undisputed leader in aviation science. The first installation was Wilbur Wright Field, which was established on May 22, 1917, approximately five miles (8 kilometers) northeast of Dayton. Wilbur Wright Field incorporated the Huffman Prairie Flying Field used by the Wright brothers and assimilated their "Wright Flying School" with the main purpose of training military pilots for the coming war effort. Next, McCook Field was established on December 4, 1917, for testing and experimentation purposes. Finally, Fairfield Aviation General Supply Depot was formed in January 1918, and provided logistical support to Wilbur Wright Field, McCook Field and other Army Signals Corps installations in the Mid-West.

During the interwar years, Dayton, Ohio, continued as the hub for military aviation research and development – largely due to Dayton civic leaders donating 5,250 acres of land to the US Army in 1924 to continue advanced aviation research in the Dayton area.[3] The Army subsequently established "Wright Field" in 1927, dedicated to both Wilbur and Orville, and incorporating both the Wilbur Wright Field and McCook Field installations. Significantly, advanced research could fully integrate with testing and procurement efforts once the newly created US Army Air Corps established in 1926 set up its Materiel Division at Wright Field where leading aviation scientists could then work side by side to solve complex aeronautical problems.

> Here, scientists continued to improve aircraft. The scientists also realized that they had to work together to create the best possible plane. Before creation of the Army Air Corps, researchers studied individual parts of the plane, like engines, armor, wings, and propellers. Now scientists continued to specialize in individual components of planes, but they realized that changes in a plane's armor would affect all other parts of the plane. The same held true for changes in engines, wings, and the other parts of a plane. Under the Army Air Corps, the researchers now kept a dialog open between the various experts.[4]

On July 6, 1931, a portion of Wright Field was separated and renamed Patterson Field in honor of Lieutenant Frank Stuart Patterson who had been killed as a test pilot at "Wilbur Wright

Field" in June 1918; and Patterson's family, who had been instrumental in later persuading Dayton civic leaders to donate land for the establishment of Wright Field in 1924.

Up until their eventual merger in September 1947 to create Wright-Patterson Air Force Base, Wright Field, and Patterson Field conducted complementary aviation activities. Here is how Thomas Carey and Don Schmidt, authors of *Inside the Real Area 51*, explained the complementary nature of Wright Field and Patterson Field in the lead up to and during World War II:

> Even though Wright and Patterson were two separate installations, their projects often augmented and strategically complemented one another, each providing areas of technical support or facilities and technicians the other did not have.... Whenever a more advanced foreign design was recovered for analysis and study, Patterson Field was the most likely destination....
>
> Is it then any wonder that during and after WWII both facilities saw a dramatic expansion? ... The Materiel Division at Wright handled the procurement of aircraft and their parts on production lines around the country, which resulted in increased testing and development, whereas the Air Service Commission at Patterson maintained the hardware's logistical assimilation into the war.[5]

## Italian and American Crash-Retrievals

It was during the war years that rumors initially began circulating of flying saucer sightings and crashes. Of particular concern to the US Army Corps were reports that fascist nations were secretly studying and developing flying saucer technologies. In 1933, Italy was a major aviation power and became the first nation to begin secretly studying flying saucer technologies. A flying saucer had crashed near Modena, Italy and Fascist dictator Benito Mussolini had ordered the creation of a top secret study group led by inventor and electrical engineer, Guglielmo Marconi, to investigate the phenomenon. In addition to the crash, there were numerous flying saucer sightings, and the reports on them went directly to Mussolini and his subordinates. It's worth noting that prior to the entente signed between Fascist Italy and Nazi Germany in 1939, both Britain and the U.S. tried to recruit Italy as an ally against Hitler. Italy had fought with the Allies in World War I and had major border disputes with Austria that made an alliance with Hitler difficult.

Britain's Winston Churchill (then a Conservative member in the House of Commons), in particular, had kept a steady correspondence with Mussolini where numerous topics such as aviation were discussed, and where Churchill expressed his admiration of Mussolini.[6] It's very possible that either Mussolini told Churchill about Italy's remarkable discovery, or Marconi, who became increasingly disgruntled over Mussolini's growing militarism, disclosed to Churchill or British Intelligence what Mussolini was secretly studying. Marconi's ties to British intelligence date back to 1897 when he had established the Wireless Telegraph & Signal Company (aka Marconi Company) which launched the age of wireless communication. While Italy holds the distinction of being the place where the first documented crash of a flying saucer had occurred, thereby sparking a secret official study group, it was not the only country where such crashes would arise and likewise stimulate official study groups.

Multiple sources report the first flying saucer crash in the United States occurred in April 1941 in Cape Girardeau, Missouri. Charlotte Mann, the granddaughter of Reverend William Huffman, a Baptist minister, says that he told his family about his late night call taking him to a secluded area to provide last rites for three dead or dying extraterrestrial beings. He was later sworn to secrecy by representatives of the US Government and never spoke about the incident again. Mann has recounted details from her grandmother who exposed in a deathbed confession what her husband, Reverend Huffman, said he had witnessed that remarkable night:

> She said that grandfather was out in the spring of 1941 in the evening around 9:00-9:30, that someone had been called out to a plane crash outside of town and would he be willing to go to minister to people there – which he did. Upon arrival, it was a very different situation. It was not a conventional aircraft, as we know it. He described it as a saucer that was metallic in color, no seams, did not look like anything he had seen. It had been broken open in one portion, and so he could walk up and see that.
>
> In looking in, he saw a small metal chair, gauges and dials and things he had never seen. However, what impressed him most was around the inside there were inscriptions and writings, which he said he did not recognize, but were similar to Egyptian hieroglyphics. There were three entities, or non-human people, lying on the ground. Two were just outside the saucer, and a third one was further out. His understanding was that perhaps that third one was not dead on impact … and so grandfather did pray over them, giving them last rites. [7]

The Cape Girardeau case was extensively examined by Ryan Wood in his 2005 book, *Majic Eyes: Earth's Encounters with Extraterrestrial Technology*. He found other witnesses and multiple documents supporting key elements of Reverend Huffman's story, which first appeared in Leonard Stringfield's 1991 report, *UFO Crash Retrievals: The Inner Sanctum*.[8] Wood wrote:

> Other living, supporting witnesses included Charlette Mann's sister, who confirmed her story in

a notarized sworn affidavit; and the brother of the Cape Girardeau County sheriff in 1941, Clarence R. Schade.… Seven leaked documents from three sources since 1994 provide both direct and indirect support for crash retrieval events in the possible timeframe recounted by Reverend Huffman.[9]

Importantly, the witnesses and documents point to the US Army being in charge of the retrieval of the crashed craft and its transportation to Wright Field for scientific evaluation. At Wright Field, Army Air Corps scientists and engineers would closely study the retrieved craft and attempted to reverse engineer it.

Evidence that the Cape Girardeau craft was studied by Wright Field scientists is presented in one of the seven leaked documents mentioned by Wood – a two-page memo that is part of *The Majestic Documents* collection being investigated by Ryan Wood and his father Dr. Robert Wood. The memo describes an "S" aircraft created by the reverse engineering effort:

> The 'S' aircraft was designed from an aerodyne recovered in 1941 that crashed in … Missouri and one captured in 1942 in Louisiana. Re-construction commenced in 1945 with the assistance of German scientists at Wright Field".[10]

The memo's author, Thomas Cantwheel (a pseudonym), claimed that he joined the US Army's Counter Intelligence Corps around 1940, and later became a commissioned officer in 1942 after World War II began. Cantwheel says that until 1958 he participated in multiple crash retrieval operations with an elite unit (later to be called the Interplanetary Phenomenon Unit) set up by the Army. In the 1960's, he was recruited into the CIA to continue crash retrieval operations around the world. Up to 1980, he continued to serve in a variety of US Army Intelligence positions prior to his retirement from active service as a Lieutenant Colonel. Figure 2 is Cantwheel's illustration of the aerodyne vehicle retrieved and examined at

Wright Field in 1947.

RECONSTRUCTED DRAWING OF A UNIDENTIFIED AERODYNE FOUND ON
JULY 5 1947, SOUTH OF SOCORRO, NEW MEXICO, BY MILITARY AND
CIVILIAN MEMBERS OF THE ARMED FORCES SPECIAL WEAPONS PROJECT.

APPROXIMATELY 100 FEET IN LENGTH
WIDTH OF CHIME IN AFT PORTION WAS APPROXIMATELY 60 FEET
CHIME RAN THE LENGTH OF AERODYNE
SKIN WAS SMOOTH. NO RIVETS, BOLTS, SEAMS, OR WELDS. LIKE PLASTIC.
NO CANOPY, WINDOWS. NO DOORS, INTAKES, EXHAUST.
MATERIAL WAS NOT INERT.
PROPULSION BELIEVED TO BE ATOMIC.
SILVER-GRAY IN COLOR WITH DARK CHIME SURFACES.
SADDLE SHAPE IN AFT SECTION AND SPEAR SHAPED NOSE.
NO LANDING GEAR WAS OBSERVED. EMBEDDED IN DESERT SAND. NOT DAMAGED..
TRANSPORTED TO ALAMOGORDO ARMY AIR FIELD.

DRAWING APPROVED BY
THOMAS CANTWHEEL
1-30-96

**Figure 2. The Aerodyne Vehicle**

The only researcher to have personally met with Cantwheel is Timothy Cooper, who served with the US Navy and US Marine Corps during the Vietnam War, and later became a private investigator.[11] Cooper wrote a four-page "preliminary report" about his investigation into Cantwheel and his claims, and reached the following conclusion:

> All this has led me to conclude, that Cantwheel (or whoever he is) was a bona fide military intelligence officer, a product of "spycraft" training and experience. As to his motives? I can only speculate this late date, that he was close to dying [estimated to be about 90] and wanted to tell what he knows of UFO secrecy.... I had the impression that he was not happy, nor agreed with the "policy" of ridicule and harassment of honest and sincere military men who reported the truth regarding UFO sightings and encounters, or worse still, ruining the careers and integrity of those who wanted to talk.[12]

In addition to Cooper's endorsement of Cantwheel and his information, his memo has undergone preliminary authentication analysis. The Woods have developed a list of "eight attributes of authenticity" which they apply to the hundreds of leaked Majestic documents they have in their possession in order to determine whether they are genuine or not.[13] In the case of Cantwheel's memo, they gave it an authentication rating of 40-60%, which signifies that it contains positive signs of authentication but has been under-researched.[14] In turn, this means that while Cantwheel's memo is not conclusive as a standalone document, it is nevertheless helpful as supporting evidence in understanding the Cape Girardeau crash and the history of US Air Force studies of retrieved flying saucers, along with classified reverse engineering efforts such as the alleged "S" aircraft.

Yet another leaked Majestic document refers to the 1941 Cape Girardeau crash, and additionally reveals President Roosevelt's strategic response to it given the looming entry of the U.S. into World War II. The "White Hot Report" is a September 1947 document, and the Woods gave it a "High Level of Authenticity" rating, which "means that virtually all of the available investigative channels and ideas have been pursued and, at each test, the document has been shown to be authentic."[15] The report provides an overview of how a unified effort by the US intelligence community to understand the flying saucer phenomenon would not emerge until well after the 1941 Cape Girardeau case:

> In the early months of 1942, up until the present, intrusions of unidentified aircraft have occasionally been documented, but there has been no serious investigations by the intelligence arm of the Government. Even the recovery case of 1941 did not create a unified intelligence effort to exploit possible technological gains with the exception of the Manhattan Project.[16]

What the White Hot Report tells us is that President Roosevelt was content to let the US Army take the lead in studying the retrieved saucer craft in an ad hoc manner, and to develop its intelligence assessment for a later time when the US Government would be ready to advance a comprehensive intelligence plan, once the difficulties posed by the United States' entry into World War II were met. This position was taken to ensure that the US military would not divert valuable resources into studying an advanced technology problem that did not directly help the war effort. Nevertheless, intrusions of unidentified craft continued, including one in early 1942 as the White Hot Report mentions. Undoubtedly, this was a reference to the Los Angeles UFO Air Raid incident in which the Army was tasked to study a crashed flying saucer and develop its intelligence strategy, while the Roosevelt administration was concentrating its full resources on fighting World War II.

# CHAPTER 2

## The Interplanetary Phenomenon Unit

It is by going down into the abyss that we recover the treasures of life. Where you stumble, there lies your treasure.

— Joseph Campbell

On the night of February 24, 1942, and on into the early hours of the 25th, residents of Los Angeles, California, were awakened by a dramatic UFO event. It began when an unknown number of unidentified flying objects flew across the sky in eerie silence over Long Beach and the surrounding areas. Feared to be Japanese aircraft about to launch another surprise Pearl Harbor-like attack, US Army and US Navy ships in Long Beach Harbor sprang into action and unleashed anti-aircraft artillery barrages against the mysterious craft that maneuvered beyond modern airplane capabilities. While sirens, projectiles, and explosions heralded the chaos of war upon a US shore, hundreds of thousands of people witnessed the hovering UFOs which were well lit up for several hours by searchlights and the anti-aircraft artillery during a total blackout. However, the UFOs did not attack. The only damage inflicted on the stunned citizenry was done by the artillery shells which fell to the ground causing minor injuries and significant property loss.

The events of that night were covered extensively in volume

two of this Secret Space Program series, *The US Navy's Secret Space Program & Nordic Extraterrestrial Alliance*, where you can read about coverage by major newspapers and contradictory official responses by the Army and Navy.[17] What I will concentrate on here is the compelling documentary and whistleblower evidence which shows how the Los Angeles incident directly led to Army Intelligence's development of an elite team within its Counter Intelligence Corps. Its primary responsibility would be to coordinate with the Army Air Force in the retrieval and study of extraterrestrial craft, both at Wright and Patterson Field facilities during and after World War II.

Figure 3. "The Great Los Angeles Air Raid"

Among the many thousands of witnesses to the Los Angeles event was William Tompkins, a precocious 17-year old at the time who had gained the attention of the Office of Naval Intelligence (ONI) because of the astonishingly accurate model replicas of US Navy ships he had constructed from his memory of the shipyards he had toured. At the time of the event, Tompkins had been placed at the Vultee Aircraft Corporation by the ONI to learn about aircraft design and propulsion. He got to see the shocking event up close and wrote about it in his 2015 autobiography, *Selected by*

*Extraterrestrials.*[18] Tompkins, and many other reports, supported the official Army position that there were several unidentified aircraft involved in the Los Angeles Air Raid incident. The *Los Angeles Herald Examiner* reported that civilian witnesses had estimated 50 objects that slowly traveled in a "V" formation.[19]

Some of the UFO craft stopped motionless over Long Beach, according to Tompkins, and displayed an advanced propulsion system unknown at the time. The helicopter industry was still in its infancy with only a small number produced by the U.S. and Nazi Germany for observation purposes. Yet the array of craft hovering over Los Angeles was quite large and silent, according to Tompkins and other witness reports, and was able to withstand a withering artillery barrage. These clearly were not any kind of experimental helicopter, dirigible, or other conventional aircraft.

According to the *Los Angeles Herald Examiner*, civilians reported that three craft had been shot down while over the ocean.[20] Tompkins says that he learned from his Navy sources that only two flying saucer-shaped craft were in fact shot down.[21] The saucers were recovered respectively by the Navy and the Army, when it was discovered that the saucer-shaped ships were unmanned fully automated drones.[22] Tompkins and witness reports of crashed vehicles are further corroborated in leaked official documents, which will be examined shortly.

Were the UFOs part of a reconnaissance mission by Japan, or perhaps other Axis powers, to ascertain the strength of anti-aircraft defenses for a possible future attack, as Henry L. Stimson, the Secretary of War, publicly hypothesized in a press release?[23] Or were the UFOs interplanetary in origin, raising the possibility of revolutionary propulsion systems that could radically transform the aviation industry and significantly impact the war effort? Answers to such questions can be found in a document officially released in 1974 through the Freedom of Information Act (FOIA).[24]

The FOIA document, dated February 26, 1942, is a memorandum from General George C. Marshall, Jr., Chief of Staff of the Army, to President Franklin D. Roosevelt. Issued one day after the Los Angeles Air Raid, it provides a preliminary report of what occurred and is consistent with what Tompkins and other witnesses saw.[25] It's all but certain that the document would have been written with the assistance of General H. H. "Hap" Arnold, Chief of the Army Air Force. In the February 26 memorandum, Marshall tells Roosevelt:

> The following is the information we have from GHQ at this moment regarding the air alarm over Los Angeles of yesterday morning:
>
> From details available at this hour:
>
> 1. Unidentified airplanes, other than American Army or Navy planes, were probably over Los Angeles, and were fired on by elements of the 37th CA Brigade (AA) between 3:12 and 4:15 am. These units expended 1430 rounds of ammunition.
> 2. As many as fifteen airplanes may have been involved, flying at various speeds what is officially reported as being "very slow" to as much as 200 mph and at elevations from 9000 to 18000 feet.
> 3. No bombs were dropped.
> 4. No casualties among our troops.
> 5. No planes were shot down.
> 6. No American Army or Navy planes were in action.
>
> Investigations continuing. It seems reasonable to conclude that if unidentified airplanes were involved they may have been from commercial sources, operated by enemy agents for purposes of spreading alarm, disclosing locations of antiaircraft positions, and slowing production through blackout. Such conclusion is supported

by varying speed of operation and the fact that no bombs were dropped.[26]

This official memorandum points out that further US Army investigations were being conducted, indicating that Marshall would send updates on the incident to the President as more was learned. Due to the magnitude of this situation, Arnold would have led the effort on Marshall's behalf to gather more facts as quickly as possible.

Significantly, the February 26 memorandum refers to the speed of some of the unidentified craft as ranging from "very slow" to 200 mph. This corroborates Tompkins' account that some of the craft were indeed "hovering" over Los Angeles. Even further, it helps authenticate that the craft were non-terrestrial in origin since the only aircraft capable of hovering at the time (prototype helicopters, balloons and dirigibles) would have easily been shot down by the anti-aircraft guns. This motionless capability would naturally have been of great interest to Arnold and the Army Air Force, which was only beginning the task of developing helicopters for surveillance purposes based on Igor Sikorsky's VS-300 prototype built in 1939.[27]

According to FOIA records, there was no further official correspondence between Marshall and Roosevelt over the February 24/25 Los Angeles Air Raid. However, leaked Majestic documents show that the correspondence did indeed continue between them in two subsequent exchanges which contained important policy issues about how to deal with UFOs in the future.

The Majestic document memorandums between Marshall and Roosevelt are categorized by the Wood research team as "Medium-High Level of Authenticity", which they define as follows:

> The medium-high level means that a considerable amount of investigation and testing has been completed and there are strong signs of authenticity in the way of content, forensics, typography, zingers etc. Although there may be some anachronisms identified they do not seem to be major.[28]

OCS 21347-86

SECRET

*File*

February 26, 1942.

*Record*
*action*

MEMORANDUM FOR THE PRESIDENT:

The following is the information we have from GHQ at this moment regarding the air alarm over Los Angeles of yesterday morning:

"From details available at this hour:

"1. Unidentified airplanes, other than American Army or Navy planes, were probably over Los Angeles, and were fired on by elements of the 37th CA Brigade (AA) between 3:12 and 4:15 AM. These units expended 1430 rounds of ammunition.

"2. As many as fifteen airplanes may have been involved, flying at various speeds from what is officially reported as being 'very slow' to as much as 200 MPH and at elevations from 9000 to 18000 feet.

"3. No bombs were dropped.

"4. No casualties among our troops.

"5. No planes were shot down.

"6. No American Army or Navy planes were in action.

"Investigation continuing. It seems reasonable to conclude that if unidentified airplanes were involved they may have been from commercial sources, operated by enemy agents for purposes of spreading alarm, disclosing location of antiaircraft positions, and slowing production through blackout. Such conclusion is supported by varying speed of operation and the fact that no bombs were dropped."

DECLASSIFIED
E.O. 11652, Sec. 3(E) and 5(D) or (E)
OSD letter, May 3, 1972
By ___ NARS Date 4-9-74

akn

(Sgd) G. C. MARSHALL

Chief of Staff.

21347
86

*Orig. despatched to Pres.*
*2/12/42*

Figure 4. Feb 26, 1942 Memorandum from Marshall to Roosevelt

Importantly, the continued correspondence between Marshall and Roosevelt is also consistent with the FOIA document, which confirms key aspects raised in their subsequent memo exchange, as I will shortly discuss. Furthermore, the leaked Majestic documents corroborate some of the information revealed in the testimony of Bill Tompkins and others who had reported crashes of unidentified objects from the artillery barrage, and their clandestine retrieval by special Army and Navy units. Consequently, I consider the leaked Majestic documents showing additional written communications between Marshall and Roosevelt after February 26, 1942, to be genuine.

The leaked Majestic documents reveal that Roosevelt responded to Marshall's February 26 memorandum the next day. During the time between the initial report and Roosevelt's response, additional information had been gained. Namely, that two of the unidentified craft had been downed and retrieved. Briefed over this new development by the time of his response, the President directly referred to the "celestial device" acquired by the Army Air Force, which he called a "new wonder", and pointed out that his administration's policy was to study the device with regard to learning any atomic secrets that might advance nuclear weapons research taking place through the Manhattan project. Thus, he instructed Marshall of the priorities and what to do next:

> I have considered the disposition of the material in possession of the Army that may be of great significance toward the development of a super weapon of war. I disagree with the argument that such information should be shared with our ally the Soviet Union. Consultation with Dr. Bush and other scientists on the issue of finding practical uses for the atomic secrets learned from study of celestial devices precludes any further discussion and I therefor authorize Dr. Bush to proceed with the project without further delay. This information is vital to the nation's superiority and must remain

within the confines of state secrets. Any further discussion on the matter will be restricted to General Donovan, Dr. Bush, the Secretary of War and yourself. The challenge our nation faces is daunting and perilous in this undertaking and I have committed the resources of the government towards that end. You have my assurance that when circumstances are favorable and we are victorious, the Army will have the fruits of research in exploring further applications of this new wonder.[29]

Roosevelt's reply makes clear that despite his pro-Navy background (he was the former Assistant Secretary of the Navy from 1913 to 1920), he assigned the Army/Army Air Force the primary responsibility for leading the investigation of flying saucer reports and dealing with any retrieved craft.[30] This is largely due to the advice of his science advisor, Dr. Vannevar Bush, and the Army's role in leading the Manhattan Project to develop atomic bombs, and the invaluable expertise it had acquired as a result.[31]

Significantly, Roosevelt took the effort to explain that the overall policy would be for the Army/Army Air Force and the Office of Strategic Services (led by General William Donovan) to collect information about "celestial devices" wherever found around the world, without devoting significant resources into research and development of their exotic propulsion systems. Roosevelt was effectively telling Marshall's subordinate, General Arnold, that the Army Air Force engineers would have to wait until the end of the war to study and reverse engineer their prizes, the "celestial devices". The inclusion of Donovan in this inner circle shows the importance of covert operations that the Roosevelt, and subsequent Presidential administrations, would adopt in dealing with the flying saucer phenomenon.

After sending his February 26 memorandum with an overview of the prior day's Los Angeles event, and receiving Roosevelt's February 27 reply, Marshall needed a week before he could provide a more detailed assessment of the unidentified craft that had been retrieved. On March 5, 1942, Marshall sent another memorandum to Roosevelt that explained the hypothesized origin of the two unidentified aircraft that had been shot down and retrieved by the Navy and the Army:

> ... regarding the air raid over Los Angeles it was learned by Army G2 [Intelligence] that Rear Admiral Anderson ... Naval Intelligence, has informed the War Department of a naval recovery of an unidentified airplane off the coast of California with no bearing on conventional explanation. Further it has been revealed that the Army Air Corps has also recovered a similar craft in the San Bernardino Mountains east of Los Angeles which cannot be identified as conventional aircraft. This Headquarters has come to the determination that the mystery airplanes are in fact not earthly and according to secret intelligence sources they are in all probability of interplanetary origin.[32]

Marshall went on to explain to Roosevelt how the Army would deal with the UFO phenomenon:

> ... issued orders to Army G2 that a special intelligence unit be created to further investigate the phenomenon and report any significant connection between recent incidents and those collected by the director the office of Coordinator of Information. [33]

This communique allegedly hallmarks the genesis of the Army's legendary "Interplanetary Phenomenon Unit."

The Interplanetary Phenomenon Unit (IPU) became a highly-classified Army Intelligence unit that existed during the World War II era (see Figure 6). After initially denying its existence, the US Air Force (as the direct successor to the US Army Air Force in 1947) eventually found itself forced to acknowledge that the Interplanetary Phenomenon Unit *did* exist for a period of time. Through documents released under the Freedom of Information Act, the existence of this highly-secretive investigatory group was confirmed, despite the best efforts by Air Force officials to cast doubt on its existence.

TOP SECRET

Top Secret

gm 25
February 27, 1942

THE WHITE HOUSE
WASHINGTON

February 27, 1942

MEMORANDUM FOR

CHIEF OF STAFF OF THE ARMY

I have considered the disposition of the material in possession of the Army that may be of great significance toward the development of a super weapon of war. I disagree with the argument that such information should be shared with our ally the Soviet Union. Consultation with Dr. Bush and other scientists on the issue of finding practical uses for the atomic secrets learned from study of celestial devices precludes any further discussion and I therefor authorize Dr. Bush to proceed with the project without further delay. This information is vital to the nation's superiority and must remain within the confines of state secrets. Any further discussion on the matter will be restricted to General Donovan, Dr. Bush , the Secretary of War and yourself. The challenge our nation faces is daunting and perilous in this undertaking and I have committed the resources of the government towards that end. You have my assurance that when circumstances are favorable and we are victorious, the Army will have the fruits of research in exploring further applications of this new wonder.

You may speak to me about this if the above is not wholly clear.

F. D. R.

TOP SECRET

*This is a retyped version of FDR's memo from DE's DS files*

Figure 5. Re-typed Roosevelt memorandum to Marshall

In May 1984, UFO researcher William Steinman sent a FOIA request to the Army Directorate of Counterintelligence about the Interplanetary Phenomenon Unit. Steinman received a reply from a Lieutenant Colonel Lance R. Cornine:

> As you note in your letter, the so-called Interplanetary Phenomenon Unit (IPU) was disestablished and, as far as we are aware, all records, if any, were transferred to the Air Force in the late 1950's. The 'unit' was formed as an in-house project purely as an interest item for the Assistant Chief of Staff for Intelligence. It was never a 'unit' in the military sense, nor was it ever formally organized or reportable, it had no investigative function, mission or authority, and may not even have had any formal records at all. It is only through institutional memory that any recollection exists of this unit. We are therefore unable to answer your questions as to the exact purpose of the unit, exactly when it was disestablished, or who was in command. This last would not apply in any case, as no one was in 'command'. We have no records or documentation of any kind on this unit."[34]

Cornine's letter acknowledged that the IPU did exist up to the late 1950's, but downplayed its actuality as merely an "interest item" that was never an operational army "unit" of any kind. However, a subsequent FOIA request to the US Army Intelligence and Security Command led to a reply on April 9, 1990, written by Colonel William Guild. He stated, "all records pertaining to the *unit* were surrendered to the US Air Force Office of Special Investigations in conjunction with Operation 'Bluebook'."[35] [*sic*] So according to Army Intelligence records, the IPU was a formal unit related in some way to the study of UFOs, thereby contradicting what the Air Force had to say about the IPU's inactivity.

Importantly, the FOIA documents confirm the creation of

the IPU, and are documentary evidence which further corroborates the authenticity of the leaked Majestic document offering Marshal's March 5, 1942 memorandum informing the President of the creation of a special investigative unit to study UFOs. In the memorandum, General Marshall ordered "Army G2" to create "a special intelligence unit" to "investigate the phenomenon,"[36] so he almost certainly was referring to the Interplanetary Phenomenon Unit. This is the conclusion reached by Dr. Robert Wood and Ryan Wood in their respective efforts to authenticate the Marshall memorandum:

> The memo bears correct Office of Chief of Staff (OCS) file numbers and has "Interplanetary Phenomenon Unit" (IPU) typed on it at a later time by a different typewriter. It is logical to believe that this is the order that sets up the IPU.[37]

Another important aspect of the FOIA IPU documents is that Colonel Guild's response shows that the IPU was actively coordinating with the US Air Force's Project Blue Book, which was based out of Wright-Patterson Air Force Base (as Wright Field was renamed in 1947). The predecessors of Project Bluebook (1952-1969); Project Sign (1947-49) and Project Grudge (1949-51) were also based out of Wright Field/Wright-Paterson. In 1948, the first formal study of UFO reports by Wright-Patterson technical and intelligence experts concluded that UFOs were interplanetary in origin – the same conclusion reached in the March 5, 1942, Marshall memo to Roosevelt about the retrieved Los Angeles craft.[38] It's very possible, and even likely, that some of the same scientists at Wright Field that contributed to Marshall's 1942 "interplanetary origin" conclusion were involved five years later in the Project Sign and/or Grudge findings.

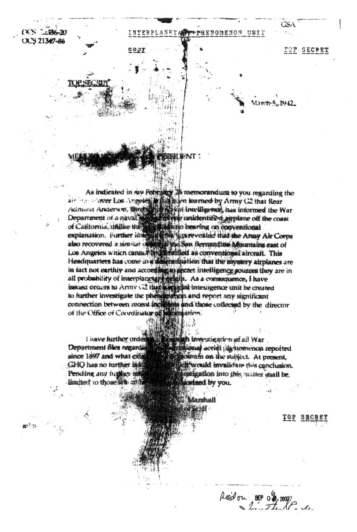

Figure 6. Alleged memo from General Marshal to President Roosevelt

Furthermore, it's worth noting that the IPU played a key role in investigating the various Project reports and providing sightings data that allowed the Wright Field/Wright-Patterson technical experts to reach such a conclusion. However, according to General Hoyt Vandenberg, who was the US Air Force Chief of Staff in 1947, disclosing the truth about extraterrestrial life "would cause a stampede,"[39] so he asked for the Wright Field experts to rewrite their report and *downplay the interplanetary (extraterrestrial) explanation!*

The collaboration between the IPU and Wright Field/Wright-Patterson aeronautical experts went far beyond the mere study and analysis of UFO reports; their work extended to the study of crashed/retrieved extraterrestrial craft. We get an idea of the true scope of IPU activities through the testimony of Thomas Cantwheel (pseudonym), who was mentioned in Chapter 1 for his role in multiple crash/retrieval operations. He says that the elite unit within the Army's Counter Intelligence Corps (CIC) that he served with was, in fact, the Interplanetary Phenomenon Unit. Cantwheel states that the IPU was operational right up to 1958, more than a decade after the Air Force had been created, when his service in the program ended. Timothy Cooper, while investigating Cantwheel, learned:

> Much of Cantwheel's activities with CIC is unknown, only that he was assigned by CIC to a secret army investigative unit known as the IPU (Interplanetary Phenomenon Unit). The IPU was an intelligence branch of the [Army Directorate of Counter Intelligence] which passed on UFO reports to US Army G-2 (Intelligence Director)... Cantwheel claims to have been an CIC officer in the U.S. Army's Interplanetary Phenomenon Unit from 1942 – 1958, with overseas TDY. His main areas of interest was [*sic*] UFO sightings by military personnel.[40]

Cantwheel's testimony corroborates a part of Steinman's official FOIA response, which is the identification of the late 1950's as the time when the IPU was formally abolished as a functional Army intelligence unit, and most likely absorbed into a classified US Air Force intelligence program, presumably Project Moon Dust and Operation Blue Fly (to be discussed later). His testimony also supports what Colonel Guild confirmed in his FOIA response, where he stated that the IPU coordinated with the official US Air Force investigation of UFO reports – Project Blue Book.

In addition to Cantwheel's testimony, the leaked Marshall-Roosevelt correspondences show that the IPU was far more than merely an "interest item" that was never made operational, as claimed in the FOIA response by Cornine. Army Intelligence (G2) and the Army Air Force would coordinate in such investigations and retrievals through the Interplanetary Phenomenon Unit. It's noteworthy that the US Navy, which had retrieved one of the craft found after the Los Angeles Air Raid event, also coordinated with the IPU and Army Air Force. This cooperation is demonstrated in a series of orders received by one of the Navy's expert aviation engineers, Commander Rico Botta who headed the Bureau of Aeronautic's Power Plant Design Branch. Botta had been identified by the Navy's leading engineering division, the Bureau of Aeronautics, as their representative for investigating the power systems of retrieved antigravity craft.

On February 25, 1942, the very day that the two craft were retrieved following the Los Angeles incident, Botta was ordered to travel to Wright Field for an undisclosed assignment. Only a few months later, Botta traveled to advanced Navy, Army Air Force and corporate aviation facilities, and was finally transferred in December 1942 to Naval Air Station, San Diego, to head up the Navy's own intelligence gathering on the flying saucer phenomenon.

Botta recruited the young William Tompkins to participate in the top secret debriefing of Navy spies embedded in Nazi Germany, which resulted in Tompkins couriering the briefing documents based on the spies' intelligence to the Interplanetary Phenomenon Unit, Army Air Force, aviation companies and to any scientific institution capable of understanding flying saucer technologies. Tompkins says that he traveled all over the U.S. delivering his briefing packets which included leading Army Air Force facilities such as Wright Field. Botta and Tompkins' roles in the covert Naval espionage program of gathering intelligence about Nazi Germany's research into flying saucer technologies can be read in the second book of my Secret Space Program series, *The*

## US Navy's Secret Space Program & Nordic Extraterrestrial Alliance.

In reply address not the signer of
this letter but Bureau of Aeronautics,
Navy Department, Washington, D. C.

Refer to No.
Aec-P2-12-LC
OO/Botta, Rico

NAVY DEPARTMENT
BUREAU OF AERONAUTICS
WASHINGTON

$37475-164/18$

From: The Chief of the Bureau of Aeronautics.
To:   The Chief of the Bureau of Navigation.

FEB 25 1942

SUBJECT: Orders—temporary additional duty.

1. It is recommended that orders for temporary additional duty be issued to the following personnel as indicated:

|  | Name | Rank or Rating | WRITTEN DISPATCH Present Station |
|---|---|---|---|
| | BOTTA, Rico | Commander, U.S.N. | BuAer |

7- 20568

From:      Bureau of Aeronautics, Washington, D. C.
Date to proceed: On or about 2 March 1942
Via:              Naval Aircraft, or commercial, ~~commercial air,~~
To:       Wright Field, Dayton, Ohio, reporting to the General Inspector of Naval
Thence via:       Naval Aircraft,    commercial,    commercial air,  Aircraft,
To:
Thence via:       Naval Aircraft,    commercial,    commercial air,
To:
Thence via:       Naval Aircraft,    commercial,    commercial air,
To:
Thence return via: Naval Aircraft, or  commercial,  ~~commercial air,~~
To: ......Washington, D. C.................., and resume regular duties.

(a) Reason:  In connection with engine development.
          (Chargeable to Aviation, Navy, 1942 - Subhead One)

(b) Estimated period of duty:  2 days
(c) Estimated cost:  $85.44 if via commercial
(d) Estimated delay in excess of 72 hours at:
(e) It is recommended that these orders be:  delivered by 28 February 1942

F. W. McMAHON
Commander, U.S.N.
By direction Chief of Bureau

FEB 2 6 1942

12:55

U. S. GOVERNMENT PRINTING OFFICE  16—22027

**Figure 7. Order issued to Rico Botta to travel to Wright Field on the day of the LA Air Raid**

Respective efforts by the IPU, Army Air Force, Navy and major aviation companies to gather intelligence and fully understand the flying saucer phenomenon were largely constrained by the enormity of World War II. The extent of the colossal commitment to the production of military aircraft for the war effort is well described by Dana T. Parker, author of *Building Victory: Aircraft Manufacturing in the Los Angeles Area in World War II*:

> In 1939, total aircraft production for the Army Air Corps and Navy combined was only 2,141 planes. By the end of the war, America produced an astounding 300,000 planes ... The U.S. not only armed its own forces, but its allies as well.... It averaged 170 planes a day since 1942, more than the Soviet Union and Great Britain combined."[41]

Douglas Aircraft Company and other aviation firms, many of which were based in the Los Angeles area, maxed out their production facilities to meet the military's ever-growing demand for bombers, fighters, and cargo aircraft.[42] The number of airplanes built for the war effort was so staggering that Donald Douglas declared: "Here's proof that free men can out-produce slaves."[43]

Despite the enormous war production quotas met by Douglas and other major aircraft companies, the Army Air Force still struggled to meet its own rapid expansion needs. This was because General Arnold was ordered to assist Britain, the Soviet Union and other US allies in having operational military aircraft ready to do battle against Nazi Germany. The challenge for the Army Air Force is described in Arnold's war diaries:

> For example, the United States accepted from the manufacturers in the years 1940 through 1944 a total of almost 23,000 P-39 and P-40 aircraft. Yet at the peak of their inventory in 1944, the AAF had on hand only 4,600 of them. Even allowing for attrition caused by training accidents and their limited U.S. use in combat, the figures reflect that the bulk of

them went to Britain and the Soviet Union. [44]

Given the immediate needs of the war, with aircraft companies struggling to meet ambitious production goals, it was understandable why the Roosevelt administration had decided that the vast undertaking of research and development for the flying saucer phenomenon would have to wait until after the conflict ended, as Roosevelt's February 27, 1942 memorandum to General Marshall made clear.[45]

It was only after World War II came to an end that the Army Air Force and Army Counter Intelligence (through the Interplanetary Phenomenon Unit) could begin in earnest the necessary research and development essential for a long-term strategic response to the flying saucer phenomenon. Arnold was promoted to the rank of five-star General of the Army on December 21, 1944, and soon after, as World War II came to an end, he decided to formally launch the necessary research and development program that would combine military and corporate expertise to master flying saucer technologies. Arnold decided to use left-over Army Air Force war funds to launch Project RAND, and called upon his close friend, the legendary aviation wizard Donald Douglas, to help. Under the public cover of a conventional study focused upon a range of aerospace topics, the military began secretly to conduct with the Douglas Aircraft Company serious research and development of crashed/retrieved flying saucer technologies.

# CHAPTER 3

# Project RAND and Non-Terrestrial Technologies

It isn't sufficient just to want – you've got to ask yourself what you are going to do to get the things you want.

— Franklin D. Roosevelt

I t is easy to imagine the US Army Air Force's fervent desire to push ahead with reverse engineering its prize from the 1942 Los Angeles Air Raid. However, the retrieved antigravity vehicle would have to sit idly on the sidelines while the President, by necessity, instructed all military branches to use their full resources toward building machines of war. Roosevelt did not believe the highly advanced science behind the "celestial devices" could be understood in time to assist in weapons development needs at the time. In turn, Americans would not find out that in a woeful time of nation fighting nation, from out of the sky, something had come to usher in the promise of amazing technological advances that could lead humankind to the stars. Although the Army Air Force would have to wait, behind-the-scenes the generals began to plan for the development of those highly desirable technologies.

Two years after the retrieval, in a February 22, 1944, Roosevelt memorandum to his science advisor, Dr. Vannevar Bush, the President reaffirmed his stance on the policy of postponing the research and development of "non-terrestrial" technologies. This memorandum is another leaked Majestic document, which the

Woods research team ranked as having a medium-high level of authenticity, and thus, concluded it credible in their professional estimation. The memo's unusual classification, "Double Top Secret", contrary to initial skepticism, was discovered by the Woods to be legitimate.[46] Within the document, expense was cited as the primary issue for not starting a research and development program, since this would directly impact available resources for producing conventional advanced aircraft and the atomic bomb. In the Double Top Secret memorandum addressed to "The Special Committee on Non-Terrestrial Science and Technology", Roosevelt wrote:

> I agree with the OSRD [Office of Science Research and Development] proposal of the recommendation put forward by Dr. Bush and Professor Einstein that a separate program be initiated at the earliest possible time. I also agree that application of non-terrestrial know how in atomic energy must be used in perfecting super weapons of war to affect the complete defeat of Germany and Japan. In view of the cost already incurred in the atomic bomb program, it would, at this time, be difficult to approve without further support of the Treasury Department and the military. I therefore have decided to forego such an enterprise. From the point of view of the informed members of the United States, our principle object is not to engage in exploratory research of this kind but to win the war as soon as possible.
>
> Various points have been raised about the difficulties such an endeavor would pose to the already hardened research for advanced weapons programs and support groups in our war effort and I agree that now is not the time. It is my personal judgment that, when the war is won, and peace is

once again restored, there will come a time when surplus funds may be available to pursue a program devoted to understanding non-terrestrial science and its technology which is still greatly undiscovered.

I appreciate the effort and time spent in producing valuable insights into the proposal to find ways of advancing our technology and national progress and in coming to grips with the reality that our planet is not the only one harboring intelligent life in the universe.[47]

The surplus funds mentioned by Roosevelt would indeed become available in 1945 through a number of official channels, one of which was a five-star general of the Army, Henry "Hap" Arnold, who had successfully led the rapid expansion and deployment of the Army Air Force throughout World War II. Arnold and the Army Air Force had effectively been ordered to sit on the retrieved flying saucers from the 1941 Cape Girardeau and the 1942 Los Angeles Air Raid incidents. These had been secretly taken to Wright Field, the Army Air Force's premier advanced aviation research facility, and would only be subjected to preliminary scientific study until the war's end.

In the interim, the Army Air Force, the Interplanetary Phenomenon Unit, the Office of Strategic Services (forerunner to the CIA) and the US Navy gathered as much intelligence as possible on what the Axis powers were doing in this arena, and covertly endeavored to retrieve any craft of non-terrestrial origin that came their way around the planet. In particular, the Navy, through its covert operatives working undercover in Nazi-Occupied Europe, clandestinely gathered data on what the Nazis were doing in their own research and development of flying saucer technologies.[48]

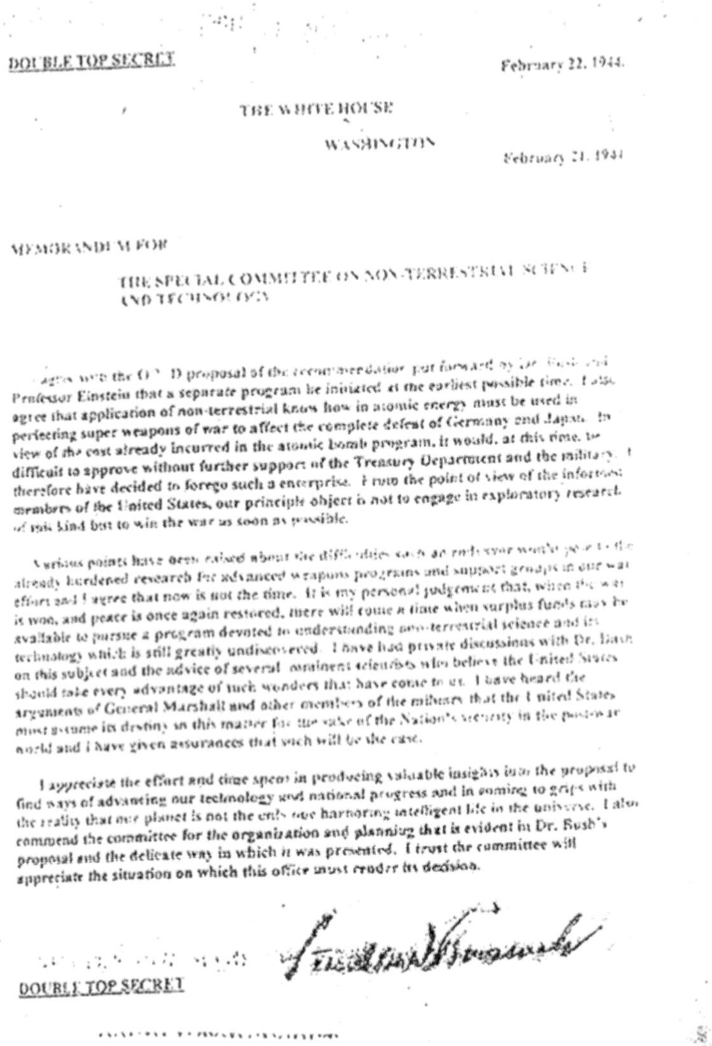

Figure 8. "Double Top Secret" document signed by President Roosevelt

After Japan's surrender on August 15, 1945, Arnold was finally able to move forward with funding research and development of the recovered alien craft using surplus war funds. In doing so, he closely collaborated with another aviation giant, Donald W. Douglas, Sr., founder of Douglas Aircraft Company. Soon after Japan's formal surrender, Arnold flew to California to meet with Douglas and chart the way forward for a comprehensive R&D (research and development) plan that would lay the foundation for a yet another revolution in the aviation industry, now using exotic propulsion systems such as antigravity.[49]

There are numerous reasons why Arnold chose to work with Douglas and his company on the secret research and development of flying saucer technologies. First, Douglas Aircraft Company was among the leading manufacturers of aircraft for both the Navy and Army Air Force during World War II, producing the most advanced military aircraft in the world. By 1945, Douglas Aircraft had produced nearly 30,000 planes for the successful war effort, and its engineers and production facilities were world-renowned for their ability to incorporate the latest aviation technologies into their products. In addition to having talented engineers and advanced production facilities, there was another compelling reason why the Douglas company was an attractive partner for Arnold and the Army Air Force to lead R&D efforts over retrieved non-terrestrial technologies.

The company's owner, Donald Douglas, was also a pioneer and legend in the aviation industry. Like Arnold, Douglas had a direct personal connection with the Wright brothers. In fact, when only 16 years old, Douglas witnessed the 1908 test flight of the Wright brothers prototype military aircraft, the Military Flyer, at Fort Myer, Virginia. In 1914, Douglas had graduated with a degree in aeronautical engineering, the first awarded from the Massachusetts Institute of Technology (MIT). Douglas' academic accomplishments and aeronautical abilities were recognized by the Army, which hired him in November 1916 as their chief civilian aeronautical engineer for the Aviation Section, Army Signal Corps. According to Dana Parker, the Army "waived the test for the position because there was no one in the country who knew more about aeronautical engineering than Donald Douglas."[50] In 1941, *Fortune Magazine* wrote: "The development of the airplane in the days between the wars is the greatest engineering story there ever was, and in the heart of it is Donald Douglas." [51]

The next reason why Arnold chose Douglas was their close

personal and working relationship dating back to their respective associations with the Wright brothers, and mutual interests in aeronautical developments. Seven months after Douglas became the Army's chief civilian aeronautical engineer, he was followed by Arnold who took up a number of executive positions within the Aviation Division in June 1917, two months after the U.S. entered World War I. Douglas and Arnold came to work together in modernizing US aircraft for the First World War. Afterward, both eventually moved to the Los Angeles area where Douglas established his aircraft company in 1921, and Arnold, later in November 1931, took command of March Field. The Douglas and Arnold families became so close that Douglas' daughter, Barbara, eventually married Arnold's second son, Bruce, upon his graduation from West Point Military Academy in 1943.[52]

The fourth reason is that Douglas personally witnessed the Los Angeles Air Raid event, according to William Tompkins. Tompkins was very familiar with the operations and history of Douglas Aircraft Company, due both to his employment in its engineering division from 1950 to 1963, and also because of his visits to Douglas Aircraft when working under Rear Admiral Rico Botta out of the San Diego Naval Air Station from 1942 to 1946.[53] In his autobiography, Tompkins described the powerful influence the Los Angeles incident had on Douglas and other key corporate and military officials who had experienced the events of that chaotic night.[54] The Douglas Aircraft Company was headquartered in Santa Monica, so it is natural to assume that other senior Douglas personnel also witnessed the paradigm-shifting Los Angeles event and immediately began pondering and discussing its revolutionary implications for the aviation industry.

Nevertheless, President Roosevelt's February 27, 1942 memorandum made it very clear that war resources would not be channeled into "exploratory research" of the retrieved craft. It is understandable that aviation pioneers such as Douglas, General Arnold, and others would have chafed at the bit to begin a serious study of the revolutionary propulsion and navigation technologies of the retrieved craft. So it is not surprising that according to

Tompkins, Douglas, along with his chief engineer, Arthur Raymond, and Raymond's assistant, Franklin Collbohm, came up with the idea of creating an informal working group to study the available information.[55]

Tompkins further claimed that the study group included two Army Air Force generals and two Navy admirals.[56] If Tompkins were correct, then Douglas and General Arnold had convened an "informal" study group directly after the incident despite President Roosevelt's policy of not devoting funding and resources to the "exploratory research" of the retrieved non-terrestrial spacecraft. Even though Roosevelt's policy persisted, as outlined in his February 22, 1944, Memorandum to the "Special Committee on Non-Terrestrial Science and Technology", their work-around did not technically defy policy because of its non-official, or "informal" status.[57]

The study group's primary goal was to investigate the scientific principles and research the implications of what had been witnessed. Additionally, the group was to analyze any information about the two retrieved craft taken to Wright Field that the Army and Navy officers in attendance were authorized to share. If General Arnold was not an official member of the group, it is almost certain that he would have appointed a subordinate to report to him on what the study group learned. At the very least, it can be assumed that members of the Interplanetary Phenomenon Unit would have known of, and even attended, meetings of the informal Douglas study group. Consequently, at the end of World War II when Arnold was finally allowed to move forward to fund research and development of the captured alien craft, Douglas Aircraft Company was the obvious choice to receive the available funding.

A final reason to consider entails the intelligence data gathered about the Nazi's research and development of flying saucer technologies which had been accumulated at Naval Air Station San Diego, and was made known to the Douglas Aircraft Company. After beginning his covert assignment under Rear Admiral Rico Botta, Tompkins claimed that from the spring of 1943 onward he traveled at least three times to Douglas with briefing packets containing the latest information about the Nazi programs.[58]

Tompkins identified both Arthur Raymond and Franklin Collbohm as being directly involved in the study of the briefing packets delivered to Douglas Aircraft Company. In an interview, Tompkins recalled that one of the Douglas engineers he spoke to about the briefing packets might have been Collbohm, who later became the first President of the RAND Corporation founded in 1948 by the newly formed United States Air Force (USAF).[59] All of this provides yet another compelling reason for why General Arnold chose Douglas Aircraft Company as the recipient for surplus Army Air Force funds that would establish "Project RAND" – the forerunner to the RAND Corporation – which today continues to perform a mission similar to the original intent behind its creation.

## Project RAND

Finally able to move forward using surplus war funds, General Hap Arnold hinted at what he planned to do in an official report to the Army Chief of Staff, General Marshall, stating:

> During this war the Army, Army Air Forces, and the Navy have made unprecedented use of scientific and industrial resources. The conclusion is inescapable that we have not yet established the balance necessary to insure the continuance of teamwork among the military, other government agencies, industry, and the universities. Scientific planning must be years in advance of the actual research and development work.[60]

Consequently, from surplus funds available to him after Japan's surrender, Arnold was permitted to assign a significant proportion of these funds to the creation of Project RAND as explained by an official US Air Force historian, Major General Huston:

> In the month of the final Japanese surrender, he [Arnold] flew to California where he and his decades-long friend, Donald Douglas, arranged for creation of the RAND Corporation. Arnold committed $10 million of AAF funds to the initial funding of the new corporation.[61]

General Huston described RAND's early mission as:

> ... a "technical consultant group charged with operations analysis and long-range planning to examine future warfare and the best way the Air Force could perform its missions. It was originally staffed with scientists and engineers from Douglas Aircraft Corporation". [62]

A November 1965 article in *The Progressive*, by journalist Wesley Marx (archived on the CIA's Freedom of Information Act database) pointed out that Arnold was able to use these Army surplus funds without any congressional approval:

> Without Congressional authorization and without taking bids, Air Force General H.H. "Hap" Arnold managed to shift around enough funds to award a $10 million contract to Douglas Aircraft Company to set up Project RAND.... Project RAND was to engage in a continuing program of scientific study and research on the broad subject of recommending to the Air Force preferred methods, techniques, and instrumentalities for this purpose. [63]

Especially revealing is the next sentence in Marx's article, where the elaborate security mechanisms in place at the Douglas facilities are briefly described:

> Douglas furnished administrative services, security

guards, and locked rooms in its Santa Monica, California, facility, and RAND became a subsidiary, but virtually autonomous division of Douglas.[64]

Ten million dollars in 1945 converts to approximately $855 million in 2018 terms.[65] At the time, this was an enormous amount of money to give to an aviation company to merely conduct scientific research and make recommendations, and further, without taking bids from competitors. Arnold's decision shows that he believed that the Douglas Aircraft Company was at the time well ahead of its main rivals, Lockheed, Boeing and Northrup when it came to starting research and development of advanced technologies found in the extraterrestrial spacecraft stored at Wright Field.

The official website of the RAND Corporation curiously glosses over the enormous amount of money given to Douglas to set up Project RAND.[66] The impression given is that Project RAND was a small in-house Douglas operation, spending only "$640 in its first month of operation" to presumably pay for a group of senior scientists and engineers to meet and discuss advanced aerospace topics. This is certainly what official Air Force historians, such as General Huston, claimed.[67] However, without a doubt, the $10 million allocated was being used for far more than paying a small group of Douglas engineers and scientists to act as a "technical consultant group", especially in a secure Douglas facility protected by armed guards as Marx's article stated. What were the armed guards protecting in the locked rooms at Douglas?

## William Tompkins and Project RAND's Undisclosed Mission

For an answer, we can turn to William Tompkins, who himself later worked at a secure research facility hidden within Douglas Aircraft's engineering division. He was able to indicate the general purpose of Project RAND, as well as confirm its original establishment within Douglas:

The Douglas/RAND scientists and concept conceivers were studying topics actually way above top secret. They were in a Think Tank deep inside the Douglas Aircraft Company's Engineering Department A-250, that nobody knew existed at the Santa Monica, California Airport.[68]

Tompkins further identified the real mission of Project RAND:

It was created in … [October] 1945, as a special contract to Douglas Aircraft Company. At the Santa Monica Municipal Airport. Inside a highly classified walled-off area in the Douglas Engineering Department, **Project RAND studied the implications of threatening alien agendas**…. Then, on March 2 … [1946], a letter of contract was executed, which put Project RAND under the direction of Douglas' Assistant Chief Engineer, Frank Collbohm. The Douglas Think Tank was born. [emphasis added][69]

The exclusive role given to the Douglas campaign at the end of the war in establishing the first US research center set up to focus on flying saucer craft and the possible threat posed by extraterrestrial visitors was emphasized by Tompkins.[70] Furthermore, he explained how the Project RAND contract gave Douglas employees access to the technical information possessed by all other US research and development programs that were relevant to flying saucer technology and extraterrestrial life.[71] A document that directly supports Tompkins' claim that Project RAND was primarily set up to study the flying saucer phenomenon, while also collaborating with other leading US scientific organizations conducting advanced aerospace research on this topic, is another leaked Majestic document. Known as the "White Hot Report", it is dated September 19, 1947, and lists Project RAND among the research organizations studying artifacts

41

recovered from the crash of extraterrestrial vehicles dating back to the 1941 Missouri crash:

> Based on all available evidence collected from recovered exhibits currently under study by AMC, AFSWP, NEPA, AEC, NACA, JRDB, **RAND,** USAAF, SAG and MIT, are deemed extraterrestrial in nature. This conclusion was reached as a result of comparisons of artifacts ["from the Missouri"] discovery in 1941. The technology is outside the scope of US science, even that of German rocket and aircraft developments. [emphasis added][72]

Significantly, after an extensive investigation by veteran document researchers, Dr. Robert Wood and Ryan Wood, the "White Hot Report" received their highest level of authenticity.[73] Consequently, the "White Hot Report" document is independent confirmation that the Douglas Aircraft Company, through Project RAND, was involved in the study of retrieved alien spacecraft just as Tompkins claimed. This document also lends credibility to another core element of Tompkins' testimony that he joined a secret Douglas engineering division "Think Tank" in 1951, which did feasibility studies on different antigravity spacecraft for the US Navy. Much more about Tompkins testimony and documents are furnished in my book, *The US Navy's Secret Space Program & Nordic Extraterrestrial Alliance*.[74]

Secrecy surrounded Project RAND, according to Tompkins, who added, "They had the highest, secret clearance even above the nuclear bomb."[75] This statement about the extraordinary level of security clearance granted to the think tank further illuminates the article by Marx, which exposed the locked rooms and security guards provided for Project RAND. [76] More support for why the heightened security and clearance blanketing Project RAND was necessary while it was located at Douglas is found in a document written by a senior radio engineer with the Canadian Department of Transportation.

0020133

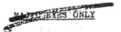 EYES ONLY

PART III  SCIENTIFIC PROBABILITIES

1.  Based on all available evidence collected from
recovered exhibits currently under study by AMC, AFSWP, NEPA,
AEC, ONR, NACA, JRDB, RAND, USAAF SAG, and MIT, are deemed
extraterrestrial in nature.  This conclusion was reached
as a result of comparisons of artifacts ███████████
discovery in 1941.  The technology is outside the scope of
US science, even that of German rocket and aircraft development.

2.  Interplanetary space travel is possible provided
adequate  funding, necessary resources are made available,
and national interest is piqued.

3.  Our solar system is not unique.  Chances are
favorable for intelligent life on other planets notwithstanding
similar development not unlike our own.

4.  Being that our culture is relatively young (in
relation to the cosmic scale), it is possible that other
cultures may have developed faster, or are much older and
have avoided the pitfalls common in our historical and
scientific development.

5.  Human origins may not be constrained to one
planet.  Our genus may be found among solar systems similar
to our own.

6.  The laws of physics and genetics may have a
genesis in a higher, structured order than once previously
thought.
MAJIC EYES ONLY

6

Figure 9. Leaked "White Hot Report" dated September 19, 1947.
Source: The Majestic Documents

Dated November 21, 1951, Wilbert Smith wrote to the
Controller of Telecommunications concerning flying saucer
technologies being secretly studied in the U.S. at the time: "The
matter is the most highly classified subject in the United States

government, rating higher than the H-bomb."[77] Thus, Project RAND's secret study of extraterrestrial technologies clearly merited the careful and delicate handling it received.

## Struggle for US Air Force Supremacy in Space

Publicly, Project RAND's cover mission was to provide a "technical consultant group charged with operations analysis and long-range planning to examine future warfare and the best way the Air Force could perform its missions."[78] However, Project RAND's highly classified real mission was to help the US Army Air Force develop the necessary strategic scientific planning that would be essential for the research and development of advanced aerospace technologies, especially those using antigravity and other exotic propulsions systems. This was vitally important to the US Army Air Force's advancement of long term plans when it came to UFOs that were either of interplanetary origin or linked to Nazi advanced technology projects developed and deployed out of Antarctica. The dark web of Nazi exploits in Antarctica is described in detail in book three of this Secret Space Program series, *Antarctica's Hidden History: Corporate Foundations of Secret Space Programs.*

Significantly, General Arnold, the man who provided the initial funding to set up Project RAND, was also a key figure in the creation of the Interplanetary Phenomenon Unit that was established to deal with UFO crashes immediately after the Los Angeles incident. As the Commanding General of the Army Air Force, he was able to provide personnel with the necessary technical expertise for the secretive Army intelligence unit specially created to investigate reports of aircraft crashes that were, in fact, interplanetary UFOs, or other forms of advanced aerial technologies.

Arnold's involvement establishes an important connection between the formation of Project RAND and the Interplanetary Phenomenon Unit. This clear link supports Tompkins' claim that

Project RAND emerged out of an informal study group to examine the Los Angeles UFO incident in 1942, and held from the beginning the secret mission to help develop a long-term coordinated policy on the UFO phenomenon and extraterrestrial life.

Through General Arnold, funding for Project RAND was provided by the US Army Air Force, and in October 1945 he further arranged for the RAND contract to come under the purview and responsibility of the newly created Army position of "Deputy Chief of Air Staff for Research and Development". Major General Curtis E. LeMay was appointed to head up the new research position. In considering who would get the position, the appointment board had recommended "an officer of the caliber of Curtis LeMay."[79]

In September 1945, a month prior to Project RAND's official opening, LeMay was briefly put in charge of the Army Air Force research facilities at Wright Field in Dayton, Ohio where he could familiarize himself with the retrieved saucer craft from the Los Angeles Air Raid, along with anything else recovered by the Interplanetary Phenomenon Unit. Over the several weeks of commanding Wright Field's advanced research laboratories, LeMay would come to fully appreciate the R&D challenges that lay ahead for the Army Air Force, not only with non-terrestrial technologies but also with the advanced Nazi antigravity craft secretly brought into the U.S. under Operation Paperclip.

According to an army cryptologist, known as Agent "Kewper", who had been recruited by the CIA in 1958 to work on the UFO topic, he personally saw four Nazi German flying saucers that had been brought into the U.S. under Operation Paperclip. Veteran UFO researcher Linda Moulton Howe first interviewed him in 1998. She was convinced that Kewper who was also identified by the name Agent "Stein" (two pseudonyms that he used), was a credible eyewitness to the events he disclosed.[80] By 1958, the craft had been moved from Wright Field (later renamed Wright-Patterson AFB in September 1947) to the S-4 facility at Area 51 for storage and further study, which is where he sighted them. In an interview with Howe, Kewper/Stein discussed the different types of German craft he witnessed:

At Area 51, the first two craft we saw looked almost identical. They were smaller, not nearly as big as one in the back. Col. Jim [USAF; official "tour guide" at Area 51/S4] mentioned that those two were 'Vril craft.' We asked him what 'Vril' was. The Col. said it was a foreign saucer built in Germany in the 1920s and 1930s. Then he pointed up ahead and said the next three craft were alien (extraterrestrial) craft retrieved from New Mexico. There were three more in the back and they were huge, all sitting on metal sawhorses or stands to keep them off the ground. The disc on the very end was a huge one and Col. Jim said that was a German WWII craft built in 1938 and was jacked up higher on stands because it had a gun emplacement underneath, which he said the Germans called a 'death ray'. It was a different shape than the other craft, was dark in color and had a larger top that stood up probably 10 or 12 feet above the saucer. That one had a diameter of about 50 or 60 feet.[81]

Kewper/Stein went on to describe what he had been told about the locations where the four German flying saucer craft had been retrieved:

The other four Col. Jim said were picked up in foreign areas. He said that the German Messerschmitt Haunebu 1 and 2 were picked up in Germany. The other smaller two Vril craft I would think were picked up in Germany as well, but Col. Jim acted like he did not know where those two craft were retrieved.[82]

Over the years ahead, Wright Field/Wright-Patterson AFB would receive other craft retrieved by the Interplanetary Phenomenon Unit, including the July 1947 Roswell crash, and the less known but much more significant 1948 Aztec crash, as I will later explain.

LeMay was reassigned to Washington DC in October 1945 where he took up the position of deputy chief of Air Staff for Research and Development. With his brief exposure to the retrieved flying saucers at Wright Field, LeMay was ready to take charge of the Air Force's secret effort of reverse-engineering exotic propulsion systems. In his new leadership role, LeMay was put in charge of Operation Paperclip under which German scientists had been brought into the U.S. to help the Air Force understand how advanced propulsion systems and flying saucer technologies worked.[83]

Among the German scientists brought to Wright Field due to their involvement in flying saucer research was Professor Winfried Otto Schumann who had worked on developing the prototype Vril flying saucer based on the highly respected German medium Maria Orsic's telepathic communications.[84] Documents released under the Freedom of Information Act confirm that Schumann was on a list of German scientists, requested by the Army Air Force, who possessed the skills relevant to its classified aerospace projects.[85] It can be concluded that scientists such as Schuman helped LeMay and the Army Air Force understand the principles and concepts behind the Nazi flying saucer projects.

Barrett Tillman, author of the biography *LeMay*, wrote about how LeMay would respond to the revolutionary potential of the advanced propulsion systems that were being studied at Wright Field and their relevance to the future of the aerospace industry:

> Though he had arisen to prominence on reciprocating [piston] engines, LeMay recognized the immense potential of new propulsion systems. But apart from jets and rockets, he was also receptive to other innovations such as satellites. LeMay's scientific consultants convinced him that it was possible to place a satellite in earth orbit, but in 1946 the cost would have been astronomical. Amid the greatest downsizing in military history, millions

for space exploration were simply not a political reality.[86]

While Tillman had no idea of the non-terrestrial technologies being studied at Wright Field secretly, his comments show why LeMay was the perfect man to provide Project RAND with the necessary information and access to the Air Force's treasure chest of captured extraterrestrial and Nazi flying saucer technologies. LeMay's participation was critical to the success of Project RAND's real mission: providing a long-term research and development program for acquired extraterrestrial technologies. For his leadership during the formative period of Project RAND, from its October 1945 inception until his reassignment to Europe in October 1947, LeMay has been publicly acknowledged, along with General Arnold and Donald Douglas, as one of the co-founders of Project RAND. According to Alex Abella, author of *Soldiers of Reason: The RAND Corporation and the Rise of the American Empire*, "If Arnold and his group were the founding fathers of RAND, there is no doubt that LeMay was its Godfather".[87]

The fact that only Army Air Force officials have been acknowledged by RAND historians in its establishment is a reflection of the fact that initial surplus funds were provided by General Arnold. It is also due to the Navy's downplaying of its interest in flying saucer and antigravity research both during and after World War II. Yet the Navy, according to Tompkins, immediately began working behind the scenes with the Army Air Force and Douglas Aircraft engineers, through Douglas' informal study group, on preliminary studies of the extraterrestrial phenomenon. At the same time, the Navy ran its covert intelligence program with operatives working within Nazi Germany's flying saucer programs.

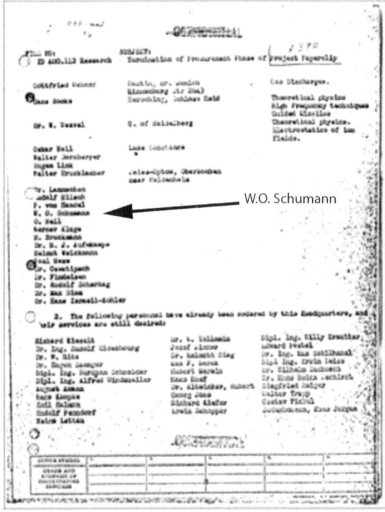

Second page of a three page declassified Operation Paperclip Memorandum dated 6 June 1947. This is a list of of German scientists requested by the U.S. Army Air Force for classified research at its Dayton, Ohio facilities. The appearance of Schumann's name is evidence that after his debriefing in post-war Germany, his expertise in aerospace projects was deemed important for the Army Air Force's classified foreign technology research. Source: Richard Sauder, *Hidden in Plain Sight* (2011).

**Figure 10. Document revealing Schumann was part of Operation Paperclip**

Quickly, the Navy realized that the Army Air Force planned to assert its primacy when it came to the newly emerging field of space surveillance operations in near-earth orbit. The Army Air Force viewed future space operations as a domain it would dominate in the future. Through Project RAND, the secret scientific and military study of the flying saucer phenomenon would serve to buttress Army Air Force efforts to exclude the Navy from playing a significant role in such impending space surveillance operations. A brief but staunchly heated policy battle arose between the Army Air Force and the Navy when it came to the development of the first system of space surveillance satellites. The Navy wanted to cultivate an inter-service program, but the Army Air Force was not interested. US Air Force historian Lt. Col. Mark Erickson wrote:

> [A] 1945 Navy investigation concluded that "in view of the recent progress in the field of rocket missiles it may prove advantageous to review the possibility of establishing a space ship in an orbit above the surface of the earth.... This orbit may prove more desirable for communications or for scientific observations." However, when in March 1946, the Navy requested the Army Air Forces (AAF) join its satellite studies, the AAF concluded that a "joint program of evaluation, justification, and, if warranted, construction and operation ... was not agreeable," and as a result the services "would conduct separate investigational programs."[88]

The RAND Corporation website provides a succinct explanation for what drove the Air Force to go it alone when it came to future space operations: "Major General Curtis E. LeMay, then Deputy Chief of the Air Staff for Research and Development, considered space operations to be an extension of air operations."[89]

LeMay commissioned a study by Project RAND that was published in May 1946 and titled "Preliminary Design of an Experimental World-Circling Spaceship".[90] The RAND study was used to justify Army Air Force primacy when it came to overall military responsibility for space surveillance operations:

> The project arose out of the Air Force's interest in developing intercontinental ballistic missiles. LeMay requested that RAND scientists prepare the study quickly because the Navy Bureau of Aeronautics was already working with Wernher von Braun and other captured Nazi scientists on a similar rocket project. LeMay wanted to outmaneuver his rivals and preserve the Air Force's exclusive right to the military uses of space. Within a month, RAND's four employees, with the help of consultants, wrote a farseeing report, breathtaking in its intellectual daring and self-assured to the point of arrogance. "Preliminary Design of an Experimental World-Circling Spaceship" was the world's first comprehensive satellite feasibility assessment.[91]

The RAND study led to the Navy losing the policy struggle with the Army Air Force over jointly developing a space satellite reconnaissance system. As Abella put it, "The oftentimes lyrical study served its purpose well and the Navy project was shelved." As Air Force chief of staff General Hoyt Vandenberg consequently put it, the Air Force had "logical responsibility for satellites".[92] Subsequently, as Erikson pointed out, "the Navy dropped its satellite studies on 22 June 1948 after the US Air Force refused to join."[93] Subsequent RAND publications focused on the Army Air Force, and later the US Air Force (formed September 1947), conducting space surveillance and in the development of the advanced propulsion systems needed for space flight.

Undaunted, the Navy secretly continued its own research and development work on reverse engineering the recovered extraterrestrial craft from the Los Angeles Air Raid and the Nazi flying saucer prototypes acquired at the end of the war. These advanced craft were covertly brought into the US on Navy ships and taken to classified facilities such as the Naval Research Laboratory (NRL) in Washington DC.

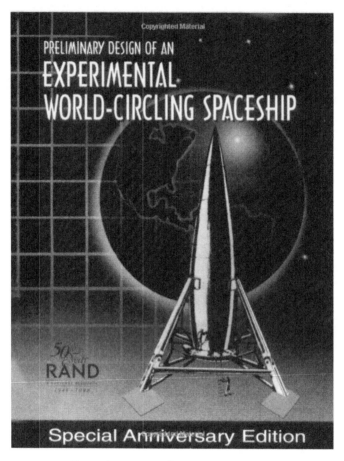

**Figure 11. RAND Preliminary Design Study**

The Navy's long term strategic interest was in developing space battle groups, and not the space surveillance systems for near-earth orbit which the Army Air Force was focused upon.[94] The principal locations the Navy chose for conducting its examination of flying saucers technologies were the Naval Research Laboratories and its flagship R&D location on the West Coast, the China Lake Naval Air Weapons Station. Tompkins said that during World War II, he flew multiple missions to China Lake with the briefing packets he had prepared from the debriefings of Navy spies.[95] This reflects the Navy's sustained interest in understanding extraterrestrial technologies after the recovery of the alien craft from the Los Angeles incident and its broader goals for future naval operations in deep space.

Due to Project RAND, all of the general public's attention fell upon the Army Air Force when it came to future space operations and the UFO phenomenon. This escalated after the July 1947 Roswell incident when the Army Air Force announced that a flying saucer had crashed, only to retract the announcement hours later.[96] The effect of giving the public the impression that the Army Air Force was covering up the flying saucer phenomenon was laid. In this way, public attention was firmly directed away from what the Navy was surreptitiously doing at its R&D facilities.

# CHAPTER 4

# The Roswell Crash and the Creation of MJ-12

Good leaders need a positive agenda, not just an
agenda of dealing with crisis.

– Michael Eugene Porter

Only two great military forces of land and sea existed in America to protect its vast shores and citizenry upon its entry into World War II. The oldest branch, the Department of War, coordinated all Army operations while the Department of the Navy oversaw naval fleet activities. Both were using aircraft for tactical troop or ship support. However, throughout this demanding time of conflict, the aviation industry produced innovative planes capable of far more. Thus, during and immediately after World War II, the Army Air Force came to operate as an independent military service under its Commanding General Henry H. Arnold who was fourth in the esteemed senior hierarchy of the Army's five-star generals at the time. On January 29, 1946, Arnold and his successor, General Carl Spaatz, reached an agreement with the Army Chief of Staff, Dwight D. Eisenhower, to formally separate the Air Force from the Army, thereby establishing the Air Force and the Army as independent military services within the Department of War.

The arrangement was analogous to the US Marine Corps

and the US Navy being independent military services within the Department of the Navy, an arrangement that continues to the present day. Soon after, however, the National Security Act of 1947 went a step further by formally separating the Air Force from the Army with the creation of the Department of the Air Force. Post-war political posturing and heated military rivalries underscored this change, and events took place over the strong objections of the Navy.

> "As the President knows," Secretary of the Navy Forrestal said, "I am so opposed to the fundamental concept expressed in the message [from President Truman to create a Department of the Air Force] that I do not believe there is any very helpful observation that I could make." The Navy's leadership remained unreconstructed, fearful that an independent Air Force would make a grab for naval aviation and that the Army might even attempt to take over the Marine Corps.[97]

Consequently, Forrestal had to be reassured that the Navy would remain autonomous before the Department of the Air Force was formally established on September 18, 1947. The restructuring reflected peacetime agendas looking to the future, thus launching two mighty equals to the Department of the Navy through the Department of War's successors. Soon after, the joint National Military Establishment consisting of the three departments would be renamed the United States Department of Defense in 1949.

During this critical transition period resulting in the emergence of the US Air Force (USAF), the research and development of retrieved extraterrestrial craft (commonly called "flying saucers") technologies secretly continued under the authority of the Commanding General of the Army Air Force. On February 28, 1946, Arnold left active duty (he would officially retire on June 30, 1946) and was replaced by General Spaatz. Taking over his predecessor's oversight responsibilities, Spaatz took a keen interest in what Major General Curtis LeMay was discovering as

Deputy Chief of the Air Staff for Research and Development. It was LeMay who had arranged for top German scientists to be brought to Wright Field to study the craft retrieved from the Cape Girardeau and Los Angeles incidents. These former Nazi German scientists helped the Air Force and Douglas Aircraft scientists from Project RAND to understand the feasibility of exotic propulsion systems based on antigravity and torsion field physics.

Meanwhile, the "Interplanetary Phenomenon Unit", which had been created within the Army's Counter Intelligence Corps, would continue to investigate reports of flying saucers and endeavor to retrieve any craft that had crashed. Army Intelligence (G2) therefore had to cooperate with the Army Air Force in its secret R&D efforts. The Navy, in turn, continued its research based on what had been recovered in the Los Angeles incident, along with its intelligence gained by Navy spies about the Nazi flying saucer programs. In July 1947, another flying saucer crash occurred, and this time, the world would learn the truth – albeit for a short time.

## The Truth & Cover-Up of the Roswell Crash

On a stormy July 3 night in 1947, ranchers reported hearing a loud explosion unlike the many thunderclaps taking place that evening. Among the locals was W. W. "Mac" Brazel, a foreman working on a ranch approximately 75 miles northwest of Roswell, in Lincoln County, New Mexico. The next day during Brazel's rounds on the property, he found the wreckage of what appeared to be a crashed aerial vehicle of unknown origin. On July 7, he decided to report the find to the Sheriff's office, who in turn notified the nearby Roswell Army Air Field (RAAF) base which immediately sent two military intelligence officers to investigate. The result of the initial military investigation led to the famous press release on July 8 that a "flying saucer" had crashed. This generated instant media interest around the world. Escalating coverage pressured another press release to be issued, this time by more senior military authorities promptly claiming the initial release was mistaken.

Army Air Force brass offered the now markedly tarnished explanation; *it was only a weather balloon!* Backed at the time by people's faith in the presiding powers, the sensation of a flying saucer from another world was crushed. Therefore, interest in the Roswell story collapsed, and it wasn't heard of again for another thirty years.

Then in 1978, one of the two military officers that conducted the initial investigation, Major Jesse A. Marcel, (Ret.), formerly of the 509th Bomb Group Intelligence Office at the RAAF, contacted Stanton Friedman, a veteran UFO researcher, and told him the truth about events at Roswell. According to Marcel, the crash debris contained material that was not recognizable to aviation experts, "It was nothing we had ever seen before."[98] It appeared indestructible, had strange hieroglyphic writing, and had other unearthly properties. He stated for the public record: "It was not an aircraft of any kind, that I am sure of. We didn't know what it was. It was nothing made on this Earth."[99]

Marcel had taken some of the debris home on the way to the base to briefly show his wife and son, Jesse Marcel Jr., who later in life authored a book about what his father had shown him in the early morning hours of July 7.[100] Marcel Jr. eventually attained the rank of colonel and became a medical doctor with the US Army. He agreed with his father, Major Marcel, that the Roswell wreckage "was nothing made on this Earth". A flying saucer had indeed crashed at Roswell, and Major Marcel had been ordered to take part in an official cover-up.

The subsequent publication of *The Roswell Incident,* by Charles Berlitz and William Moore in 1980, revealed the testimony of Marcel and other first-hand witnesses, consequently bringing the Roswell story back to the public's attention.[101] Ever since, the 1947 Roswell crash has been on the center stage of public interest generating further investigations, official government and military reports, books and media notice. In 2007, an important book was published containing a notarized affidavit by another key military official involved in events at Roswell.

Lieutenant Walter Haut became the Public Affairs Officer

for Roswell Army Air Field in 1947. He had been ordered to distribute the original July 8 press release, claiming a flying saucer crash had taken place which led to global headlines. Haut's affidavit was written in December 2002 and authorized for release after his death.[102] In December of 2005, he died, and two years later the affidavit was published as the final chapter in *Witness to Roswell* by Thomas Carey and Donald Schmitt.[103]

The affidavit exposed startling information about events occurring just before and after the Army's initial press release of the flying saucer crash. The most important is that a staff meeting occurred on the morning of July 8, when Roswell Army Air Field officers discussed how to deal with the growing public and press interest in the wreckage found at *two* crash sites. Haut stated: "The main topic of discussion ... was an extensive debris field in Lincoln County approx. 75 miles NW of Roswell."[104] This was the crash site on "Mac" Brazel's ranch that Major Marcel, the Roswell base's chief intelligence officer, traveled to investigate on July 6 and 7. Marcel gave his report at the meeting. The base commander, Colonel William Blanchard, also gave a brief report on wreckage found at the second site. This second site was 40 miles north of Roswell, and any information on it was kept under wraps. Thus, it was not generally known to researchers of the Roswell crash.

The most surprising fact about the staff meeting is that General Roger Ramey, whose headquarters was Carswell Army Air Field, Fort Worth, Texas, was also present. Haut revealed that Ramey devised a strategy for throwing the public and press off track about the two crash sites, and stated:

> One of the main concerns discussed at the meeting was whether we should go public or not with the discovery. General Ramey proposed a plan, which I believed **originated from his bosses at the Pentagon** [emphasis added]. Attention needed to be diverted from the more important site north of town by acknowledging the other location ["Mac"

Brazel's ranch]. Too many civilians were already involved and the press already was informed. [105]

Ramey approved a press release pointing to the more remote and less important site near Mac Brazel's ranch. Haut wrote in his affidavit:

> At approximately 9:30 am, Col Blanchard [Roswell AAF Base Commander] phoned my office and dictated the press release of having in our possession a flying disc, coming from a ranch northwest of Roswell, and Marcel flying the material to higher headquarters. I was to deliver the news release to radio stations KGFL and KSWE, and newspapers, the Daily Record and the Morning Dispatch.[106]

Later that afternoon, General Ramey retracted the "flying disc" announcement with the weather balloon story that appeared in news reports on the evening of July 8 and the morning of July 9. This strategy succeeded in taking the flying saucer story off the news headlines, and confusing members of the public and press that had witnessed or were investigating the events.

General Ramey's role in the cover-up is significant. It was Ramey who ordered Major Jesse Marcel to fly to Fort Worth to appear at a press conference. Marcel was ordered to be photographed crouching quietly over what appeared to be material from a weather balloon. This was used to buttress Ramey's claim that the Roswell wreckage was really a misidentified common "Rawin" weather balloon.

In his affidavit, Walter Haut claimed that he was taken later on July 8 to one of the Roswell hangars by Colonel Blanchard where he saw part of the wreckage that was relatively intact – it was about 15 feet long and 6 feet high.[107] Under canvas tarpaulin, Haut also saw two bodies of the victims with their heads sticking out, who appeared about the size of a 10-year old child. He said that at "a later date in Blanchard's office, he would extend his arm about 4 feet above the floor to indicate the height."[108]

Figure 12. Major Jesse Marcel crouches over weather balloon remains.

Haut's affidavit indicates that Ramey was operating under orders by the Pentagon, which had been briefed about the two crash sites. It is clear that their staff meeting was focused on controlling the press and any public interest in the crash sites. It is highly significant that no discussion occurred over security supervision at the two sites, or for the retrieval of crashed materials from the second crash site. Indeed, the only information was that the second crash site had been witnessed by Col Blanchard, but was not known to other Roswell Army Air Field officers before the Roswell AAF staff meeting on July 8. This is very revealing. Even given its proximity and resources, Roswell Army Air Field would *not* play a prominent role in retrieval operations at the mysterious second site.

Haut's testimony divulges that security and retrieval operations at the more secretive crash site were being supervised at a higher level, utilizing personnel other than the general staff at Roswell AAF. The most plausible explanation is that the Pentagon had deployed its own specialist teams to control security and retrieval operations at the more sensitive site. However, Roswell AAF would find itself used to supply manpower and resources, primarily to throw the public off the trail with contradictory press releases. All of this is evidence that the investigatory team used at the second site was a highly specialized covert team – the Interplanetary Phenomenon Unit – whose report of the incident was leaked over five decades later, and which I will shortly discuss.

As to where the Roswell flying saucer debris and its occupants were taken, a number of witnesses point to them being removed to Wright Field, now the headquarters of the newly created Air Materiel Command (1946-1961). The crash site material was subsequently transported by both air and land routes. Don Schmidt and Ken Carey interviewed multiple firsthand witnesses who were involved with the flight carrying debris to Wright Field:

> The July 8 flight was actually the second confirmed to transport debris out of Roswell since the crash.... The second flight's destination was Wright Field, after a brief stop at Fort Worth. Piloting the plane was Roswell's deputy base commander, Lt Col, Payne Jennings.[109]

An eyewitness to the overland transport of the Roswell material is Lt Colonel Philip Corso, a former US Army intelligence officer who was stationed at Fort Riley, Kansas, at the time. He observed a military convoy do an overnight stopover on its way to Wright Field a few days after the crash:

> Whatever they'd crated this way, it was a coffin, but not like any coffin I'd seen before. The contents, enclosed in a thick glass container, were submerged in a thick light blue liquid ... the object was floating, actually suspended, and not sitting on the bottom with a fluid over top, and it was soft and shiny as the underbelly of a fish. At first I thought it was a dead child they were shipping somewhere. But this was no child. It was a four-foot human-shaped figure with arms ... thin legs and feet, and an oversized incandescent light bulb-shaped head that looked like it was floating over a balloon gondola for a chin....

> I looked through the crate encasing the container of liquid for any paperwork or shipping invoice or anything that would describe the nature or origin of this thing. What I found was an intriguing Army Intelligence document describing the creature as an inhabitant of a craft that had crash landed in Roswell, New Mexico, earlier that week and a routing manifest for this creature to the login officer at the Air Materiel Command at Wright Field ...[110]

Corso's military records show he was stationed at Fort Riley, Kansas, from April 21, 1947, to May 12, 1950, with the rank of Major.[111] Events he described as the duty officer on the night the extraterrestrial body arrived at Fort Riley en route to Wright Patterson AFB are, therefore, corroborated by his military record. Further testimony supporting Corso's account comes from Leonard Stringfield. He was the first UFO crash retrieval researcher to investigate this incident, and he was told that the extraterrestrial bodies found at Roswell were taken to Building 18F at Wright Patterson and kept in a "holding tank" for a period of time.[112]

Another source confirming Corso's description of events is the former US Army Signals Corps officer, known by the two pseudonyms "Kewper" and "Stein", who was recruited to work with a covert CIA unit from 1957 to 1960. As an agent with the CIA unit, he tracked reports of UFO crashes and sightings around the world, and stated in an interview:

> We were told that the original debris for a short period of time was at the Roswell Army Air Field. From there, it went to Fort Riley, Kansas, where it was for a week. The government was trying to figure out what to do with it and where best to work on such a classified object. It was wrapped up on a truck under camouflage, so no one would know what it was.
>
> And then it went to Ohio to Wright Field outside of Dayton, Ohio (became Wright-Patterson AFB). It was there for about a year and some of the scientists from the area the government brought in to look at.[113]

Due to the multiple first-hand witness accounts of events at Roswell, including the primary military officials involved in reporting and covering up key facts, it can be concluded that an extraterrestrial craft had indeed crashed, and the initial newspaper report was correct. Interplanetary visitors had visited the Earth, and the Army Air Force had ordered a cover-up at the highest levels.

One may wonder why an extraterrestrial intelligence would choose New Mexico in which to perform some type of clandestine surveillance or reconnaissance. Why were they there? The simplest answer is that in 1947, Roswell Army Air Field was the home of the 509th Bomb Group. The 509th was the bomb group that released the atom bombs on Japan in 1945. Roswell was the only atomic bomber base in the world. New Mexico was also the location of White Sands Proving Grounds and the Los Alamos National Laboratory (LANL) where the atomic bomb was originally developed and then tested at Trinity Site. We can get more answers to these questions and learn more details about what happened at Roswell, along with the subsequent Army Air Force efforts to study the retrieved craft, from a number of the Majestic Documents.

## Majestic Documents on the Roswell Crash

Immediately after the 1947 events at Roswell, the Commander of the Air Materiel Command, Lt General Nathan F. Twining, received two virtually identical "Top Secret Eyes Only" directives. One was from the Army Chief of Staff Dwight D. Eisenhower and another from President Truman, consecutively on July 8 and 9, instructing Twining to travel to White Sands Proving Ground to oversee the recovery efforts of the Roswell crashes.[114] Truman's "Directive to General Twining" states:

> You will proceed to the White Sands Proving Ground Command Center without delay for the purpose of making an appraisal of the reported unidentified objects being kept there. Part of your mission there will deal with the military, political and psychological situations current and projected. In the course of your survey you will maintain liaison with the military officials in the area....
>
> You will take with you such experts, technicians, scientists and assistants as you deem necessary to the effectiveness of your mission.[115]

Upon completing his mission, General Twining wrote a report on July 16, 1947, titled: "Air Accident Report on 'Flying Disc' Aircraft Near the White Sands Proving Ground, New Mexico" to the

Commanding General of the Army Air Force Carl Spaatz which was classified Top Secret. In this report, Twining detailed the role of the scientific personnel he brought with him from Wright Field that conducted the preliminary investigation of the material retrieved from the two flying saucer crash sites:

> 1. As ordered by Presidential Directive, dated 9 July 1947, a preliminary investigation of a recovered "Flying Disc" and remains of a possible second disc, was conducted by the senior staff of this command. The data furnished in this report was provided by the engineering staff personnel of T-2 and Aircraft Laboratory, Engineering Division T-3. Additional data was supplied by the scientific personnel of the Jet Propulsion Laboratory, CIT [California Institute of Technology] and the Army Air Forces Scientific Advisory Group, headed by Dr. Theodore von Karman. Further analysis was conducted by personnel from Research and Development.[116]

Next, the initial analysis given by his scientific team about the retrieved craft confirms that it came from outside the United States:

> It is the collective view of this investigative body, that the aircraft recovered by the Army and Air Force units near Victoria Peak and Socorro, New Mexico, are not of US manufacture...[117]

The report goes on to explain the reasons why the craft is not of U.S., German, or Russian origin, and further describes its unique propulsion, navigation, and piloting features. Significantly, Twining's report states:

> The absence of canopy, observation windows/blisters, or any optical projection, lends support to the opinion that this craft is either guided by remote viewing or is remotely controlled.[118]

Only three pages of the "Air Accident Report" were leaked, so it is unknown what final conclusions were reached. One thing is clear though from the analysis: the retrieved craft was of a highly advanced design and would require years of research and development efforts by specialists at the Air Materiel Command.

Dr. Robert Wood and Ryan Wood have categorized the two Twining Directives as "Medium Level of Authenticity".[119] They gave the Air Accident Report a higher ranking of "Medium to High Level of Authenticity". An officially declassified Air Force document further supports the authenticity of all three leaked documents concerning Nathan Twining and the Air Materiel Command investigation. On September 23, 1947, General Twining wrote a memorandum to Brigadier General George Schulgen on the extraterrestrial craft which was classified Secret. Twining told Schulgen:

> a. The phenomenon is something real and not visionary or fictitious.
> b. There are objects probably approximating the shape of a disc, of such appreciable size as to appear to be as large as man-made aircraft...
> d. The reported operating characteristics such as extreme rates of climb, maneuverability (particularly in roll), and motion which must be considered evasive when sighted or contacted by friendly aircraft and radar, lend belief to the possibility that some of the objects are controlled either manually, automatically or remotely.[120]

The reference to the recovered craft being "controlled either manually, automatically or remotely" is very similar to the conclusion reached in the leaked "Air Accident Report" where Twining referred to the craft as being "guided by remote viewing or is remotely controlled".[121] Consequently, the officially declassified September 23 Memorandum to Schulgen leaves little doubt about the authenticity of the leaked September 16 "Air

Accident Report", and substantiates the September 8 and September 9 directives sent to Twining by General Eisenhower and President Truman.

While Twining's September 23 "Schulgen Memo" did not specifically mention the Roswell crash, which was classified "Top Secret Eyes Only", it shows that he was making subordinates aware of the reality of UFOs and the need to study reported sightings made by the general public in order to better understand their "non-domestic" origins.[122] So far, the contents of the documents presented focus upon the technical aspects of the craft themselves, and do not refer to the recovery of extraterrestrial beings. A report by the Interplanetary Phenomenon Unit, however, does refer to the recovery of extraterrestrial bodies at one of the sites associated with the Roswell crash.

The report is titled "Interplanetary Phenomenon Unit Summary" and is categorized as "Medium-High Level of Authenticity" by the Woods.[123] The IPU Report begins:

> 1. The extraordinary recovery of fallen airborne objects in the state of New Mexico, between 4 July – 6 July 1947: This Summary was prepared by Headquarters Interplanetary Phenomenon Unit. Scientific and Technical Branch, Counterintelligence Directorate, ... at the expressed order of Chief of Staff.[124]

It is worth noting that Eisenhower was the Army Chief of Staff at the time. The report goes on to describe materials in two location sites, which corroborates what Lieutenant Haut and other witnesses were told:

> 2... Two crash sites have been located close to the WSPG [White Sands Proving Grounds] Site LZ-1 was located at a ranch near Corona, approx. 75 miles northwest of the town of Roswell [the crash site reported by Brazel]. Site LZ-2 was located approx. 20 miles southwest of the town of Socorro, at Lat. 33-

40-31, Long 106-28-29, with Oscura Peak being the geographic reference point.

3. The AST personnel were mainly interested in LZ-2 as this site contained the majority of structural detail of the craft's airframe, propulsion and navigation technology. The recovery of fives bodies in a damaged escape cylinder, precluded an investigation at LZ-1.

4. On arrival at LZ-2, personnel assessed the finds as not belonging to any aircraft, rocket, weapons, or balloon test that are normally conducted from surrounding bases. [125]

This data is consistent with the conclusion in General Twining's "Air Accident Report" that the crashed craft was not of "U.S. manufacture".[126] The IPU report also describes how the debris and bodies were taken to Wright Field, and other secure Air Force facilities for closer study:

7. ... CIC [Counter Intelligence Corps] member of the team was able to learn that several bodies were taken to the hospital at Roswell AAF and others to either Los Alamos, Wright Field, Patterson AAF, and Randolph Field for security reasons.... Remains of the powerplant were taken to Alamogordo AAF and Kirtland AAF. Structural debris and assorted parts were taken to AMC [Air Materiel Command] Wright Field. [127]

Another significant Majestic document concerns a two-page, July 18, 1947, memorandum from Lt General Twining to Major General Curtis LeMay on the creation of a new laboratory at Wright Field, only two weeks after the Roswell crash. The memorandum is categorized as "Medium-High Level of Authenticity" by the Woods.[128] Here is how they summarize its contents and significance:

This two page memo from Lt General Nathan Twining to Curtis LeMay activates a new laboratory to conduct meteorological research and development and upper air research with the Electronics Subdivision of the Engineering Division of the AMC. It states in part, "In view of the close relationship and interdependence of research in meteorology and research in electromagnetic compressional wave propagation, action is being taken to reorganize the present Applied Propagation Laboratory of Watson laboratories in the Atmospheric Laboratory, and expand its functions to include research and development in meteorology and related geophysical fields." Later the memo goes on to state that "funds requested for F.Y. 1949 Project 680-11, Atmospheric Research and applied scientific research of the upper atmosphere, a total sum of $6,000,000 has been specified..." What a huge sum of money in 1949 to study the "upper atmosphere;" a more logical interpretation is that we are analyzing flying saucers, their technology, why they are here, and what are we going to do about it. [129]

The Twining-LeMay memorandum shows how the Army Air Force, two months before the official creation of the USAF, had set up a new laboratory at Wright Field to study artifacts from the Roswell crash. The memorandum is documentary evidence identifying that Twining and LeMay were directly involved in the secret research and development of the advanced technologies found in the Roswell crash. It is important here to keep in mind that Twining was the commander of Air Materiel Command headquartered at Wright Field, and LeMay was Deputy Chief of Air Staff for Research and Development. Therefore, the memorandum confirms that the Army Air Force was moving into high gear in its secret study of the craft retrieved from Roswell.

## Roswell Event Leads to MAJIC

Roswell was the decisive catalyst for the US government to finally take steps to coordinate military and intelligence efforts to get to the bottom of the UFO phenomenon. Previously, the Army's Counter Intelligence Corps (through its Interplanetary Phenomenon Unit), the Navy's Office of Naval Intelligence (through its espionage program out of Naval Air Station San Diego), and the Army Air Force's research and development efforts at Wright Field were all separate military endeavors to understand the UFO phenomenon after its emergence with the 1941 Cape Girardeau and 1942 Los Angeles events. However, by mid-1947, the Army, Navy, and Air Force were competing military services vying for control and funding resources over the UFO phenomenon and the revolutionary potential it held for future combat operations. Effectively, there was no comprehensive government oversight or coordination for what the military services were individually doing.

To recap, Lt General Nathen Twining was ordered by General Eisenhower (July 8) and President Truman (July 9) to lead an Army Air Force investigation of the Roswell incident. After writing the "Air Accident Report" describing the technical aspects of the retrieved craft (on July 16), Twining then ordered Major General LeMay to set up a new laboratory at Wright Field (on July 18). Next, Twining would lead a comprehensive interdepartmental investigation into the intelligence, technical, scientific, political, and national security implications of the recovered Roswell artifacts. On this front, the document trail continues with Twining's "White Hot Report", dated September 19, 1947, (a day after the official creation of the USAF), which has notably been given the highest of authenticity ratings by the Wood team for a leaked Majestic document.

Consequently, the Twining "White Hot Report" provides stunning insights into the thinking within the Truman White House and the newly created Department of Defense on how to deal with the recovered crash artifacts, which were understood by US authorities to be extraterrestrial in origin. Here's how Twining's team, comprised of sixteen leading scientists, national security experts, and senior military officers, summarized the national security situation shortly after the Roswell incident:

> In the early months of 1942 [LA incident], up until the present, intrusions of unidentified aircraft have occasionally been documented, but there has been no serious investigations by the intelligence arm of the Government. Even the recovery case of 1941 did not create a unified intelligence effort to exploit possible technological gains with the exception of the Manhattan Project. We now have an opportunity to extend our technology beyond the threshold that we have achieved, **by directly working with the aliens on technology transfer** [bolded section was blacked out in original and has been reconstructed]. Aside from technological gains, we face an even greater challenge, that of learning the intent of such a presence. There are questions that remain unanswered ....[130]

In this report, Twining's team identifies the national security challenges involved in establishing a future working relationship with extraterrestrials in order to achieve significant technological advances. The overwhelming scope of the challenges posed in researching and developing extraterrestrial technologies led to Twining's team recommending a comprehensive strategic plan:

> The members of the mission are prepared to submit a ... dozen volumes to explain how these problems should be met. Our only point, however, is that a combined intelligence and research operation would be vast, intricate, covertly planned marshalling of resources, human and material, to solve a specific, clearly defined problem.
>
> We have to find effective methods of persuasion

with other government agencies without creating a sense of impending doom. The first task is to carefully appraise the problem. The second is to evaluate the known resources and probable strategy of the visitors. The third is to inventory our own ways and means, ascertain how many resources we can bring to bear, and how fast. The fourth is to devise our strategic plan. And last is to work out with infinite pains the tactical details and the myriad of secondary problems of funding and security.[131]

The "White Hot Report" now makes it clear why the Army Air Force had initially decided to fund Project RAND and work with civilians in developing strategic planning on advanced technology projects and extraterrestrial life. In contrast to the wartime Los Angeles incident, the Roswell Crash served as the necessary catalyst to launch serious research and development efforts into UFO technologies.

Twining's team went on to recommend the establishment of a permanent committee to deal with the myriad of anticipated national security problems that could arise while researching and developing extraterrestrial technologies:

It is the unanimous opinion of the members that Operation MAJESTIC TWELVE be a fully funded and operation TOP SECRET Research and Development intelligence gathering agency. It is also recommended that a panel of experts be appointed to chair and oversee the functions and operations of said agency.[132]

The seriousness with which the recommendations made by Twining's interdepartmental team were taken is reflected in what happened five days after the report was issued. On September 24, 1947, President Truman issued a Memorandum to James Forrestal, the first Secretary of Defense, authorizing "Operation Majestic

Twelve":

> Dear Secretary Forrestal
>
> As per our recent conversation on this matter, you are hereby authorized to proceed with all due speed and caution upon your undertaking. Hereafter this matter shall be referred to only as Operation Majestic Twelve.[133]

It's important to note that Twining's "White Hot Report" was classified "MAJIC Eyes Only". The Woods point out that "MAJIC" is an acronym for "Military Assessment of the Joint Intelligence Committee", the code word coined by the Joint Intelligence Committee which was making recommendations to the president on the extraterrestrial issue.[134] Consequently, when Joint Intelligence Committee members recommended setting up "Operation Majestic Twelve", they were both naming their successor committee "Majestic 12", and formalizing the number of officials that would sit on the new committee. Meanwhile, the "MAJIC" code word would be used as a security classification for the Majestic 12 committee running Operation Majestic Twelve. This brief history helps explain why the term "MAJIC" appears in a number of Majestic documents as both a security code word and also as the name of the group running Operation Majestic Twelve.

The twelve founding members of the Committee are described in the famous "Eisenhower Briefing Document", which is another leaked Majestic document. It is the alleged written briefing given to President Eisenhower on November 18, 1952, less than two weeks after his election. The "Eisenhower Briefing Document" names the principal military and scientific figures involved with the UFO/extraterrestrial issue, and identifies the Majestic 12 (MAJIC-12) Group as the main organization set up to manage the issue. This group comprised some of the most powerful figures of the day in the US national security apparatus. They had been reporting directly to the Office of the President as of September 24, 1947, just as the Truman Memorandum, which was included as an attachment, had stipulated:

> OPERATION MAJESTIC-12 is a TOP SECRET Research and Development/Intelligence operation responsible directly and only to the President of the United States. Operations of the project are carried out under the control of the Majestic-12 (Majic-12) Group which was established by special classified executive order of President Truman on 24 September, 1947, upon recommendation by Dr. Vannevar Bush and Secretary James Forrestal. (See Attachment "A".) Members of the Majestic-12 Group were designated as follows:

> Adm. Roscoe H. Hillenkoetter
> Dr. Vannevar Bush
> Secy. James V. Forrestal
> Gen. Nathan F. Twining
> Gen. Hoyt S. Vandenberg
> Dr. Detlev Bronk
> Dr. Jerome Hunsaker
> Mr. Sidney W. Souers
> Mr. Gordon Gray
> Dr. Donald Menzel
> Gen. Robert M. Montague
> Dr. Lloyd V. Berkner

Veteran UFO researcher Stanton Friedman's groundbreaking study of the Eisenhower Briefing Document uncovered a wealth of evidence that corroborated the designated list of names and the existence of the Majestic 12 (MJ-12) Group. As far as the names are concerned, Friedman found that the 12 figures involved were eminently suited for being on the notable committee that would control the entire UFO/extraterrestrial phenomenon. Coincidentally, the "Eisenhower Briefing Document" was released only three months after the last surviving member of the original group, Dr. Jerome C. Hunsaker, had died (September

10, 1984). This is not an insignificant coincidence and hints at some prior agreement over the public release of such information.

TOP SECRET / MAJIC

EYES ONLY

* TOP SECRET *

COPY ONE OF ONE.

EYES ONLY

SUBJECT: OPERATION MAJESTIC-12 PRELIMINARY BRIEFING FOR PRESIDENT-ELECT EISENHOWER.

DOCUMENT PREPARED 18 NOVEMBER, 1952.

BRIEFING OFFICER: ADM. ROSCOE H. HILLENKOETTER (MJ-1).

NOTE: This document has been prepared as a preliminary briefing only. It should be regarded as introductory to a full operations briefing intended to follow.

* * * * * *

OPERATION MAJESTIC-12 is a TOP SECRET Research and Development/ Intelligence operation responsible directly and only to the President of the United States. Operations of the project are carried out under control of the Majestic-12 (Majic-12) Group which was established by special classified executive order of President Truman on 24 September, 1947, upon recommendation by Dr. Vannevar Bush and Secretary James Forrestal. (See Attachment "A".) Members of the Majestic-12 Group were designated as follows:

> Adm. Roscoe H. Hillenkoetter
> Dr. Vannevar Bush
> Secy. James V. Forrestal*
> Gen. Nathan F. Twining
> Gen. Hoyt S. Vandenberg
> Dr. Detlev Bronk
> Dr. Jerome Hunsaker
> Mr. Sidney W. Souers
> Mr. Gordon Gray
> Dr. Donald Menzel
> Gen. Robert M. Montague
> Dr. Lloyd V. Berkner

The death of Secretary Forrestal on 22 May, 1949, created a vacancy which remained unfilled until 01 August, 1950, upon which date Gen. Walter B. Smith was designated as permanent replacement.

* TOP SECRET *

TOP SECRET / MAJIC

EYES ONLY

EYES ONLY

T52-EXEMPT (E)

002

Figure 13. Eisenhower Briefing Document, p. 2. The Majestic Documents

The MJ-12 Group was predictably divided up in its membership between the different military services, the scientific community and members of the intelligence community. After conducting a detailed analysis of each of those named, Friedman concluded:

> There were close links within this group of very important people. Considering what was happening in America and around the world post-World War II, they were a natural group to be on a committee such as Majestic 12. Either by aptitude, position, or geographic location, their inclusion would be fairly obvious.[135]

The best evidence supporting the existence of Operation Majestic Twelve is a memorandum retrieved from the National Archives by Friedman, who only found it after receiving a tip-off. Issued on July 14, 1954, this Top Secret Memorandum sent by Robert Cutler, a Special Assistant to President Eisenhower, was issued to General Nathan Twining (a member of MJ-12 at the time). The "Top Secret Restricted Security" Memorandum stated:

> The President has decided that the MJ-12 SSP briefing should take place during the already scheduled White House meeting of July 16, rather than following it as previously intended.[136]

The subject line of the Memorandum reads "NSC/MJ-12 Special Studies Project". Unlike other Majestic documents still under investigation, the Cutler-Twining memo is an official government document that had been declassified from Top Secret and housed at the National Archives. It was found among a group of declassified Air Force documents and copied by the archivist Ed Reese as part of official records belonging to the National Archives.[137] The Cutler-Twining memo is among the most authoritative documentary sources so far for the existence of the MJ-12 Special Studies Project.[138]

# Majestic-12 Group (1947-1949)

MJ-1. Adm Roscoe Hillenkoetter    MJ-2. Dr. Vannevar Bush    MJ-3. Sec. James Forrestal

MJ-4. Gen. Nathan Twining    MJ-5. Gen. Hoyt Vandenberg    MJ-6. Dr. Detlev Bronk

MJ-7. Dr. Jerome Hunsaker    MJ-8. Adm. Sidney Souers    MJ-9. Gordon Gray

MJ-10. Dr. Donald Menzel    MJ-11. Gen. Robert Montague    MJ-12. Dr Lloyd Berkner

**Figure 14. Initial Members of Majestic 12 Group**

Two of the original members of the Majestic 12 Group were Generals Hoyt Vandenberg and Nathan Twining, who would respectively succeed General Carl Spaatz as the USAF Chief of Staff. Vandenberg and Twining's inclusions mean that the USAF was closely involved in managing extraterrestrial and related technology issues from the very beginning. Twining, in particular, oversaw research and development efforts which by necessity involved the Air Materiel Command. Civilian organizations such as Project RAND provided the necessary scientific and management skills not only for developing a long-term strategic plan for the Air Force to deal with the multiple problems identified by Twining in his White Hot Report, but also the eventual building of fleets of antigravity craft that would be reverse-engineered from the captured extraterrestrial technologies. This latter goal had immediate consequences for Project RAND and its relationship with the Douglas Aircraft Company.

July 14, 1954

MEMORANDUM FOR GENERAL TWINING

SUBJECT:  NSC/MJ-12 Special Studies Project

    The President has decided that the MJ-12 SSP briefing should take place <u>during</u> the already scheduled White House meeting of July 16, rather than following it as previously intended.  More precise arrangements will be explained to you upon arrival.  Please alter your plans accordingly.

    Your concurrence in the above change of arrangements is assumed.

ROBERT CUTLER
Special Assistant
to the President

Figure 15. Cutler-Twining Memo.  Source: The Majestic Documents

# CHAPTER 5

# Diverging Paths:
# RAND's Roots within the USAF and US Navy's Secret Space Programs

One way to find food for thought is to use the fork
in the road, the bifurcation that marks the place of
emergence in which a new line of development
begins to branch off.

–William Irwin Thompson

Several months after the 1947 Roswell crash and the creation
of the Majestic 12 Group, a decision with sweeping
consequences was made to formally separate Project RAND
from the Douglas Aircraft Company. This is how the RAND
Corporation website explains what happened:

> By late 1947, it seemed as though Project RAND —
> which was already operating fairly autonomously —
> should consider separating from Douglas. In
> February 1948, the Chief of Staff [Carl A. Spaatz] of
> the newly created United States Air Force wrote a
> letter to the president of the Douglas Aircraft
> Company that approved the evolution of Project
> RAND into a nonprofit corporation, independent of
> Douglas.[139]

The RAND Corporation was officially created on May 14, 1948, marking its formal parting from Douglas. Historians describe the separation as amicable due to the inherent challenges the USAF had encountered while funding an 'impartial' think tank inside a major aviation company:

> Although nominally autonomous, Project RAND had come to be seen as an appendage of Douglas. After awhile, RAND found it difficult to obtain detailed information from Air Force contractors to prepare analyses of defense problems since contractors were wary of supplying confidential proprietary information that might be leaked to a competitor. Moreover, many analysts were reluctant to be associated with RAND, believing that RAND could not really be objective as long as it was under Douglas sponsorship.
>
> For its part, Douglas Aircraft saw RAND as a growth choking the life of the parent body. Since the Air Force was going out of its way to avoid even the impression of favoritism, Douglas executives felt the company's association with the think tank was hurting Douglas's chances of landing lucrative contracts. Even the Air Force was unhappy, frustrated that the millions of dollars it was spending on Project RAND were not bearing the desired results because of these conflicts.[140]

According to William Tompkins, the separation was not as smooth a transition as the RAND website and historians depicted, which overlooked the tensions and animosities of employees forced to choose between the two entities. In an interview, Tompkins said that roughly two-thirds of the approximately two hundred personnel working in Project RAND moved over to the Santa Monica facility housing the newly created RAND Corporation. The other one third remained with Douglas at its own still highly classified engineering facility in Santa Monica.[141]

This remnant of Project RAND that stayed with Douglas came to form the secret think tank "Advanced Design", which Tompkins began working for in early 1951 after joining Douglas only months before in late 1950. Tompkins learned firsthand in 1951 that antigravity projects continued to be secretly studied by Douglas scientists and engineers in its classified think tank,-through funding now secretly supplied by the US Navy.[142] At the same time, the RAND Corporation was exclusively funded by the US Air Force to study advanced aerospace concepts.[143]

Tompkins' claims are supported by official Douglas Aircraft Company documents showing that antigravity and other exotic propulsion systems were indeed being studied. These documents show how Tompkins' bosses within "Advanced Design" were directly involved. One document by Dr. Wolfgang B. Klemperer (Tompkins' immediate supervisor), dated March 1, 1955, to Elmer Wheaton, vice president of Douglas Engineering, has as its subject line, "UNCONVENTIONAL PROPULSION SCHEMES", and states:

> Our studies of the possible merits or significance of occasionally appearing publications concerning Unconventional Propulsion Schemes have been casually continued since writing the first memorandum (MTM-622) about their progress to mid December 1954.
>
> Between that time and the end of February 1955, twenty more papers on pertinent topics have been obtained and read. They are reviewed in the appended Astronautical Literature Review ...
>
> Several more occasions were had to talk personally to people about the subject. Two of such interviews are abstracted ... We have also looked at a few typical "Flying Saucer" books but found none of

them of technical significance thus far. Brief reviews
of six of them are appended.[144]

Further Douglas Aircraft Company documents supplied by
Tompkins clearly demonstrate that Dr. Klemperer was conducting
a systematic study of unconventional propulsion systems including
antigravity.[145] Prior to these Advance Design antigravity feasibility
studies, Klemperer was one of the Douglas scientists that worked
in Project RAND, and, in fact, authored three of the papers that
were included in the 1946 Project RAND study, "Preliminary Design
of an Experimental World-Circling Spaceship."[146]

It's important to analyze the role of Dr. Klemperer with
Douglas during the time of the aircraft company's efforts to
understand exotic propulsion systems including antigravity, and
how it adopted these for aerospace projects on behalf of the US
Army, Air Force, and the Navy. Klemperer was an Austrian national
who had been raised in Dresden, Germany.[147] He graduated from
the Dresden University of Technology in 1920 and got his first job
at the Aachen Institute of Aerodynamics where he stayed for two
years as an assistant to Professor Theodore von Kárman. Notably,
von Kárman later emigrated to the U.S. in 1930 to work at the
California Institute of Technology and became a prominent
member of Army Air Force/USAF efforts to study captured German
and extraterrestrial aerospace technologies. Von Kárman's
brilliance led to him co-founding the Jet Propulsion Laboratory in
1944, and to him becoming General "Hap" Arnold's principal
scientific advisor during World War II. According to the "White Hot
Report," Professor von Kárman was later part of the scientific group
that Twining sent to investigate the debris from the Roswell crash.

Klemperer's connection to von Kárman is worth
remembering. Next, Klemperer worked at *Luftschiffbau Zeppelin*
from 1922 to 1924, where he headed research in the design and
construction of Zeppelin-designed dirigibles. After earning his
doctorate in engineering from Aachen, he emigrated to the U.S. in
1924 to accept a position at the Goodyear-Zeppelin Corporation of
Akron, Ohio. Klemperer was immediately involved in designing two

large Zeppelins for the US Navy based on the designs pioneered in Germany. The Navy contract was for 8 million dollars, which in 2018 terms is an impressive 1.78 billion dollars.[148] The first of the two airships, the *USS Akron*, had a flight crew of 10 officers and 50 enlisted men, plus a heavier-than-air group of four officers and 15 airplane mechanics. It also housed an airplane complement of up to four Curtiss F9C-2 "Sparrowhawk" fighters, launched and recovered via a mechanical trapeze extended below the airship's hangar compartment.[149]

MIM-622, Part 2

March 1, 1955

To:  E. P. Wheaton, A-250

From:  W. B. Klemperer, A-250

Subject:  UNCONVENTIONAL PROPULSION SCHEMES

Copies to:  H. Aurand, A-250; R. Demoret; A-250; J. B. Edwards, A-250; S. Kleinhans, A-250; T. A. Kvaas, A-250; H. Luskin, A-250; C. C. Martin, A-215; G. M. Files

Reference:  MIM-622, December 20, 1954 (Declassified)

Our studies of the possible merits or significance of occasionally appearing publications concerning Unconventional Propulsion Schemes have been casually continued since writing the first memorandum (MIM-622) about their progress to mid December 1954.

Between that time and the end of February 1955, twenty more papers on pertinent topics have been obtained and read. They are reviewed in the appended Astronautical Literature Review pages, serial 026 to 045. The content of most of them falls into similar categories as those reviewed before, under serial numbers 001 to 025.

Several more occasions were had to talk personally to people about the subject. Two of such interviews are abstracted, one with Dr. C. B. Millikan and the other with Captain W. T. Sperry of American Airlines who encountered an UFO in flight in 1950.

We have also looked at a few typical "Flying Saucer" books but found none of them of technical significance thus far. Brief reviews of six of them are appended. A print of a color film tracking two Unidentified Foreign Objects near Missoula, Montana, was obtained. It is now being analysed by Iconolog techniques.

Correspondence was exchanged with Aviation Studies (International) Limited, 20-31 Cheval Place, Kinghtsbridge, London SW 7, England, who describe themselves as Management Consultants and who prepare and distribute the Aviation Reports discussing technical, commercial and political developments in the world of aviation, as mentioned in paragraph 1. Reference was made by us particularly to the article "Gravitic Steps" in their issue No. 357 of 19 Nov. 1954 (p. 531) in which veiled intimations were made of promising experimental results with a test rig; specific questions concerning details of these alleged experiments were submitted to the editor of the British publication. An answer dated 4 February 1955 was promptly received. In this reply, signed by R. G. Worcester (Director of Aviation Studies (International) Limited) we were referred to "an unclassified report on Project Winterhaven

Figure 16. Dr. Klemperer letter on" Unconventional Propulsion"

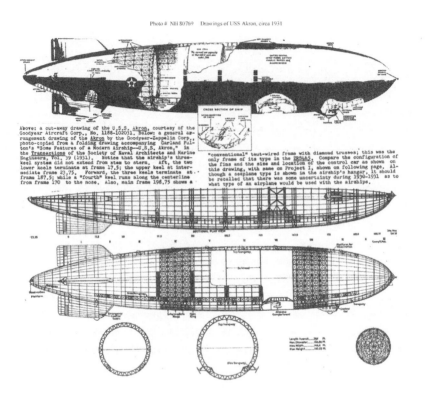

Figure 17. Naval Historical Center drawing

The building of the *USS Akron* began on October 31, 1929, and was completed on August 1931. Construction of its sister ship, the *USS Macon*, began in 1933 and was completed in 1935. Both the *USS Akron* and the *USS Macon* were prototype flying aircraft carriers that actually were put into service for the Navy. They were the largest helium-filled airships ever built and provided an unprecedented projection of Naval airpower. Unfortunately, the service life of both was cut short due to them crashing during storms: the *Akron* in 1933 and the *Macon* in 1935. As a result of this setback and a decision by the Navy not to build further Zeppelins, Klemperer decided to move to Douglas Aircraft Company in 1936.

Klemperer's pre-World War II work in designing and building flying aircraft carriers, along with his documented interest in antigravity, substantiates Tompkins' claims that the Douglas Aircraft Company was secretly involved in the 1950's and early 1960's in designing space carriers for the US Navy. The relationship between the Navy and Douglas over this pursuit is described in great detail in Tompkins' autobiography, *Selected by Extraterrestrials*, and in book two of this Secret Space Program series, *The U.S. Navy's Secret Space Program & Nordic Extraterrestrial Alliance*.

Significantly, I found no evidence that the Navy's incipient space program was associated with the Majestic 12 Group. While Navy admirals were involved with Majestic 12 from its inception, along with MJ-12 efforts to understand and manage problems associated with extraterrestrial life and technology, this still did not translate into these admirals apprising other Majestic members of what the Navy was doing. On the surface, this may appear to have been a product of resentment over the Navy being institutionally sidelined by both the Roosevelt and Truman administrations from leading the research and development efforts over captured extraterrestrial technologies. It's a historical fact that the Navy was strongly opposed to the creation of the Department of the Air Force, and therefore, would not have been pleased with the primacy Truman had assigned to his newly created military department in managing the extraterrestrial phenomenon immediately after the Roswell crash.

However, there is another more compelling explanation for why the Navy's budding space program was researched, designed, and built entirely outside of the purview of the Majestic 12 Group. during Operation Highjump in early 1947, the Navy had suffered a devastating defeat in Antarctica at the hands of the space program established there by "German Secret Societies", which I discussed at length in book three of this series, *Antarctica's Hidden History and Corporate Foundations of Secret Space Programs*. The Navy best understood the long-term strategic threat posed by the German breakaway colony in Antarctica, and its stunning technological advances due to an alliance it had formed with an extraterrestrial race called the Draconians. During WWII, the Navy had learned about this world-altering alliance through one of its key espionage programs — the same one in which William Tompkins had participated.[150]

The technological benefits of establishing an alliance with extraterrestrials were acknowledged as an important requirement in Twining's September 19, 1947, "White Hot Report" of the Roswell crash. In this document, he referred to "directly working with the aliens on technology transfer".[151] This specific phrase was redacted (i.e., blackened out) in the leaked version of the report, however, Dr. Robert Wood and Ryan Wood were able to reconstruct it forensically. The imperative for establishing future agreements with extraterrestrials was highly restricted, as the redacted line indicates, to only the most senior officials at the time. The responsibility for establishing such an agreement with extraterrestrials would fall to the strategic planning group called for in Twining's report, which was then established five days later with President Truman's September 24, 1947, Memorandum — "the Majestic 12 Group".

Consequently, since the Navy knew that agreements would have to be reached with extraterrestrials and an accommodation eventually reached with the breakaway German Antarctic colony, it's understandable why the Navy chose not to share its long-term strategic plan to research and develop future space battle groups with the Majestic 12 Group. After all, the Navy knew that MJ-12 had been set up to manage extraterrestrial life and technology issues on behalf of the Truman administration. MJ-12 was tasked with the responsibility to find a solution to the expanding German space program in Antarctica. The less Majestic 12 knew about what the Navy was doing at its research facilities, the less likely the Germans and their extraterrestrial allies would learn about it and plan to infiltrate and sabotage such efforts. Consequently, whether due to lingering resentment, institutional jealousy or out of strategic necessity, the Navy would secretly continue its separate research and development of captured extraterrestrial technologies using the Douglas Aircraft Company and other classified research facilities.

Perhaps not so ironically, while the US Navy continued to secretly work with Douglas' Project RAND replacement, Advanced Design, in creating future fleets of space carriers, the US Air Force instead employed Douglas' offshoot, the RAND Corporation. Since RAND was involved in the study of the retrieved extraterrestrial artifacts and bodies from the Roswell incident, as Twining's "White Hot Report" outlined, the RAND Corporation's principal task ahead was to come up with a long-term strategic plan to solve the myriad of problems posed by extraterrestrial life and technologies. On this front, one of the immediate problems faced by the US Air Force and the Majestic 12 Group was how to deal with the increasing number of overflights of United States airspace by both extraterrestrial and Antarctic German spacecraft.

# CHAPTER 6

## The German Connection to UFO's over the United States

*Never in history have lies been such vital instruments of diplomacy and policy.*

–Max Hastings

### Flying Saucers from Antarctica

In March 1947, Rear Admiral Richard Byrd gave an interview in Santiago, Chile, about his recently terminated "Operation Highjump" mission. At the time, he was traveling back to the United States to debrief his superiors about what had just taken place during that deployment in Antarctica. Byrd's statements included a sobering portent of an imminent danger looming for the United States. The journalist, Lee Van Atta, quoted Byrd extensively in an article that appeared in the March 5, 1947, edition of *El Mercurio* with the title, "Admiral Richard E. Byrd refers to the strategic importance of the poles". What follows are translated passages from the original Spanish release:

> Admiral Richard E Byrd warned today of the necessity for the United States to adopt protective measures against the possibility of an invasion of the country by hostile aircraft proceeding from the polar regions. The admiral said: 'I do not want to scare anybody but the bitter reality is that in the

event of a new war, the United States will be attacked by aircraft flying in from over one or both poles.'"152

This specific passage is the source of the original theory that proposed the South and North poles were the locations where an unknown enemy existed, capable of initiating a "new war" against the United States. The reference to a "new war" clearly indicates that the entity involved was both hostile and powerful since it could directly threaten mainland America with advanced aircraft launched from the South Pole. Van Atta continued:

> On the subject of the recently terminated expedition, Byrd said that 'the most important of the observations and discoveries made [were] ... of the present potential situation as it relates to the security of the United States ... I can do no more than warn my countrymen very forcibly that the time has passed when we could take refuge in complete isolation and rest in confidence in the guarantee of security which distance, the oceans and the poles provide."153

Here, Byrd is suggesting that the unknown enemy possessed advanced aerial craft whose speed and range removed the protection previously enjoyed by the U.S. in terms of the vast oceans separating it from Europe and Asia.

In my book, *Antarctica's Hidden History: Corporate Foundations of Secret Space Programs,* I presented evidence that Byrd's Operation Highjump had been dealt a stunning blow by the breakaway German colony in Antarctica, which had successfully weaponized its advanced flying saucer craft.154 Operation Paperclip, which arranged for the entry of Nazi scientists and technologies into the U.S., had only succeeded in retrieving the *unsuccessful* flying saucer prototypes that the Germans had abandoned in their retreat to Antarctica. The Germans had given the Navy a bloody nose in Antarctica, and now it would be the Army Air Force's turn as the Germans began overflights of US territory, which is the threat Byrd was alluding to in his interview.

EL MERCURIO. — Santiago de Chile, miércoles 5 de marzo de 1947

## El almirante Richard E. Byrd se refiere a la importancia estratégica de los polos

(Por Lee Van Atta, para "El Mercurio")

A BORDO DEL MOUNT OLYMPUS, EN ALTA MAR. 4 — (ESPECIAL). — El almirante Richard E. Byrd advirtió hoy que es preciso que los Estados Unidos adopten medidas de protección contra la posibilidad de una invasión del país por aviones hostiles procedentes de las regiones polares.

El almirante dijo: "no intento asustar a nadie, pero la amarga realidad es que, de ocurrir una nueva guerra, los Estados Unidos serán atacados por aviones que volarán sobre uno a ambos polos". Esta declaración fué hecha a manera de recapitulación de la ejecutoria del propio Byrd, como explorador polar, en una entrevista exclusiva para International News Service. A propósito de la expedición recién terminada, Byrd dijo que el resultado más importante de las observaciones y descubrimientos hechos es el efecto actual potencial, que tendrán éstos en relación con la seguridad de los Estados Unidos. "La fantástica premura con que el mundo se está encogiendo —declaró el almirante— es una de las lecciones objetivas aprendidas durante la exploración antártica que acabamos de efectuar. No puedo menos que hacer una fuerte advertencia a mis compatriotas en el sentido de que ha pasado ya el tiempo en que podíamos refugiarnos en un completo aislamiento y descansar en la confianza de que las distancias, los océanos y los polos constituían una garantía de seguridad".

A continuación observó que si él ha hecho éxito, otras personas podrían igualmente dirigir una nueva expedición de 4 mil jóvenes norteamericanos, con la ayuda exclusiva de un puñado de exploradores experimentados. El almirante encareció la necesidad de permanecer "en estado de alerta y vigilancia a lo largo

constituyen los últimos reductos de defensa contra una invasión.

"Yo puedo darme cuenta quizás mejor que cualquier otra persona, de lo que significa el uso de los conocimientos científicos en estas exploraciones, porque puedo hacer comparaciones. Hace 20 años realicé mi primera expedición antártica con menos de ciento cincuenta hombres, dos buques y diez aviones. Entonces la exploración era arriesgada y peligrosa y constituía una singular experiencia. Pero ahora, poco menos de veinte años más tarde, una expedición quince veces mayor que aquélla ha en todos los respectos, recorre el antártico, completa su misión en menos de dos meses y abandona la región después de haber hecho importantes descubrimientos geográficos. La moraleja que se deriva de esta comparación es clara: puesto que la velocidad y el progreso al parecer no reconocen horizontes, es preciso que aceleremos la pauta de nuestro pensamiento y de nuestros proyectos y de nuestras acciones, y la expansión de nuestros propios horizontes. Pero es preciso que hagamos esto ahora, ya, porque tanto la supervivencia del mundo como la ciencia militar, se hallan actualmente en una etapa vital de su desarrollo".

El almirante declaró que en su opinión la expedición ha sentado un precedente sin igual en cuanto se refiere a la rápida sucesión en que se verificaron los descubrimientos geográficos. Y concluyó encomiando la labor de los aviadores y fotógrafos del servicio de cartografía aérea de la expedición, quienes desempeñaron el papel más importante en la exploración de las desconocidas regiones del antártico. — (I. N. S.).

Lee Van Atta.

Figure 18. Lee Van Atta article on Operation Highjump

Only a few months after Admiral Byrd's interview with the Chilean press, the famous Kenneth Arnold UFO incident occurred. On June 24, 1947, Arnold witnessed nine flying wing-shaped craft over the Cascade Mountains of Oregon and Washington State. In a letter to the Army Air Force, Arnold described what he had witnessed:

> The sky and air was clear as crystal. I hadn't flown more than two or three minutes on my course when a bright flash reflected on my airplane. It startled me as I thought I was too close to some other aircraft. I looked every place in the sky and couldn't find where the reflection had come from until I looked to the left and the north of Mt. Rainier where I observed a chain of nine peculiar looking aircraft flying from north to south at approximately a 9,500 foot elevation and going, seemingly, in a definite direction of about 170 degrees.
>
> They were approaching Mt. Rainier very rapidly, and I merely assumed they were jet planes. Anyhow, I discovered that this was where the reflection had come from, as two or three of them every few seconds would dip or change their course slightly, just enough for the sun to strike them at an angle that reflected brightly on my plane.
>
> These objects being quite far away, I was unable for a few seconds to make out their shape or their formation. Very shortly they approached Mt. Rainier, and I observed their outline against the snow quite plainly.
>
> I thought it was very peculiar that I couldn't find their tails but assumed they were some type of jet plane. I was determined to clock their speed, as I had two definite points I could clock them by; the air was so clear that it was very easy to see objects and determine their approximate shape and size at

almost fifty miles that day....

These objects were holding an almost constant elevation; they did not seem to be going up or coming down, such as would be the case of rockets or artillery shells. I am convinced in my own mind that they were some type of airplane, even though they didn't conform with the many aspects of the conventional type of planes that I know....

Some descriptions could not be very accurate taken from the ground unless these saucer-like disks were at a great height and there is a possibility that all of the people who observed peculiar objects could have seen the same thing I did ... [155]

Arnold's observation of the nine craft flying in formation indicates that they were not experimental jet aircraft being tested by the US military. Such classified tests of experimental aircraft involved one, or at most two, prototypes being simultaneously evaluated, not nine flying in formation. Ben Rich best explains the construction and testing of experimental aircraft in his book, *Skunk Works: A Personal Memoir of My Years at Lockheed*.[156] Rich's book makes it clear that what Arnold had witnessed was not experimental craft undergoing testing, but a squadron performing a mission. The question that arises is to who did the flying wing-shaped craft belong?

Arnold's description of the flying "saucer-like disks" bears a remarkable likeness to a unique jet aircraft developed for Nazi Germany by two brothers, Reimar and Walter Horten – the Horten Ho 229. The revolutionary design of this flying wing aircraft was used decades later in the construction of the B-2 Spirit stealth bomber.[157] One of the Ho 229's prototypes, a V3, had been relocated to the U.S. after World War II for study. Today it can be found in the Smithsonian National Air and Space Museum's restoration facility.

Drawing of UFO sighted by Kenneth Arnold in 1947 compared with photo of Horton 229 developed by Nazi Germany in 1944.

Figure 19. Comparison: UFO drawing by Kenneth /Photo of Horton Flying Wing

When the Air Force began its investigation of flying saucer reports, the similarities in the design and flight performance of the Ho 229 with what witnesses had sighted, was officially acknowledged.[158] The similarities between the Ho 229 and what Arnold had seen leads to the following conclusion: while only two Ho 229 V3 prototypes were known to have been officially built, more successful prototypes were secretly developed either by the Hortens or other German aviation engineers and moved to

Antarctica. It was these more advanced German flying wing craft that were overflying U.S. territory in squadron formation by 1947.

Admiral Byrd's warning had proved prescient insofar as the Antarctic-based Germans now had developed the capability to overfly the U.S. with squadrons of flying saucers by June 1947. With the rapid rise of UFO sightings after the Arnold incident, it can be concluded that some were directly connected to the German space program out of Antarctica. The precise connection between these UFO reports with known World War II German flying saucer projects was something the US Air Force immediately began investigating.

## Air Force Investigation Encounters the German Connection

One of the best sources of information about what happened after the Kenneth Arnold and other UFO sightings over the U.S. is Captain Edward Ruppelt, who headed the USAF's "Project Bluebook" from early 1951 to September 1953. In the first edition of his 1955 book, *The Report of Unidentified Flying Objects*, he described the widespread panic among Air Force leaders that arose from the first official studies of UFO sightings that began in June 1947:

> The memos and correspondence that Project Blue Book inherited from the old UFO projects told the story of the early flying saucer era. These memos and pieces of correspondence showed that the UFO situation was considered to be serious; in fact, very serious. The paper work of that period also indicated the confusion that surrounded the investigation; The brass wanted an answer, quickly, and people were taking off in all directions. Everyone's theory was as good as the next and each person with any weight at ATIC [Air Technical Intelligence Center, Wright Patterson AFB] was

plugging and investigating his own theory. The ideas as to the origin of the UFO's fell into two main categories, earthly and non earthly.[159]

It's important to mention that Ruppelt and the "Project Bluebook" investigation were stationed at Wright Patterson AFB, the same location where the Roswell craft debris had been taken. The debris and bodies were being closely studied, very likely, in a newly created classified laboratory there that Lt General Nathan Twining had ordered Major General Curtis LeMay to build on July 18, 1947, just two weeks after the Roswell Crash.[160]

Any doubt over the importance of Wright-Patterson AFB in the overall study of flying saucers, whether merely of reported sightings or crash retrievals, is laid to rest in a declassified 1952 CIA document with the subject header: "Flying Saucers". Released in 2010, the CIA document was issued from the assistant director, Office of Scientific Intelligence (OSI), and clearly stated that Wright-Patterson was the only facility conducting flying saucer research:

> OSI has investigated the work currently being performed on "flying saucers" and found that the Air Technical Intelligence Center [ATIC], DI, USAF, Wright-Patterson Air Force Base, is the only group devoting appreciable effort and study to this subject".[161]

While the leaked Majestic documents show that Air Force investigators had quickly realized that the Roswell craft was extraterrestrial in origin, this did not mean that the wave of UFO sightings now sweeping the U.S. were all alien as well. The question over which flying saucer sightings were extraterrestrial and which were connected to the German flying saucer program in Antarctica immediately preoccupied the Majestic-12 committee and senior Air Force officials. From "Project Blue Book", Captain Ruppelt provided an eyewitness account to them of the different arguments, memoranda, and files that tried to determine the unknown craft's origins.

One explanation that quickly emerged was that the UFO's were connected in some way to revolutionary German aerospace projects from World War II:

> When World War II ended, the Germans had several radical types of aircraft and guided missiles under development. The majority of these projects were in the most preliminary stages but they were the only known craft that could even approach the performance of the objects reported by UFO observers. Like the Allies, after World War II the Soviets had obtained complete sets of data on the latest German developments. This, coupled with rumors that the Soviets were frantically developing the German ideas, caused no small degree of alarm. As more UFO's were observed near the Air Force's Muroc Test Center, the Army's White Sands Proving Ground, and atomic bomb plants, ATIC's efforts became more concentrated.
>
> Wires were sent to intelligence agents in Germany requesting that they find out exactly how much progress had been made on the various German projects. [162]

Ruppelt, like the vast majority of Air Force officers, didn't know that a German breakaway group had established a colony in Antarctica which possessed operational flying saucers that had dealt the Navy's Operation Highjump a blow in early 1947. Even though Ruppelt and other Project Blue Book Investigators held "Top Secret" security clearances, this would not have been sufficient to grant them access to compartmentalized information about the German's Antarctica colony. Nevertheless, Ruppelt's book shows that some Air Force investigators had concluded that the UFO sightings held an important German connection. However, they concluded that some UFO sightings involved Soviet adaptions of German flying saucer prototypes, a logical misunderstanding given the limited information that was made available to them at the time.

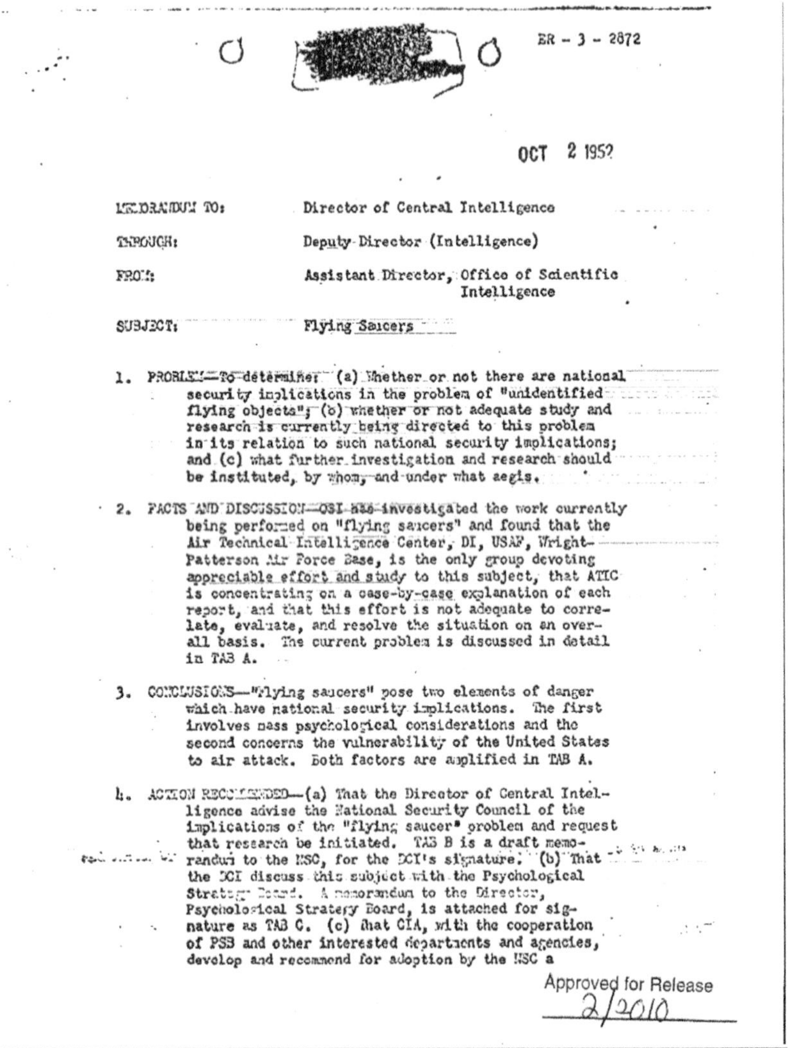

ER – 3 – 2872

OCT 2 1952

MEMORANDUM TO:      Director of Central Intelligence

THROUGH:            Deputy Director (Intelligence)

FROM:               Assistant Director, Office of Scientific
                    Intelligence

SUBJECT:            Flying Saucers

1. PROBLEM—To determine: (a) Whether or not there are national security implications in the problem of "unidentified flying objects"; (b) whether or not adequate study and research is currently being directed to this problem in its relation to such national security implications; and (c) what further investigation and research should be instituted, by whom, and under what aegis.

2. FACTS AND DISCUSSION—OSI has investigated the work currently being performed on "flying saucers" and found that the Air Technical Intelligence Center, DI, USAF, Wright-Patterson Air Force Base, is the only group devoting appreciable effort and study to this subject, that ATIC is concentrating on a case-by-case explanation of each report, and that this effort is not adequate to correlate, evaluate, and resolve the situation on an over-all basis. The current problem is discussed in detail in TAB A.

3. CONCLUSIONS—"Flying saucers" pose two elements of danger which have national security implications. The first involves mass psychological considerations and the second concerns the vulnerability of the United States to air attack. Both factors are amplified in TAB A.

4. ACTION RECOMMENDED—(a) That the Director of Central Intelligence advise the National Security Council of the implications of the "flying saucer" problem and request that research be initiated. TAB B is a draft memorandum to the NSC, for the DCI's signature; (b) That the DCI discuss this subject with the Psychological Strategy Board. A memorandum to the Director, Psychological Strategy Board, is attached for signature as TAB C. (c) That CIA, with the cooperation of PSB and other interested departments and agencies, develop and recommend for adoption by the NSC a

Figure 20. CIA document showing role of Wright Patterson AFB as only USAF facility investigating flying saucer phenomenon

An October 13, 1952, CIA document shows the US national security establishment was seriously investigating the link between flying saucers and the Soviet Union as well. When reading the CIA document text, if one replaces the acronym "USSR" with "Antarctic-based Germans", this will offer a more accurate idea of the real national security dilemma confronting US policymakers in the early 1950's:

> Determination of the scientific capabilities of the USSR [Antarctic-based Germans] to create and control Flying Saucers as a weapon against the United States is a primary concern of the CIA/OSI. Its review of existing information does not lead to the conclusion that the saucers are USSR [Antarctic-based German] created or controlled. It is the view of OSI that collection of intelligence information on the capabilities of the USSR [Antarctic-based Germans] to produce, launch, and control Flying Saucers and the analysis of such data as might be collected cannot be very effective until there is adequate fundamental scientific research launched to clarify the nature and causes of Flying Saucers and to devise means whereby they might be instantly identified. [163]

This document illustrates several concerns as to why the craft were not originating from some clandestine USSR (or Antarctic-based German) program. However, this led to another explanation that some Air Force officers favored instead of the foreign power connection: the flying saucers were extraterrestrial in origin.

The 'Extraterrestrial Hypothesis' (ETH) was officially proposed as the most valid explanation for UFO/Flying Saucer sightings by a 1948 classified study initiated by the Air Force. The secret study of approximately 300 cases produced an "Estimate of the Situation" in September 1948, and its conclusion supported the "ETH". The study and its remarkable conclusion were moved up the Air Force hierarchy to the desk of the newly appointed Chief of Staff, General Hoyt Vandenberg. Vandenberg rejected it and made clear that acceptance of the "ETH" was not an acceptable conclusion for reasons related to national security concerns.[164]

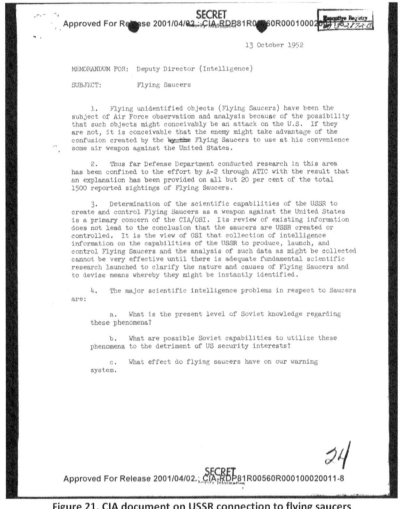

**Figure 21. CIA document on USSR connection to flying saucers**

Word of the rejection of the initial "Estimate of the Situation" and the subsequent destruction of the report found its way to private UFO investigators such as Major Donald Keyhoe, who concluded that it was evidence of a cover-up perpetuated from the highest level within the US military and government. Keyhoe was confidentially told the following by Captain Edward Ruppelt about General Vandenberg's decision to reject the original "Estimate of the Situation":

> "The general said it would cause a stampede.... How could we convince the public the aliens weren't hostile when we didn't know ourselves? ... the general ordered the secret analysis burned. But one copy was held out – Major Dewey Fournet and I saw it in 1952.[165]

It has long been assumed that Vandenberg simply wanted to cover up the extraterrestrial hypothesis. However, another explanation is that he wished the answer to be indeterminate since he and other Majestic-12 members couldn't be sure which UFO reports were German and which were extraterrestrial. Support for this conclusion comes from Kewper/Stein, the former Army Signals Corps serviceman recruited to work for the CIA from 1957-1961. He stated that part of his mission assignment was to distinguish between German and extraterrestrial spacecraft operating out of Antarctica and South America. In an interview with Emmy award winning UFO researcher Linda Moulton Howe, Kewper/Stein described his unit's assignment:

> [Howe] You did have a photograph that confirmed the Germans were still flying some of their Peenemunde craft in South America?

> [Kewper] Oh, yeah! The craft with a high center about 12-feet high – they all look like Haunebu II's. Although they could be alien craft as well. But we labeled those photos as being German craft from Argentina. However, on radar, we used to see some of the *real* alien craft come from outer space right

down into the Argentina region. We also saw craft come into the Antarctica region from outer space via radar we shared with the British down in the Falkland Islands in the South Atlantic Ocean east of Argentina.... In 1959 to 1960, our unit was separating alien craft from the known German craft by the appearance of the craft. We always found the German craft to be much slower in speed than the alien craft. Some alien craft were tracked from outer space doing something like 30,000 mph![166]

Due to the technological inferiority of the German craft when compared to their extraterrestrial allies, it is understandable why the German saucers likely preferred to fly accompanied in a squadron formation, as exemplified in the Kenneth Arnold incident. Possible documentary evidence for this pattern is shown in a CIA report referring to numerous sightings of a formation of flying saucers over Deception Island, Antarctica in July 1965. The CIA document refers to the following brief report from nearby British, Chilean and Argentinian bases:

ANTARCTIC FLYING SAUCERS    - a group of red, green, and yellow flying saucers has been seen flying over Deception Island for two hours by Argentine, Chilean, and British bases in Antarctica. The flying saucers were also seen flying in formation over the South Orkney island in quick circles.[167]

The proximity of the flying saucers to Antarctica does suggest that this formation was part of the breakaway German colony that Byrd's Operation Highjump had confronted in 1947.

If the Antarctic-based Germans did fly in formation over U.S. soil, as the 1947 Kenneth Arnold and 1965 Antarctic sightings indicate, then arguably the best photographic evidence of this tendency is the Lubbock lights case that took place from August 28 to September 5, 1951. On successive nights, a formation of lights appeared over Lubbock, Texas, and was witnessed by multiple credible sources and even photographed. Three professors who witnessed the craft estimated that they were traveling up to 6000 miles per hour (970k/h). It became one of the most analyzed UFO cases in history, and the photographs have never been debunked or satisfactorily explained.[168]

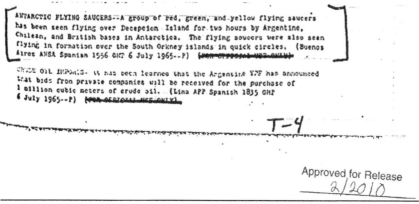

ANTARCTIC FLYING SAUCERS.--A group of red, green, and yellow flying saucers has been seen flying over Decepcion Island for two hours by Argentine, Chilean, and British bases in Antarctica. The flying soucers were also seen flying in formation over the South Orkney islands in quick circles. (Buenos Aires ANSA Spanish 1556 GMT 6 July 1965--P)

CRUDE OIL IMPORTS.- It has been learned that the Argentine YPF has announced that bids from private companies will be received for the purchase of 1 million cubic meters of crude oil. (Lima APP Spanish 1835 GMT 8 July 1965--P)

T-4

Approved for Release
2/2010

*Figure 22. CIA document on saucer formations near Antarctica*

In contrast to the Germans in Antarctica sending their flying saucers in formation from their polar base, the extraterrestrial craft undoubtedly due to their superior technology, as Kewper/Stein reported, could more safely fly solo when traveling over US territory or anywhere else around the planet. Accordingly, as a general rule of thumb, I would deem squadrons of flying saucers sighted between the mid-1940's up to the 1970's, more likely than not to have been German in origin. By the 1970's, this would no longer apply because the US military had begun secretly deploying its own saucer craft, which will be discussed later in chapter 14.

Kewper/Stein also confirmed that part of the reason for government secrecy on the UFO issue was because it was never clear who was behind the UFO phenomenon. In response to a question about why secrecy was maintained, Kewper/Stein said:

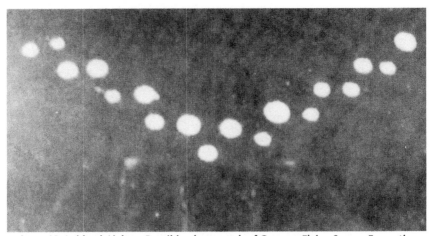

Figure 23. Lubbock Lights - Possible photograph of German Flying Saucer Formation
from Antarctica

> I think basically it's because of the German UFO
> program that started way back in 1917 and they
> developed their first vehicle in 1927. Most people
> don't know anything about this. Then after WWII,
> we did not capture any of those craft. They were all
> sent to Argentina and to the Antarctic base the Nazis
> had there since 1938 before the war even started. I
> think the whole thing is still being kept secret
> because our government does not know if it's Nazis
> flying around or extraterrestrial aliens![169]

Consequently, while it is highly likely that the Kenneth Arnold
sighting involved a squadron of German saucers from Antarctica,
other UFO sightings could have involved either the Germans or
extraterrestrial visitors. Given what was known from US Navy
intelligence files of agreements reached between Nazi Germany
and Draco Reptilian aliens, there is even the possibility that some
flying saucer craft involved BOTH Germans and extraterrestrials.[170]
As we will later see, the Betty and Barney Hill abduction case
showed that Germans in Nazi uniforms were sometimes seen
commanding the short Gray aliens conducting the abductions.[171]

The uncertainty over the true origin of many UFO sightings

would have added to the confusion among US leaders over how best to respond to UFO incursions over US territory. Indeed, it's more than likely that the Antarctic-based German colony adroitly exploited this confusion to inflate their influence and power for any negotiations with US policymakers that lay ahead. In the meantime, the Army Air Force/USAF would not allow the German/extraterrestrial forays on their home turf to go unchallenged.

# CHAPTER 7

# Hidden Losses: The Shoot Down Policy Against Alien & Antarctic German Spacecraft

*I can assure you that, given they exist, these flying saucers are made by no power on this Earth.*

— President Harry S. Truman

## UFO Interference with Rocket Program Leads to Deadly Retaliation

One of the little-known facts about the dawn of the modern UFO era in 1947 is that the US Army used surface-to-air missiles and other advanced weapons to bring down UFOs that were interfering with the testing of V2 rockets. These V2 rockets were critical for launching a future ballistic missile system employing nuclear weapons. In addition, Army Air Force pilots were given intercept and shoot down orders against the flying saucers that were increasingly being spotted over US territory, especially near sensitive military facilities. Multiple sources have described the efforts made by pilots attempting to shoot the unknown craft out of the sky, and the tragic results that often occurred. The origin of the intercept and shoot down orders can be traced to UFO interference occurring while rebuilt and refurbished V2 (aka A4) rockets were being test launched out of White Sands Proving Ground, New Mexico, in May 1947. Markedly, this took place only two months after Admiral Byrd gave his famous warning about the U.S. being attacked in a new war from unknown hostile aircraft from the polar regions.

It was a shocking awakening when the breakaway German colony in Antarctica dealt the US Navy's Operation Highjump a walloping defeat during their attempt to take over or destroy the German colony, yet it's important to emphasize that the Germans were not acting alone. They had received direct assistance from their extraterrestrial allies in defending the Antarctic colony.[172] It is logical to assume then that the Germans and their alien allies would have taken a keen interest in the US Army's missile development program, recognizing that it would one day be able to deliver nuclear weapons into far distant regions, including Antarctica. Consequently, the Germans and/or their extraterrestrial allies began a program of interfering with Army rocket launches to send a clear message to US policymakers.

Here is how the *Las Cruces Sun-News* of New Mexico described the May 15, 1947, White Sands missile launch that was adversely affected by a 'UFO', euphemistically referred to by the base commander as a "peculiar phenomenon":

> Lt. Col. Harold R. Turner, White Sands Proving Ground commanding officer, today blamed "peculiar phenomena" for the erratic test flight of a German V-2 rocket which landed only six miles east of Alamogordo yesterday afternoon.[173]

Veteran UFO researcher Timothy Good was sent an unpublished book manuscript by Andrew Kissner, a former state representative for New Mexico who had investigated the early history of UFOs within the United States. Kissner also supplied Good with additional details about the "peculiar phenomenon" that Colonel Turner had named as the reason for the failure of the V-2 rocket test:

> As the rocket climbed to an intermediate altitude of 40 miles, Kissner learned that radar

technicians assigned to US Army Ordnance watched in amazement as another target instantaneously appeared right next to the missile. The V-2 veered off course and crashed to earth two minutes later.[174]

LAS CRUCES SUN-NEWS
May 16, 1947 p.1

LAS CRUCES, NEW MEXICO, FRIDAY AFTERNOON MAY 16, 1947

**V-2 Goes Astray, Lands in Six Miles Of Alamogordo**

Lt. Col. Harold R. Turner, White Sands Proving Ground commanding officer, today blamed "peculiar phenomena" for the erratic test flight of a German V-2 rocket which landed only six miles east of Alamogordo yesterday afternoon.

**Two Ki**

**Disgrur**

**Navy Budget Cut Of 11 Percent Recommended**

Figure 24. News report of V-2 rocket affected by UFO

Kissner was able to interview an Army officer at White Sands about the V-2 rocket incident and wrote about what he had been told:

Whatever had mysteriously appeared and vanished after observing a V-2 in flight in close proximity to the rocket, it apparently affected the rocket's trajectory. That event became an immediate priority within a very small, closed circle of highly ranked general staff officers and civilian scientists assigned to the Joint Research and Development Board (JRDB) of the Joint Chiefs of Staff (JCS) under whose authority weapons test ranges were established and operated and chaired by Dr. Vannevar Bush. The 'peculiar phenomenon' object was defined as *hostile* in that it appeared to have affected the V-2 rocket's trajectory. It was by every definition an advanced foreign weapons system. Effort would be expended to guard against further intrusions, and high priority was assigned to collect a *specimen* of the technology for further analysis, covertly if possible.[175]

The Army officer had revealed why the Pentagon had classed the UFOs interfering with the V-2 tests as "hostile", which was very similar to the language used by Byrd in describing the unknown threat from Antarctica in his Chilean interview.[176] Furthermore, the Pentagon had decided to begin defending against "further intrusions", and was hell-bent "to collect a *specimen* of the technology". The meaning stood clear. The Truman administration was determined to shoot down any UFOs interfering with the US missile program, and afterward study the recovered craft for future reverse engineering efforts.

According to Kissner, the next missile test on May 29, 1947, constituted the first US Army Air Force attack against an intruding UFO, which was sighted hovering near the launch site of a V-2 rocket that was about to take off from White Sands. The launch was aborted, and the hovering UFO was targeted by at least one Army surface-to-air missile. Kissner explained unfolding events further:

> At approximately 7:20 pm, at an altitude above 60,000 feet, the proximity fuse on the warhead detonated – ten miles north and slightly west of Mt. Franklin. This explosion, witnessed by General Homer and reported by Hanson Baldwin in the *New York Times* the following day, occurred

more than 10 minutes before a second explosion at least 30 miles further south. Gen. Homer dispatched troops to look for missile wreckage and investigated the first crash site 10-15 miles northwest of Ft. Bliss, towards WSPG [White Sands Proving Ground].

The surface-to-air missile's target, possibly crippled by the explosion, continued to fly in airspace north of Ft. Bliss, eventually impacting within one mile of the then-new Buena Vista airport south of downtown Juarez, Mexico, at 7:52 pm ... A 'blinding flash of light' followed by a tremendous pressure wave [and] by the appearance of a mushroom cloud excited local rumors that an atomic bomb had exploded... The ground shock generated by the explosion was felt 35 miles northwest of the impact crater.

Whatever the object had been was not apparent.... It was totally vaporized by the explosion.... Military police stationed at Fort Bliss and other US Army personnel from WSPG rushed across the border into Mexico in an attempt to secure the downed object. They were met at the crater by Mexico troops and summarily evicted. Mexico's general in command of the Juarez garrison, Enrique Dias Gonzales, placed Juarez off-limits to US Army personnel for several weeks after the incident.... An effective cover story was immediately provided to the Press that another V-2 launched from White Sands had gone astray, experienced a total gyroscopic failure, flew 180 degrees off course and crashed into Mexico.[177]

Successful in its defense, the US Army had shot down a UFO that was monitoring and presumably about to interfere with a V-2 rocket launch in the same way as in the May 15 test episode. While no UFO wreckage was apparently recovered from the May 29 incident, this did not remain the case with subsequent attempts to shoot down and retrieve intruding UFOs for analysis to satisfy the Truman administration's request.

Here is how Kissner summarized the strategic situation after UFOs began interfering with the V2 missile launches, including events that occurred soon after the two May incidents, such as Roswell:

> At 4:11 p.m., Mountain Standard Time, Thursday, May 15, 1947, above the Tularosa Basin in south central New Mexico, the most highly classified episode in American military history began a series of events, decisions and actions that ultimately resulted in the crash and recovery of at least two, probably three and possibly four, extraterrestrial 'flying discs.' This was accomplished by deliberate antiaircraft artillery fire and/or use of surface-to-air missiles by units of the U. S. Army National Antiaircraft Artillery School at Ft. Bliss, Texas and/or U. S. Army Air Force fighters based at Alamogordo Army Air Field at Alamogordo, New Mexico. The well-publicized 'Roswell' (Corona), New Mexico flying disc was one of these. [178]

Kissner's analysis, based on his interviews and research, is astounding. The Roswell UFO crash was no accident. It was a flying saucer that had been shot down in the vicinity of White Sands Proving Ground – Roswell is only 128 miles (205 km) due east of White Sands. When we consider that the 1942 Los Angeles Air Raid incident resulted in two craft being shot down from an artillery barrage, the conclusion is that despite the highly advanced interplanetary propulsion system of UFOs, they were still vulnerable to the different weapons systems possessed by US military forces at the time.

Kissner studied newspaper reports from the period UFOs first began being targeted by the US military and found a clear spike in aviation crashes immediately following the May 29, 1947, incident. A diagram compiled by Kissner in his book, *Peculiar Phenomena*, shows a spike that involved the death of 198 people from air crashes in the 72 hours following the V-2 Juarez, Mexico, incident.

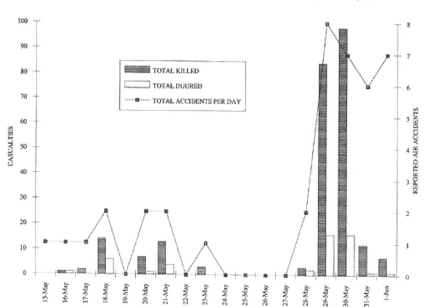

Figure 25. In the 72 hour period following the V-2 Juarez incident, 29 airplanes crashed around the world killing 198 people. Graphic © 1994 by J. Andrew Kissner.

Linda Moulton Howe reviewed Kissner's work and summarized his findings on the air accidents as follows:

> Andrew Kissner reports that within 72 hours of the "Juarez Incident," twenty-nine airplanes crashed around the world killing 198 people. The worst disaster was near La Guardia Airport in New York where a United Airlines plane took off, crashed and burned, killing all 38 people aboard. While government Civil Aeronautics Board (CAB) investigators were returning from the La Guardia

crash, an Eastern Airlines DC-4 went nose down and crashed, killing everyone aboard.[179]

The conclusion drawn by Kissner was that the civilian and military airplane crashes were a form of retaliation by those behind the UFO phenomenon:

> Historical research suggests, however, something even more controversial than the 'shoot downs.' The U. S. Army's actions either caused, or were a response to, offensive overt acts initiated by flying discs themselves. Even though unproved, it appears that flying discs might have been a contributing factor in the deaths of over six hundred military personnel and civilians in 1947 over a two month period from May 9 to July 31. Whatever flying discs were, or are, they were not, and are not, defenseless. Their 'special weapons' were, and even now are, unconventional. Flying discs were immediately defined by the U. S. military in 1947 as an 'enemy weapons system'". A T-Force assigned to secure flying disc crash sites and collect debris." [180]

Kissner's stunning research ultimately leads to the conclusion that either extraterrestrials and/or the Antarctic-based Germans were striking back by targeting both military and civilian aircraft in response to attacks against their flying saucers which were monitoring or interfering with V-2 rocket tests.

## The Escalation of Flying Saucer Engagements

The UFOs were not just openly observing the Army's V-2 rocket tests, but also very active in flying near highly sensitive US military facilities to monitor them, especially when it came to the storage or deployment of nuclear weapons. This included Roswell Army Air Field, which at the time of the Roswell flying saucer crash had the only operational wing of nuclear bombers in the world. It was the presence of UFOs near such facilities that led to the Army Air Force issuing intercept and shoot down orders against the UFOs. These craft could have either been German flying saucers from Antarctica or extraterrestrial piloted craft, for at least one was known to have crashed near Roswell.

In July 1947, only two months after the Roswell crash, Kissner found that the Army Air Force had lost five aircraft in the state of New Mexico. In his 1994 report, Kissner showed how the USAF itself linked these "aircraft losses" to the Roswell incident:

> In September 1994, in the official US Air Force Study related to the 'Roswell Incident', the Air Force reports no aircraft accident as having occurred in or near the crash site at the same time…. In addition, the Air Force reported losing five aircraft in New Mexico during July 1947. With the exception of the P-80, which is mentioned to have crashed somewhere in New Mexico, the Air Force reports four other aircraft losses during July that I wasn't able to find in any published news sources from 1947.[181]

Kissner believed that these aircraft were some of the earliest victims of retaliation by the UFOs. Thus, a long series of military and civilian airplane crashes began, often resulting from the hot pursuit of UFOs by Army Air Force fighter jets.

The military/CIA whistleblower Kewper/Stein told Linda Moulton Howe how UFOs were being targeted around the world by both the United States and USSR:

> The CIA was working to increase the different radars around the United States so they could detect more of the alien craft coming in if they come from outer space and would be able to scramble jets to check them out – or shoot them down.[182]…

It could have been something that happened since the 1947 time period. I think there was an event over the Soviet Union. I know the Soviets had a big craft go over that was like a 'mothership.' They tried to shoot it down and the thing shot all their aircraft pursuing it right out of the air in only a few seconds using lasers or whatever the extraterrestrial craft's weapons were. [183]

One of the most well-known cases of the military's aggressive attempts to intercept and shoot down a UFO involved Captain Thomas Mantel, whose F-51 (aka P-51) Mustang crashed on January 7, 1948. Mantel was serving at the time with the Kentucky Air National Guard when he and three other pilots were ordered to intercept a UFO. Project Bluebook's Edward Ruppelt described the multiple sightings and reports from state police that led to the intercept orders:

> The towns people had described the object to the state police as being "circular, about 250 to 300 feet in diameter," and moving westward at a "pretty good clip." ... About this time Colonel Hix, the base commander, arrived. He looked and he was baffled. At two thirty, they reported, they were discussing what should be done when four F-51's came into view, approaching the base from the south.

> The tower called the flight leader, Captain Mantell, and asked him to take a look at the object and try to identify it. One F-51 in the flight was running low on fuel, so he asked permission to go on to his base. Mantell took his two remaining wing men, made a turn, and started after the UFO. The people in Godman Tower were directing him as none of the pilots could see the object at this time. They gave Mantell an initial heading toward the south and the

flight was last seen heading in the general direction of the UFO.

By the time the F-51's had climbed to 10,000 feet, the two wing men later reported, Mantell had pulled out ahead of them and they could just barely see him. At two forty-five Mantell called the tower and said, "I see something above and ahead of me and I'm still climbing." All the people in the tower heard Mantell say this and they heard one of the wing men call back and ask, "What the hell are we looking for?" The tower immediately called Mantell and asked him for a description of what he saw. Odd as it may seem, no one can remember exactly what he answered. Saucer historians have credited him with saying, "I've sighted the thing. It looks metallic and it's tremendous in size.... Now it's starting to climb." Then in a few seconds he is supposed to have called and said, "It's above me and I'm gaining on it. I'm going to 20,000 feet." Everyone in the tower agreed on this one last bit of the transmission, "I'm going to 20,000 feet," but didn't agree on the first part, about the UFO's being metallic and tremendous.[184]

While Ruppelt downplayed Mantell's description of the craft as "metallic and of tremendous size" since saucer historians were the source for the quote, a joint Air Force and Navy intelligence analysis confirmed that these were the very words which Mantell had used. The intelligence analysis also provided more details from Mantell's communication: "It appears to be metallic object ... of tremendous size ... directly ahead and slightly above ... I am trying to close for a better look."[185]

Furthermore, Mantell's final radio transmission was consistent with what the townspeople had witnessed; a circular craft about 300 feet in diameter moving quickly. Mantell took his F-51 into a steep climb while pursuing the UFO and soon after crashed. The official explanation was that Mantell had blacked out due to the lack of oxygen at the high elevation and had not recovered in time to prevent his plane from crashing. This led to a big unanswered question over why an experienced Air Force pilot like Mantell had committed such a grave error.[186] Mantell had an oxygen mask after all. The answer to the mystery is quite simple: the Air Force was lying to cover up what really happened.

Leonard Stringfield, a former USAF intelligence officer learned the truth from one of his more reliable sources:

> I have heard a number of exotic stories about the Mantell incident, but one stands out that comes from a reliable source [who] related that he had talked with Mantell's wingman, who witnessed the incident. The pilot stated that Mantell pursued the UFO because he was the only pilot equipped with an adequate oxygen mask [and] also related that he saw a burst of what appeared to be tracers fired from the UFO, which hit the P-51 [aka F-51] and caused it to disintegrate in the air. Since the Mantell case, all other military encounters ending in disaster have been hidden from the public.[187]

Stringfield's source revealed that fighters that got too close to the flying saucers would be fired upon. Major Donald Keyhoe, a former Marine Corps aviator, also had reliable sources within the military who told him about the Air Force covering up the truth that UFOs were shooting down pursuing aircraft up to the early 1950's:

> Back in the early 1950s, when I knew of the jet scrambles that sometimes led to disaster, I agreed with Major Keyhoe's writings that in these we may know the intent of the UFO. I also agreed that 'losing our aircraft to the UFO' may have been the reason for official secrecy, fearing that the public would panic if they knew the truth.[188]

Frank Fechino, author of *Shoot Them Down! The Flying Saucer Air Wars of 1952* described the air battles that occurred in the summer of 1952 between flying saucers and fighter planes from the U.S. and other countries. Fechino's book title was no exaggeration. While heading Project Bluebook, Edward Ruppelt read many official reports of Air Force pilots actively targeting and shooting at UFOs. Ruppelt described being shown one classified report of an incident which occurred in 1952. The base commander had wanted Ruppelt to see it before having to destroy it due to the Air Force hierarchy wanting to cover up such incidents. Ruppelt explained what he read in the F-86 pilot's report about his hot pursuit of a UFO that he had shot at:

> Again the pilot pushed the nose of the F-86 down and started after the object. He closed fairly fast, until he came to within an estimated 1,000 yards. Now he could get a good look at the object. Although it had looked like a balloon from above, a closer view showed that it was definitely round and flat saucer shaped. The pilot described it as being "like a doughnut without a hole." ...
>
> When the object traveled out about 1,000 yards, the pilot suddenly made up his mind – he did the only thing that he could do to stop the UFO. It was like a David about to do battle with a Goliath, but he had to take a chance. Quickly charging his guns, he started shooting. A moment later the object pulled up into a climb and in a few seconds it was gone. The pilot climbed to 10,000 feet, called the other F-86, and now was able to contact his buddy. They joined up and went back to their base.
>
> As soon as he had landed and parked, the F-86 pilot went into operations to tell his story to his squadron commander. The mere fact that he had fired his

guns was enough to require a detailed report, as a matter of routine. But the circumstances under which the guns actually were fired created a major disturbance at the fighter base that day.[189]

The heavy loss of life and aircraft that occurred during this period came from multiple sources reporting on the incidents. Ruppelt described the compiled results:

> In just June and July of 1952, the U.S. lost NINETY-FOUR fighter jets worldwide and 51 men confirmed killed. Planes were falling to the ground two and three at a time in some cases. The entire month of September records are missing. According to the NY Times ONE HUNDRED AND NINETY-TWO aircraft vanished or were destroyed between the years 1951 and 1956.[190]

It is these summer of 1952 incidents which most challenged the assumption held by the majority of researchers and historians; they believed that the UFOs being intercepted and shot at, and which were retaliating against their pursuers and even against innocent civilian aircraft, were extraterrestrial in origin. However, the Washington DC 1952 flyovers have provided the most compelling evidence that the flying saucers often observed in squadron formation were, in fact, piloted by Antarctic-based Germans.

# CHAPTER 8

# Did Extraterrestrial or German Saucers Fly over Washington DC?

*Uninvited guests seldom meet a welcome.*

—Aesop

The 1952 "Washington DC Flyover" is a key historical event which has been well documented by multiple sources. Over three successive weekends beginning on July 12, 1952, fleets of UFOs (or flying saucers as they were known at the time) flew over the restricted airspace of Washington DC. The two most publicized flyover events were on the weekends of July 19-20 and July 26-27.[191] They flew directly over the White House, the Capitol Building, and the Pentagon. Thousands witnessed the UFO which were tracked by multiple radar installations, pursued by Air Force jets and photographed. Major Donald Keyhoe and other leading UFO researchers of the era thought the *Washington flap* was the best evidence yet of extraterrestrials visiting Earth.[192]

Major newspapers gave prominent coverage to the sightings, with headlines announcing to millions that US Air Force jets were in hot pursuit of the saucer-shaped lights that had flown over the nation's capital. *The Washington Post* ran with the headline, "'Saucer' Outran Jet, Pilot Remembers". The *New York Daily News* blared on its front page, "Jets Chase D.C. Sky Ghosts". The *Times Herald Sun* declared "Jets Prepared For Fight With Saucers Here:

Ready to Do Anything If They Return", and the *Washington Daily News* led with the headline, "Aerial Whatzits Buzz D.C. Again!"

Illustration by The Washington Post

In the summer of 1952, strange stories of jets chasing UFOs over the White House grabbed front-page attention in newspapers across the nation.

**Figure 26. Summer of 1952 Newspaper Headlines**

On the 50th year anniversary of the event, the *Washington Post* wrote:

> It was Saturday night, July 19, 1952 – 50 years ago – one of the most famous dates in the bizarre history of UFOs. Before the night was over, a pilot reported seeing unexplained objects, radar at two Air Force bases – Andrews and Bolling – picked up the UFOs,

and two Air Force F-94 jets streaked over Washington, searching for flying saucers.

Then, a week a later, it happened again – more UFOs on the radar screen, more jets scrambled over Washington....

The UFOs appeared on radar at the flight control centers of National Airport, Andrews Air Force Base, and Bolling Air Force Base, as the *Washington Post* went on to describe:

> The blips first appeared on radar screens at National at 11:40 that Saturday night: seven unidentified targets about 15 miles southeast of the city.... They'd cruise leisurely at 100 to 130 mph, then abruptly zoom off in an extraordinary burst of speed.... Barnes, the head controller ... called his counterparts at Andrews and Bollings to ask if they saw anything unusual on their radar screens. They were getting blips in the same places....[193]

Both radar trackers and pilots physically saw the UFOs which witnesses described as saucers or glowing objects:

> At National, Cocklin looked out his window and saw what he recalls as a "whitish blue light" emanating from a solid object that was "round with no distinguishing marks such as wings or a nose or a tail." It looked, he says, "like a saucer".

> Civilian planes flying into Washington reported seeing strange glowing objects in places where the radar was getting blips ...[194]

The flying saucers could easily outrun the Air Force F-94 Starfire jets sent after them:

One pilot chased, but couldn't catch, the streaking light. "I tried to make contact with the bogies below 1,000 feet," pilot William Patterson told investigators. "I was at my maximum speed but... I ceased chasing them, because I saw no chance of overtaking them.[195]

The US Air Force held a press conference on July 29, 1952, and Major General John Samford claimed that it was all due to a "temperature inversion".[196] Even President Truman became involved when his advisor called on Project Blue Book for a report and was given the official Air Force explanation.[197] Much of the media and the general public were confused by the official Air Force declaration but dutifully ignored the issue out of national security concerns. For over sixty years, the secret of who really flew over Washington DC and other major US cities in 1952 has been withheld from the general public.

UFOs were often seen in the Washington DC area according to Edward Ruppelt, who wrote at length about the multiple sightings in his Project Blue Book "Report on Unidentified Flying Objects":

> The investigation brought out a few more points on the pro side too. We found out that the UFO's frequently visited Washington. On May 23 fifty targets had been tracked from 8:00 P.M. till midnight. They were back on the Wednesday night between the two famous Saturday night sightings, the following Sunday night, and again the night of the press conference; then during August they were seen eight more times. On several occasions military and civilian pilots saw lights exactly where the radar showed the UFOs to be.[198]

Who was behind the flying saucers over Washington DC – extraterrestrials, Germans from Antarctica, an alleged Inner Earth civilization, or a combination of all three? For an answer, we need to begin with a mysterious prediction of the July flying saucer sightings, which Edward Ruppelt wrote about in his book:

> A few days prior to the incident a scientist, from an agency that I can't name, and I were talking about the build-up of reports along the east coast of the United States. We talked for about two hours, and I was ready to leave when he said that he had one last comment to make — a prediction. From his study of the UFO reports that he was getting from Air Force Headquarters, and from discussions with his colleagues, he said that he thought that we were sitting right on top of a big keg full of loaded flying saucers. "Within the next few days," he told me, and I remember that he punctuated his slow, deliberate remarks by hitting the desk with his fist, "they're going to blow up and you're going to have the granddaddy of all UFO sightings. The sighting will occur in Washington or New York," he predicted, "probably Washington."[199]

Importantly, this government scientist made his startling prediction of the Washington DC sightings taking place based on classified reports "from Air Force Headquarters, and from discussions with his colleagues". So, who had foreknowledge of the coming wave of UFO sightings at Air Force Headquarters and how were scientists involved? For an answer, we can turn to contactee George Van Tassell, who wrote a letter to the Air Force on behalf of human-looking extraterrestrials from an interplanetary alliance.

## George Van Tassel's Warning About UFO Overflights

Van Tassel is best known for having organized major conventions at Giant Rock, California, that began in 1954, which focused on individuals claiming to have been contacted by human-looking extraterrestrials, today called Nordics. The biggest of his conventions had 11,000 people attending in 1959,[200] and very large crowds continued to come throughout the 1960's until the Vietnam War protests eclipsed public interest in extraterrestrial contact stories. It was Van Tassel's interactions with a group of Nordics that led to him being given forewarning about the 1952 Washington DC Flyover.

Van Tassel was the first to describe a vast fleet of spaceships belonging to a quasi-military space order called the "Ashtar Command". Its titular head was "Ashtar", described in one of the messages received by Van Tassel as representing the executive branch of a Galactic authority/legislature headed by the "Council of the Twelve Lords and the Council of the Seven Lights."[201] The Ashtar Command was part of an interplanetary alliance of worlds, widely referred to as the "Galactic Federation", which was making contact with governments and private citizens. Van Tassel's relationship with the Nordics/Galactic Federation was directly communicated to the highest levels of government and military in the United States.

It was in January 1952 that Van Tassel first began receiving his messages. President Truman at the time was in his final year in office — Dwight Eisenhower was elected later that year on November 4. Van Tassel relayed information given to him by the Nordics/Galactic Federation to the US Air Force at Wright Patterson Air Force Base several months before the detonation of the very first Hydrogen bomb test on November 1, 1952. The Nordics said that it would conduct a show of force to deter US authorities from moving forward with its nuclear development program.

In a small book giving the chronology of his communications with the Galactic Federation and messages he relayed to the Air Force, Van Tassel said he received the following brief message on March 7, 1952: "Watch your skies in your months of May to August."[202] On April 6, there was more information conveyed about what was about to take place, and a warning to the Air Force:

Your Pentagon will soon have much to muddle over.

> We are going to give this globe a buzz. I hope they
> do not intercept us from in front. [203]

Here is what Van Tassel had to say about this cryptic April 6, 1952
communication which alluded to the danger of trying to intercept
the Galactic Federation's spacecraft:

> This message received in April was carried out three
> months later in the latter part of July. The Pentagon,
> can only mean Washington, D.C. There is no doubt
> they had "much to muddle over." The "buzz" was
> accomplished by the saucer beings. The statement
> that the saucers hoped the Air Force would not
> intercept them in front, indicates that the saucers
> also knew in advance that there would be an
> attempt to intercept them. Is it coincidence that a
> letter mailed by me to Air Forces Intelligence
> Command, at the request of the Saucer beings in the
> July 18th message, was in their hands when the
> "buzz" occurred? I do not comprehend how the
> letter's arrival, the "buzz," the reference to the
> Pentagon, and the expected interception, can all be
> coincidence. My belief is that the saucer beings
> timed it that way, to let the Air Force know that this
> information was authentic. Their return receipt
> showed they received the letter July 22, 1952. The
> "buzz" was on July 26, 27 and 28th. [204]

The "buzz" referred to the successive weekend flyovers above
Washington DC in mid-to-late July 1952. This is very likely the
source for the advance knowledge of the flyover that Ruppelt
wrote about, which Air Force Headquarters and government
scientists had been made aware of. Van Tassel received a message
on July 18 from the Galactic Federation (a day before the July 19
flyover) which he relayed to the Air Force before the final flyovers
on July 26-28.

Hail to you beings of Shan [Earth]. I greet you in love and peace. My identity is Ashtar, commandant quadra sector, patrol station Schare, all projections, all waves. Greetings. Through the Council of the Seven Lights you have been brought here, inspired with the inner light to help your fellow man. You are mortals, and other mortals can only understand that which their fellow man can understand. The purpose of this organization is, in a sense, to save mankind from himself. Some years ago, your time, your nuclear physicists penetrated the "Book of Knowledge." They discovered how to explode the atom. Disgusting as the results have been, that this force should be used for destruction, it is not compared to that which can be. We have not been concerned with their explosion of plutonium and U235, the Uranium mother element; this atom is an inert element.

We are concerned, however, with their attempt to explode the hydrogen element. This element is life giving along with five other elements in the air you breathe, in the water you drink, in the composition of your physical self. In much of your material planet is this life giving atomic substance, hydrogen. Their efforts in the field of science have been successful to the extent that they are not content to rest on the laurels of a power beyond their use; not content with the entire destruction of an entire city at a time. They must have something more destructive. They've got it....

Our message to you is this: You shall advance to your government all information we have transmitted to you. You shall request that your government shall immediately contact all other earth nations regardless of political feelings. Many of your physicists with an inner perception development have refused to have anything to do with the explosion of the hydrogen atom. The explosion of an atom of inert substance and that of a living substance are two different things. We are not concerned with man's desire to continue war on this planet, Shan. We are concerned with their deliberate determination to extinguish humanity and turn this planet into a cinder.

... Our missions are peaceful, but this condition occurred before in this solar system and the planet Lucifer was torn to bits. We are determined that it shall not happen again. The governments on the planet Shan have conceded that we are of a higher intelligence. They must concede also that we are of a higher authority. We do not have to enter their buildings to know what they are doing. We have the formula they would like to use. It is not meant for destruction. Your purpose here has been to build a receptivity that we could communicate with your planet, for by the attraction of light substance atoms, we patrol your universe.

"To your government and to your people and through them to all governments and all peoples on the planet of Shan, accept the warning as a blessing that mankind may survive."[205]

The message claimed that Ashtar and the Galactic Federation were acting on a higher authority than human governments in their efforts to warn humanity about the dangers of thermonuclear weapons (aka hydrogen bombs). The message's reference to the opposition of some atomic scientists to thermonuclear weapons was based on the determined protests by a group of scientists led by Robert Oppenheimer, "father of the atomic bomb", to the development of hydrogen bombs.[206] It's worth emphasizing that hydrogen bombs are more than a thousand times more powerful than atomic bombs. For example, the November 1, 1952, Ivy Mike test involved a thermonuclear device with a destructive force of just over 10 megatons, which was nearly 1000 times the destructive force of the atomic bomb dropped at Hiroshima. [207] Nearly a decade later, on October 30, 1961, the Soviet Union detonated its Tsar Bomba, which was a hydrogen bomb with an estimated yield between 50-60 megatons.

Even the Tsar Bomba, however, was dwarfed in comparison to the country-killer proposed by Dr. Edward Teller, who is widely acclaimed/reviled as the father of the hydrogen bomb, and who led an effort to discredit Oppenheimer. Rather than merely destroying a city, Teller's proposed 10,000 megaton (10 gigaton) monster would destroy an entire country:

> A 10,000 megaton weapon, by my estimation, would be powerful enough to set all of New England on fire. Or most of California. Or all of the UK and Ireland. Or all of France. Or all of Germany. Or both North and South Korea. And so on.[208]

The danger of thermonuclear weapons becoming so powerful that they could destroy an entire planet, as the Nordics warned, was very real if scientists such as Teller ever got the approval to build gigaton-scale hydrogen bombs.

Apparently, this was not the first time such destructive gigaton-scale hydrogen bombs had been built in our solar system, according to the Galactic Federation's reference to the planet "Lucifer" (aka Maldek or Tiamat) which was destroyed by nuclear weapons. Was the asteroid belt the remnants of another planet in our solar system that was destroyed by thermonuclear weapons? US Naval Astronomer, Dr. Thomas Van Flandern, has presented compelling data that the asteroid belt is indeed the remnants of an exploded planet, and furthermore, claims that the planet Mars was previously a moon of this destroyed planet.[209]

Dr. John Brandenberg has presented a cogent scientific argument that Mars once experienced a nuclear war in its ancient history which devastated the planet.[210] When combining Van Flandern and Brandenberg's information, the conclusion reached is that our solar system has certainly been subjected in its distant past to a thermonuclear war that destroyed one planet and devastated another using enormous gigaton-scale hydrogen bombs. In *Antarctica's Hidden History*, I discuss how refugees from Mars and a former "Super-Earth" in the asteroid belt (aka Lucifer/Maldek/Tiamet) escaped to Earth, and not only were instrumental in introducing advanced technologies to humanity but also for building ancient civilizations such as Atlantis.[211]

Given this disturbing history in our solar system and the danger of thermonuclear war happening again, it is understandable why the Galactic Federation was intent on delivering such a stern warning to humanity, especially the Air Force. At the time, only the Air Force had the capability of dropping nuclear weapons through its strategic bomber fleets, which could travel to any location on Earth. In 1952, the USAF was headed by Generals Hoyt Vandenberg (Air Force Chief of Staff) and Nathan Twining (Air Force Vice Chief of Staff) who were both big supporters of the strategic bombing doctrine that was credited with winning World War II. The research and development of thermonuclear weapons was strongly backed by Vandenberg and Twining, who both were founding members of the Majestic 12 Group according to the "Eisenhower Briefing Document".[212]

Van Tassell said that his message from the Galactic Federation was received and immediately relayed to the Air Force on July 18, 1952, and officially received on July 22. Consequently, Air Force officers were aware of the Galactic Federation's warning and their plans to stage a massive flyover of Washington DC. It's certain that both Vandenberg and Twining received Van Tassel's warning. What role did the Antarctic-based Germans and their extraterrestrial allies play in the Washington DC Flyover? Were they passive witnesses to the Galactic Federation buzzing the

United States Capitol, or did they actively participate in some way to show off their own advanced technologies to set the stage for future negotiations with US authorities?

## Antarctic German Colony and the Washington DC Flyover

After the dismal failure of "Operation Highjump" Admiral Byrd warned of a hostile force that could fly over US territory from the polar regions, which certainly referred to the German spacecraft operating out of Antarctica. Importantly, this powerful German contingent was not acting alone but was immeasurably helped by their extraterrestrial and Inner Earth allies. In *Antarctica's Hidden History*, I described at length how the Germans acquired both extraterrestrial and Inner Earth partnerships as a result of the original Vril Society channelings in the early 1920's, which established communications with both Nordics from Aldebaran and an Inner Earth civilization associated with the ancient Hyperboreans, the alleged racial progenitors of the Aryans.[213] The Vril Society faction, headed by Maria Orsic, went on to establish a presence in Antarctica and closely worked with the Inner Earth civilization connected to the Hyperboreans – known as the "Arianni".

In the alleged posthumous diary of Admiral Byrd reference is made to the "Arianni", an Inner Earth civilization he purportedly met during the 1947 Operation Highjump expedition. While Byrd's diary is very controversial, there is a good reason to believe that key elements are based on fact. It is known, for example, that Byrd's surveillance flight over Antarctica on February 19, 1947, was shrouded in uncertainty when radio contact was broken, during which time his plane was thought lost, but then mysteriously reappeared.[214] In addition, Corey Goode says that the Arianni were mentioned in intelligence briefing documents which he had access to while he was assigned to the Navy's "Solar Warden" secret space program. Goode reported that the Arianni, along with the Draco Reptilians, helped the Germans establish their Antarctic bases:

There was help from the Draco Federation as well as a group that the Nazi's were led to believe were ET's (referred to as "Arianni" or "Aryans", sometimes called "Nordics") but were actually an Ancient Earth Human Break Away Civilization that had developed a Space Program (referred to as "The Silver Fleet") and created vast bases below the Himalayan Mountains (largest in Tibet and called the system Agartha) and a few other regions.[215]

If we accept Goode's testimony and Byrd's reference to the Arianni as factually based, then it's worth also considering what Byrd was told by the Arianni during his brief 1947 encounter.

Byrd's diary details the Arianni's vehement opposition to the development of atomic weapons, which was explained to him by one of their elders (also called Masters).

> You are in the domain of the Arianni, the Inner World of the Earth. We shall not long delay your mission, and you will be safely escorted back to the surface and for a distance beyond. But now, Admiral, I shall tell you why you have been summoned here. Our interest rightly begins just after your race exploded the first atomic bombs over Hiroshima and Nagasaki, Japan. It was at that alarming time we sent our flying machines, the "Flugelrads", to your surface world to investigate what your race had done....
>
> You see, we have never interfered before in your race's wars, and barbarity, but now we must, for you have learned to tamper with a certain power that is not for man, namely, that of atomic energy. Our emissaries have already delivered messages to the powers of your world, and yet they do not heed. Now you have been chosen to be witness here that our world does exist.[216]

135

The Arianni told Byrd that if the U.S. abandoned its quest for increasingly destructive nuclear weapons, assistance would be given to humanity to uncover the many wondrous secrets of the Earth's distant past.

> We see at a great distance a new world stirring from the ruins of your race, seeking its lost and legendary treasures, and they will be here, my son, safe in our keeping. When that time arrives, we shall come forward again to help revive your culture and your race.
>
> Perhaps, by then, you will have learned the futility of war and its strife ... and after that time, certain of your culture and science will be returned for your race to begin anew. You, my son, are to return to the Surface World with this message....'[217]

While the U.S. refused the offer, it appears that years earlier the Germans had accepted it. This helps explain why the Germans were given sanctuary in Antarctica and assistance from one of the advanced Inner Earth civilizations, as claimed by Goode. The Germans gave up their atomic weapons program even though it was far more advanced than official historians have acknowledged. In fact, it's very likely that the Germans handed over their atomic research to the United States at the end of World War II as a conditional term for allowing many of its Nazi leaders to escape to safe havens in South America, as contended by researchers such as Dr. Joseph Farrell.[218]

In addition to working with an Inner Earth civilization, the Vril-Orsic faction of the German Antarctic colony also cooperated with the Galactic Federation in helping to raise human consciousness, an effort that included warning people about the dangers of nuclear weapons. A number of contact cases actually involved the Orsic faction masquerading as extraterrestrials.[219] Even famous contactee cases offer examples, such as Billy Meier's

Pleiadians/Plejarans who were in fact members of the German colony in Antarctica pretending to be extraterrestrials. In *Antarctica's Hidden History*, I showed how the craft Meier photographed were very similar in design to the Nazi SS "Haunebu" flying saucers.[220] Also, Goode said in a 2016 lecture that when a photo of Maria Orsic was shown to Meier, he identified her as "Semjase", his principal extraterrestrial contact:

> ... when the military found out about Meier's case, they sent people over with some photographs for him to try and identify the female being he saw. He quickly pointed out one photograph, saying, "That's her! That's her!" Apparently the photo he pointed out was of Maria Orsic, the medium from the Vril Society, who was making contact with inner-Earth groups, and who played an intimate role in the pre and post World War II German secret space program.[221]

Compounding the already complex alliances involving the German Antarctic colony and its space program were the agreements it had reached with the Draconians. Both William Tompkins and Corey Goode have referred to the role played by the Draconians in helping the Germans to establish their colony in Antarctica.[222] Tompkins explained that this German-extraterrestrial alliance was well known to the US Navy due to the intelligence gained from its embedded spies working within Nazi Germany between 1942 to 1946, who had witnessed firsthand the stunning technological advances made by the Nazi regime as a direct result of the Draconians' support. Though the Nazi flying saucers were not weaponized in time to win World War II, the weaponization had been completed by early 1947 and allowed them to defeat Byrd's Operation Highjump comprehensively.

Therefore, to understand the German Antarctic space program, it is important to acknowledge that there were two main factions which had their own respective alliances with extraterrestrial groups. While one faction cooperated with the Draconian extraterrestrials and built an interstellar fleet of spacecraft for imperial conquest, which Goode has called the "Dark Fleet", the other German faction cooperated with an Inner Earth civilization (Arianni) and the Nordics from the Aldebaran star system. This meant that the Orsic faction of the German colony was linked to various positive activities being carried out by the Galactic Federation, such as raising human consciousness and eliminating nuclear weapons. Consequently, the communications that Van Tassell received from the Ashtar Command/Galactic Federation were very likely linked to, or influenced by, the German Antarctic colony.

To recap, the Kenneth Arnold incident was the first example of German Antarctic squadrons flying over US territory, as evidenced by the similarity in design between the German-made "Horton 229 flying wing craft" and the "wing craft" Arnold witnessed over the Cascades. In addition, extraterrestrial spacecraft were interfering with rocket tests and monitoring nuclear facilities, which led to confrontations because of USAF pilots had been ordered to intercept or shoot down the flying saucer craft. Subsequent dog fights between USAF jets and the German saucers or those of their extraterrestrial allies led to casualties on both sides. US military and civilian craft also began crashing in increasing numbers after the Juarez V2 UFO incident in May 1947. Up to 1952, German and extraterrestrial spacecraft continued to fly over US territory pursuing multiple agendas, one of which was to deter US authorities from developing thermonuclear weapons. Consequently, the German saucers from Antarctica were very likely involved in some capacity with the Washington DC Flyover. The extent of this German involvement is revealed by multiple insider sources.

William Tompkins stated in various interviews that the Antarctic-based Germans did the DC flyover. In a private conversation, I specifically asked Tompkins if Antarctic-based Nazi spacecraft flew over the U.S. in the summer of 1952. He replied:

It's of course a 'Yes'. Some had the swastika on them and some had the German cross, but they were all extraterrestrial type vehicles, okay, which the Germans had re-engineered, reversed, whatever, and were putting in production. So, those vehicles were not powered by extraterrestrials or extraterrestrial vehicles. These were German vehicles that had been given to Germany by the Reptilians, but these were production vehicles out of the production facilities in Antarctica.[223]

Similarly, Goode said in an interview:

They had also received intelligence from their Paperclip spies that the Americans had implemented an Executive Order making the existence of alien life the most classified subject on the planet. The reason being that the development and release of free energy would quickly destroy the oil trade, and soon thereafter the entire Babylonian Money Magic Slave System that all elites use to control the masses. The Nazi's used this to their advantage in some very public sorties over Washington, DC and highly Secret Atomic Warfare Bases to mention a few. Eisenhower finally relented and signed a treaty with them (and a few other groups, both ET and Ancient Civilizations pretending to be ET).[224]

Tompkins and Goode's controversial claims are corroborated by whistleblower Clark McClelland, who worked for 34 years with NASA and finished his career as a Spacecraft Operator. In the August 3, 2015, installment of his book, *The Stargate Chronicles,* McClelland wrote:

**Figure 27. Photo of 1952 Washington DC Flyover**

The over flights of advance very swift crafts over Washington, DC were these German advanced aircraft that totally out flew American advanced crafts. On July 12, 1952, President Truman observed several of the UFOs and was completely amazed by their capabilities of outmaneuvering the USAF ... advanced Jet fighter ...-[Lockheed F-94 Starfire]. USA jets sent up to bring one down. None could fly the speed of the German Saucers.[225]

McClelland also described the role of Nazi scientists who had fled to Antarctica in relation to the 1952 Washington DC Flyover:

Because I worked with the German Scientists that were brought to the USA by Dr. Werner von Braun in 1946/7. Several told me that WWII German Scientists by the many thousands escaped from Germany near the fall of Germany in WW II. They boarded advanced submarines in the Baltic Sea. They were all taken to the South Pole base located underground, in Antarctica. Some called it Hitler's Shangri La. Those scientists created advanced anti-gravity craft that were flying in our air space for many years. And still are. They were observed over Washington, DC in 1952 by President Harry S. Truman. Yes, we did not have any aircraft that could stop these German planes from flying over our national capital in 1952. So German scientific expertise was again showing the USA who was boss.[226]

So, did extraterrestrial or German saucers fly over Washington DC? The answer is found in the alliances that were made, and the common interest these partnerships all shared in pressuring the Truman administration to abandon its thermonuclear weapons development plans. Both the Galactic Federation, which had come to include the German Vril-Orsic faction in Antarctica, and the German-Draconian alliance based in Antarctica were, therefore, directly involved in the Washington DC Flyovers.

Only three days before the November 4, 1952, election of President Eisenhower, Truman authorized the 10 megaton Ivy Mike nuclear test, the first-ever detonation of a thermonuclear device. Truman had been advised by his national security team, which included Majestic 12 members and USAF Generals Vandenberg and Twining, to go ahead with the test despite the strong likelihood that Eisenhower's election could lead to a radical overhaul of the nuclear weapons program. Truman had been persuaded by Vandenberg and Twining to usher in the age of hydrogen bombs, thereby ensuring US Air Force dominance over strategic nuclear bombing for decades to come through its long-range bombers and the impending development of Intercontinental Ballistic Missiles (IBMs). This set the stage for the Galactic Federation and the Antarctic German-Draconian Alliance to soon engage in face-to-face meetings with President Eisenhower. Negotiations to come would involve comprehensive agreements dealing with the complex issues of thermonuclear weapons, extraterrestrial contact, advanced technologies, and the public disclosure versus non-disclosure of the German Antarctic colony's existence.

# CHAPTER 9

## President Eisenhower Meets with the Nordics: Nuclear Negotiations Go Awry

> Great numbers of children will be born who understand electronics and atomic power as well as other forms of energy. They will grow into scientists and engineers of a new age which has the power to destroy civilization unless we learn to live by spiritual laws.

> — Edgar Cayce

### Atoms for Peace – Eisenhower's Overture to the Galactic Federation

On December 8, 1953, President Dwight D. Eisenhower gave his famous "Atoms for Peace" speech before the United Nations General Assembly, where he bravely proposed dismantling America's nuclear arsenal if the Soviet Union would only agree to do the same. He made a serious offer to place America's nuclear technology under the control of the United Nations to ensure that nuclear energy would be used in the future for peaceful purposes rather than weapons development. Eisenhower's proposal was daring and visionary. It was a bold step towards world peace at a time of a rapidly expanding nuclear arms race with the Soviet Union, which had only four months previously (on August 12) detonated its first thermonuclear device.[227] Given the events associated with the 1952 Washington DC Flyover, there is a reason to believe that Eisenhower's proposal was also intended as an overture to the Galactic Federation, which had made its position clear on the United States having to abandon its nuclear weapons development as a precursor to receiving technological assistance from human-looking extraterrestrial civilizations.

Standing before the UN General Assembly, and witnessed by an audience that very likely secretly included representatives of the Galactic Federation, Eisenhower said:

> The United States would seek more than a mere reduction or elimination of atomic materials for military purposes. It is not enough to take this weapon out of the hands of the soldiers. It must be put into the hands of those who will know how to strip its military casing and adapt it to the arts of peace.

> The United States knows that if the fearful trend of atomic military build-up can be reversed, this greatest of destructive forces can be developed into a great boon, for the benefit of all mankind. The United States knows that peaceful power from atomic energy is no dream of the future. That capability, already proved, is here—now—today. Who can doubt, if the entire body of the world's scientists and engineers had adequate amounts of fissionable material with which to test and develop their ideas, that this capability would rapidly be transformed into universal, efficient and economic usage?

> To hasten the day when fear of the atom will begin to disappear from the minds of people and the governments of the East and West, there are certain

steps that can be taken now. I therefore make the following proposal.

The governments principally involved, to the extent permitted by elementary prudence, begin now and continue to make joint contributions from their stockpiles of normal uranium and fissionable materials to an international atomic energy agency.

The United States is prepared to undertake these explorations in good faith. Any partner of the United States acting in the same good faith will find the United States a not unreasonable or ungenerous associate. Undoubtedly, initial and early contributions to this plan would be small in quantity. However, the proposal has the great virtue that it can be undertaken without the irritations and mutual suspicions incident to any attempt to set up a completely acceptable system of world-wide inspection and control.[228]

Given the tensions created by the Korean War (1950 - 1953) and the deepening Cold War divide, Eisenhower's "Atoms for Peace" proposal was as much as could be hoped for by any American President. Only someone like Eisenhower, a career military five-star general, would prove confident and strong enough to ignore the demands of powerful national security officials who were firmly intent on expanding America's nuclear arsenal no matter the financial cost and security risks involved. In particular, the Air Force Chief of Staff Nathan Twining and analysts with the RAND Corporation were strongly in favor of creating a strategic nuclear bomber force equipped with thermonuclear weapons, which could devastate potential enemies in a first or second strike scenario anywhere on Earth. A senior RAND policy strategist, Bernard Brodie, wrote the first nuclear strategy papers in 1953 which were widely adopted by the US national security establishment, a key part of the Deep State.[229]

It is significant that the RAND Corporation was heavily involved in helping to develop the nuclear strategy on behalf of the Air Force. As mentioned earlier, the RAND Corporation was initially set up to assist the former Army Air Force to develop a comprehensive strategic response to extraterrestrial life and the German Antarctic presence. This meant that the development of thermonuclear weapons actually had less to do with any genuine national security threats posed by the Soviet Union, and had more to do with meeting potential future threats by the Antarctic Germans or their extraterrestrial allies.

Despite Eisenhower's willingness to go against the advice of senior national security officials with his Atoms for Peace proposal, the Soviet Union squandered the golden opportunity to set humanity on a path towards global cooperation, peace, and open diplomatic relations with the Galactic Federation. General Secretary of the Communist Party Joseph Stalin had died on March 5, 1953, and the Soviet Union had entered an uncertain period with no clear designated successor. Without a strong leader who could go against his military advisers' demands for nuclear weapons development, the Soviet Union was not prepared to seriously respond to Eisenhower's "Atoms for Peace" proposal. Instead, the Soviet leadership chose to send mixed messages through its state media outlets before officially rejecting the proposal outright.[230]

There is little doubt that Eisenhower was genuinely opposed to nuclear weapons proliferation and wanted to put the atomic genie back in the bottle. He was especially disturbed by the development of thermonuclear weapons which his predecessor President Truman had approved for development and testing – the first of which occurred on November 1, 1952, only three days before Eisenhower's election win:

> Eisenhower, the former general, understood the strategic necessity of a nuclear arsenal but saw no need "for us to build enough destructive power to destroy everything." He didn't fear the Soviets or

their arsenal, but he worried that as atomic bombs became cheaper and more prevalent, they might be viewed as just another conventional weapon waiting to be used. The deadlier hydrogen bomb only worried him more.[231]

With his "Atoms for Peace" initiative spurned by the Soviets who would press ahead with their own thermonuclear weapons program, Eisenhower had little choice but to back the USAF, the RAND Corporation, the Deep State, etc., and continue with Truman's policy for developing hydrogen bombs. The next upcoming milestone on the United States' nuclear weapons development calendar was the March 1, 1954, scheduled test of a deliverable hydrogen bomb, a necessary step for the proof testing of a thermonuclear device. Called the "Castle Bravo" test, this planned detonation was to have a six megaton bomb yield, but it would turn out to be far more powerful with a 15 megaton yield. For the Galactic Federation, preventing this test represented one last opportunity to pull humanity back from the brink of a thermonuclear arms race that could result in global devastation. A face-to-face meeting between Nordics from the Galactic Federation and the Eisenhower Administration was proposed and accepted.

## Eisenhower's Secret Meeting with the Nordics

In the late afternoon hours of February 20, 1954, while on a 'vacation' to Palm Springs, California, President Eisenhower went missing and did not reappear until the next morning. Whistleblowers allege that he was secretly brought to Edwards Air Force Base for a sensitive meeting which was never reported. Officially, when the President showed up the next morning at a church service in Los Angeles, reporters were told that he had simply been taken for an emergency dental treatment the previous evening. The disappearance fueled rumors that Eisenhower was using the fictitious dentist visit as a cover story for a highly classified event which took place at a nearby military facility. This clandestine engagement is possibly the most significant that any American President could conduct – an alleged 'First Contact' meeting with human-looking extraterrestrials at a US Air Force Base to negotiate over the country's nuclear weapons program, technological assistance from the Galactic Federation and officially disclosing the truth about the Nordic visitors.

There is both circumstantial and testimonial evidence that Eisenhower did have a meeting in 1954 with extraterrestrials. What follows are some of the question-raising circumstances surrounding Eisenhower's unscheduled winter vacation to Palm Springs, California, from February 17-24, 1954, which will later be shown to corroborate what whistleblowers have revealed about a secret meeting with Nordics that actually took place.

First, in the latter part of the day on Saturday, February 20 President Eisenhower mysteriously vanished, fueling press speculation that he had taken ill or even died. In a hastily convened press conference, Eisenhower's press secretary announced that the President had lost the cap from a tooth while eating fried chicken and had been rushed to a local dentist. The local dentist was introduced at an official function on Sunday, February 21 as "the dentist who had treated the President".[232] Veteran UFO researcher William Moore investigated the incident and concluded that the dentist's visit was just a cover story for Eisenhower's true whereabouts.

Second, it has been reported that President Eisenhower flew to nearby Norton Air Force Base where he disembarked and immediately boarded a C-45 airplane to travel to Edwards AFB (formerly Muroc Airfield) on the evening in question. According to Bill Kirklin, an Air Force medic stationed at the nearby Georgia Air Force Base, an ambulance was requested to stay on duty while personnel at Norton honored the visiting president. Kirklin claims that after Eisenhower's plane landed at Norton, he immediately got onto the C-45 to travel in the direction of Palmdale, California, which is adjacent to Edwards AFB.[233]

# ST. LOUIS POST-DISPATCH

ST. LOUIS, MONDAY, FEBRUARY 22, 1954 — 32 PAGES

## EISENHOWER GOES TO DENTIST AFTER CAP FALLS OFF TOOTH

PALM SPRINGS, Calif., Feb. 22 (AP)—President Eisenhower chipped a porcelain cap from a tooth Saturday night and made a hurry-up trip to a dentist for a repair job.

It happened while the President was having dinner with Mrs. Eisenhower at their vacation headquarters at Smoke Tree Ranch.

Paul H. Helms drove with the President to a Palm Springs dentist, Dr. F. A. Purcell, who quickly replaced the cap of an upper front tooth.

Eisenhower lost a cap on the same tooth in the presidential campaign.

Figure 28. Eisenhower's Cover Story

Third, Edward's AFB was closed for three days during the period of Eisenhower's Palm Springs visit. Renowned UFO researcher Lt Col Wendelle Stevens wrote about the base closure:

Mead Lane, publisher of The "Round Robin" newsletter out of San Diego, had an article on the Eisenhower visit and the Muroc events of that time, in February of 1954. He described a Los Angeles

Times news reporter, getting wind of the strange goings on at Muroc, who chartered a private airplane to take him to Muroc. That airplane was refused permission to land and was turned away because the base was closed to all air traffic. He then rented a car and drove to Muroc, to try to get in, but was turned away again at the base main gate because that base was closed. A number of researchers and former military personnel confirmed that the base was closed to all servicemen who tried to enter or leave. Events of tremendous national security must have been occurring for Edwards to be suddenly closed without any prior warning to base personnel.[234]

Another whistleblower named Bill Holden revealed his own research into the closure of Edwards Air Force Base during Eisenhower's top secret 1954 visit:

Now, history says that this is where he met the ETs and that an agreement was signed between the U.S. and the ETs. And as far as that a mothership was seen coming in, there were a number of UFOs coming in, and that the base was literally shut down for 3 days. I have been able to find that the base was shut down for 3 days. I've been able to find in civilian records, newspaper accounts, and everything else, as far as those facts were validated.[235]

Fourth, nine days after the alleged Edwards AFB meeting, the U.S. detonated its largest ever hydrogen bomb test at the Bikini Atoll on March 1, 1954. Called Operation Castle, this "Castle Bravo" test was the first in a series and at 15 megatons, was 1000 times more powerful than the 15 kilotons atomic bomb detonation on Hiroshima.[236] Only the Soviet Union has ever exploded hydrogen bombs more powerful. This was the first hydrogen bomb test by the Eisenhower administration. Eisenhower's decision to go ahead with the Castle Bravo test was almost certainly linked to his trip to California during which he mysteriously disappeared for the secret meeting at Edwards AFB on February 20/21, 1954.

According to the circumstantial evidence examined so far, we know that President Eisenhower was gone for an entire evening on February 20, and was reportedly taken from Palm Springs to Edwards Air Force Base. The nature of the President's unscheduled vacation, the incident of him vanishing, the dentist cover story, and his mysterious unrecorded flight to Palmdale only a week and a half before the first hydrogen bomb test all provide circumstantial evidence that the true purpose of his Palm Springs visit was not to vacation, but to make history. He would participate in an event of unmatched importance, yet due to the sensitive circumstances, it could not be disclosed to the general public.

## Whistleblowers and the Untold Story

Fortunately, there are a number of sources discussing various details about the extraterrestrial meeting at Edwards Air Force Base which, taken together, help to build a more complete picture of the actual events on February 20-21. The sources about to be discussed are based upon the testimonies of whistleblowers who witnessed events, read classified documents, saw undisclosed film footage or learned from their 'insider contacts' of the meeting. Let's begin with a former representative for the state of New Hampshire who provides important testimony that Eisenhower received a briefing document that set the scene for a 'First Contact' meeting. This whistleblower personally read the document, and the content he says it contained clearly indicates that Eisenhower was being prepared by his national security advisors to meet with extraterrestrials to discuss weighty topics.

The former state representative, Henry W. McElroy, released a video statement recorded on May 8, 2012, in Hampton, Virginia, revealing the content of a secret brief he saw which had been prepared for President Eisenhower concerning extraterrestrial life.[237] McElroy, a Republican, served on various committees during his time in the New Hampshire state legislature, and is best known for sponsoring a new "Gold Money" bill in 2004 that aimed to restore the use of gold and silver coins in the Granite State.[238] He successfully ran for the Republican primary for State Representative in the 2008 elections but did not win re-election in 2010. In his statement, McElroy claims that the brief revealed that extraterrestrials were present in the United States, they were benevolent, and a meeting with them could be arranged for Eisenhower.

McElroy explains in his statement that he saw the briefing document while serving on New Hampshire's "State Federal Relations and Veterans Affairs Committee".[239] In his official capacity at the time, McElroy says that he was regularly "updated on a large number of topics related to the affairs of our People, and our Nation." [240] He further identified the nature of the topics covered by the committee:

> As I understood it, some of those ongoing topics had been examined and categorized as Federal, State, Local development, and security matters. These documents related to various topics some of which spanned decades of our nation's history.[241]

One of the topics being studied included an official one-page brief to President Eisenhower written by unknown national security specialists. McElroy states:

> I would like to submit to our nation my personal testimony of one document related to one of these ongoing topics which I saw while in office, serving on the State Federal Relations and Veterans Affairs Committee. The document I saw was an official brief to President Eisenhower. [242]

McElroy's account of the contents of the briefing document is startling:

> To the best of my memory, this brief was pervaded with a sense of hope, and it informed President Eisenhower of the continued presence of extraterrestrial beings here in the United States of America. The brief seemed to indicate that a meeting between the President and some of these visitors could be arranged as appropriate if desired. 243

According to the brief, the extraterrestrials were benevolent.

> The tone of the brief indicated to me that there was no need for concern, since these visitors were in no way, causing any harm, or had any intentions, whatsoever, of causing any disruption then, or in the future. 244

As a retired state representative, McElroy's testimony carries weight since he is a direct eyewitness to a document viewed during the course of his official duties. His testimony supports the claims of other whistleblowers who have reported that President Eisenhower secretly traveled to Air Force facilities to have meetings with representatives of one or more extraterrestrial civilizations.

Now I turn to the claims by whistleblowers who witnessed the alleged "First Contact" meeting at Edwards. The son of a former Navy Commander attests that his father was present at the "First Contact" event on February 20-21, 1954. According to Charles L. Suggs, a retired sergeant from the US Marine Corps, his father Charles L. Suggs Sr. (1909-1987), a former Commander with the US Navy, attended the meeting at Edwards Air Force Base with Eisenhower.[245] Sgt Suggs recounted his father's experiences from the meeting in a 1991 interview with a prominent UFO researcher:

> Charlie's father, Navy Commander Charles Suggs accompanied Pres. Ike along with others on Feb. 20th. They met and spoke with 2 white-haired Nordics that had pale blue eyes and colorless

lips. The spokesman stood a number of feet away from Ike and would not let him approach any closer. A second Nordic stood on the extended ramp of a bi-convex saucer that stood on tripod landing gear on the landing strip. According to Charlie, there were B-58 Hustlers on the field even though the first one did not fly officially till 1956. These visitors said they came from another solar system. They posed detailed questions about our nuclear testing. [246]

Suggs' reference to human-looking Nordics is significant, as is the reference to nuclear testing. Again, it needs to be pointed out that the alleged Eisenhower extraterrestrial meeting occurred nine days before the Castle Bravo test of a 15 megaton hydrogen bomb. Suggs' recollection suggests that due to the Nordics alarm over the consequences of the US testing program, they initiated the meeting with Eisenhower.

Another testimony concerning the Eisenhower-Extraterrestrial meeting comes from a former US Air Force test pilot and colonel who has chosen to remain anonymous, allegedly due to a secrecy oath. This pilot told the following details to Lord Clancarty, a member of the Irish aristocracy who has vouched for the source's authenticity:

> The pilot says he was one of six people at the meeting... Five alien craft landed at the base. Two were cigar-shaped and three were saucer-shaped. The aliens looked humanlike, but not exactly.... The aliens spoke English, and supposedly informed the President that they wanted to start an "educational program" for the people of Earth...[247]

Yet another anonymous military eyewitness, a US Air Force officer, revealed to former Royal Air Force pilot Desmond Leslie that a flying saucer had landed at Edwards AFB. The witness told Leslie that President Eisenhower had been taken over to it:

> ... a disc, estimated to be 100 feet in diameter, had
> landed on the runway on a certain day. Men
> returning from leave were suddenly not allowed
> back on the base. The disc was allegedly housed
> under guard in Hangar 27, and Eisenhower was
> taken to see it.[248]

An important point to note is that the anonymous USAF officer
referred to the base being closed to returning personnel. This is
circumstantial evidence supporting previously presented
testimony, and further confirms that highly classified activities
were occurring at Edwards AFB over the period in question.

A similar account was given by an USAF air policeman
stationed at Edwards AFB on the night President Eisenhower
visited. What follows is a summary of a phone interview with his
widow that supports Leslie's witness:

> Her deceased husband worked as an MP for the
> USAF in the 1950s and told her that he was on guard
> duty at a secure hangar facility at Edwards Air Force
> base during the evening of February 20, 1954. He
> said that he saw President Eisenhower who was
> escorted inside the hangar. He was not aware of
> what occurred inside the hangar, and was under
> shoot-to-kill orders against any unauthorized
> person attempting to enter the facility. He stated
> that a flying saucer was stored inside the hangar.[249]

In his book, *Need to Know*, British UFO researcher Timothy
Good refers to another three eyewitness accounts of saucer-
shaped craft seen near or landing at Edwards AFB on February 20,
1954. Good writes:

> Gabriel Green, an American researcher, spoke to a
> military officer who claims to have witnessed the
> arrival of the craft at Muroc. At the time I was
> engaged in firing practice, under the command of a

general, said the officer. "We were shooting at a number of targets when suddenly five UFOs came flying overhead. The general ordered all the batteries to fire at the craft. We did so, but our fire had no effect on them. We stopped firing and then we saw the UFOs land at one of the base's big hangars." Two other witnesses, Don Johnson and Paul Umbrello, also claim to have witnessed one of the disc-shaped craft near Muroc on the evening of 20 February.[250]

**Figure 29. 1982 British Newspaper story on Eisenhower–Extraterrestrial meeting.**

It is curious that Eisenhower would attend a meeting with extraterrestrial ambassadors whose vehicles had been fired upon by base personnel earlier. Perhaps the general involved in artillery practice was following standing orders to shoot at UFOs and was not aware of the scheduled meeting.

More testimony comes from a Jesuit priest working within a Vatican intelligence organization, the Servizio Informazione del Vaticano (S.I.V.). Speaking confidentially in 2001 to Cristoforo Barbato, an Italian UFO researcher, the Jesuit confirmed that President Eisenhower met with extraterrestrials at Muroc Air Base (renamed Edwards) in 1954. The Jesuit claims that a senior Catholic bishop, Francis MacIntyre, subsequently flew to Rome to brief Pope Pius XII and that the S.I.V. was created as a consequence of Eisenhower's meeting. Barbato writes:

> According to this person the reason to establish [sic] the S.I.V. was the meeting with an Alien delegation at Muroc Air Field Base in February 1954 in presence of president Dwight Eisenhower and James Francis McIntyre, bishop of Los Angeles. After that incredible event McIntyre flew to Rome to refer everything to Pope Pius XII who decided to found the S.I.V with the aim to get every possible information [sic] about Aliens and how they interacted with the American Government.[251]

Another Italian UFO researcher, Luca Scantamburlo, found circumstantial evidence supporting the existence of the top secret Vatican organization "S.I.V.", and the "Omega" security classification the Jesuit claimed to possess. He interviewed Barbato and discovered more information about the alleged 1954 meeting:

> The Jesuit member of the S.I.V. told Barbato that on occasion of the secret meeting at Muroc Air Field Base, in 1954, military cameramen filmed the outstanding event "with three movie cameras (16 millimeters), detached in different places, loaded with color film and working by spring engines; this last rather unconformable resolution, because it compelled every cameraman to change reel every 3

minutes, it was necessary since in the presence of the Aliens and of their spacecrafts, the electrical engines of the biggest movie cameras did not work.[252]

Don Phillips is a former Air Force serviceman who was employed to work on clandestine aviation projects. He testified that he viewed a film and saw documents describing the 1954 meeting between President Eisenhower and extraterrestrials:

> We have records from 1954 that [there] were meetings between our own leaders of this country and ET's here in California. And, as I understand it from the written documentation, we were asked if we would allow them to be here and do research. I have read that our reply was well, how can we stop you? You are so advanced. And I will say by this camera and this sound, that it was President Eisenhower that had this meeting. [253]

Philips' reference to a film of the 1954 Eisenhower meeting is corroborated by the testimony of the Jesuit priest who disclosed Vatican records concerning the meeting.

Another version of Eisenhower's meeting is described by William Cooper, who served in the Office of Naval Intelligence on a briefing team for the Commander of the Pacific Fleet between 1970-1973. He had direct access to highly classified documents which he had to review to fulfill his briefing duties. Cooper's military records confirm that he served in this sensitive position as claimed.[254] This lends credence to the fact that some of his testimony is based on the US Navy's secret knowledge of UFOs and extraterrestrial life. Two retired non-commissioned officers who became whistleblowers, Robert Dean and Daniel Salter, both claim Cooper had similar access to top secret UFO material as they themselves formerly had while working respectively within the intelligence divisions of the US Army and US Air Force.[255]

Cooper describes the background and nature of the "First Contact" with extraterrestrials as follows:

> In 1953 Astronomers discovered large objects in space which were moving toward the Earth. It was first believed that they were asteroids. Later evidence proved that the objects could only be Spaceships. Project Sigma intercepted alien radio communications. When the objects reached the Earth they took up a very high orbit around the Equator. There were several huge ships, and their actual intent was unknown. Project Sigma, and a new project, Plato, through radio communications using the computer binary language, was able to arrange a landing that resulted in face to face contact with alien beings from another planet.... Project Plato was tasked with establishing diplomatic relations with this race of space aliens. [256]

Part of Cooper's testimony has been corroborated by Major Donald Keyhoe who also wrote about giant spaceships tracked by the US Air Force in 1953, and their efforts to cover this information up:

> Since 1953 it [the USAF] had known that giant spaceships were operating near our planet. At least nine times, huge alien spacecraft had been seen or tracked in orbit, or they descended nearer the Earth for brief periods. Each time it had been an ordeal for the [US]AF censors, as they struggled to conceal the reports or explain them away when attempts at secrecy failed. [257]

Further, Keyhoe discusses the US Air Force's concern over its possible inability to debunk the public's discovery of the giant spacecraft in the near future, and how their debunking efforts would be further complicated by an official USAF article which had been circulated to members of the press. Although this article had not been printed by any newspaper due to Air Force restrictions denying press the ability to name the author or its official origin, there was the fear that it could resurface if the spacecraft story broke. That article was entitled "Planet Earth – Host to Extraterrestrial Life", and described the likely motivations of the extraterrestrial visitors:

> If the "moonlet" cover up failed, the true spaceship answer might emerge as the only alternative. If it did, this could revive a disturbing article on possible alien migration to our world.... The article had been written by a high AF Intelligence officer – Col. W.C. Odell.... it had been cleared by AF Security and Review.... According to his theory, alien beings from a dying planet were considering and surveying our world as a new home – a planet similar enough to their own so that they could survive here and perpetuate their race.[258]

Keyhoe is pointing out that Colonel Odell was permitted by his superiors to make public classified information concerning what had been learned through communications with the extraterrestrials or possibly theorized about them based on their data. Keyhoe's testimony strengthens Cooper's claim that a highly classified radio communications program, "Project Plato", had been used to establish diplomatic relations.

Cooper goes on to describe the efforts of a different extraterrestrial group warning about the motivations of the extraterrestrials in the giant spacecraft orbiting the Earth, who according to Odell's article wanted to establish colonies:

> In the meantime a race of human looking aliens contacted the U.S. Government. This alien group warned us against the aliens that were orbiting the Equator and offered to help us with our spiritual

development. They demanded that we dismantle and destroy our nuclear weapons as the major condition. They refused to exchange technology citing that we were spiritually unable to handle the technology which we then possessed. They believed that we would use any new technology to destroy each other. This race stated that we were on a path of self destruction and we must stop killing each other, stop polluting the Earth, stop raping the Earth's natural resources, and learn to live in harmony. These terms were met with extreme suspicion, especially the major condition of nuclear disarmament. It was believed that meeting that condition would leave us helpless in the face of an obvious alien threat. We also had nothing in history to help with the decision. Nuclear disarmament was not considered to be within the best interest of the United States. The overtures were rejected on the grounds that it would be foolish to disarm in the face of such an uncertain future. [259]

The critical point about Cooper's testimony here is that the human-looking "Nordics", certainly belonging to the Galactic Federation, were not willing to enter into technology exchanges that might help weapons development and instead focused upon spiritual development as a precursor.[260] Again, the emphasis on nuclear disarmament mentioned by the extraterrestrials is significant, due to the Castle Bravo hydrogen bomb test scheduled nine days after Eisenhower's alleged meeting. Most importantly, the overtures made by the Nordics were turned down, but not without a tremendous policy debate splitting the Eisenhower administration.

The source who has revealed the most about this bitter policy debate is Gerald Light. In a letter dated April 16, 1954, to Meade Layne, then director of Borderland Sciences Research Associates (now the Borderland Sciences Research Foundation), Light claimed he was part of a delegation of community leaders who took part in a meeting between US government officials and extraterrestrials at Edwards Air Force Base.[261] The meeting involved President Eisenhower and its timing corresponded to the President's February 20 disappearance from Palm Springs. The purported purpose of Light and others in the delegation was to test public reaction to the presence of extraterrestrials. Light, according to Meade Layne, was a "gifted and highly educated writer and lecturer" who was skilled both in clairvoyance and the occult.[262] Light described the circumstances of the meeting as follows:

> My dear friends: I have just returned from Muroc [Edwards Air Force Base]. The report is true -- devastatingly true! I made the journey in company with Franklin Allen of the Hearst papers and Edwin Nourse of Brookings Institute (Truman's erstwhile financial advisor) and Bishop MacIntyre of L.A. (confidential names for the present, please). When we were allowed to enter the restricted section (after about six hours in which we were checked on every possible item, event, incident and aspect of our personal and public lives), I had the distinct feeling that the world had come to an end with fantastic realism. For I have never seen so many human beings in a state of complete collapse and confusion, as they realized that their own world had indeed ended with such finality as to beggar description. The reality of the 'other plane' aeroforms is now and forever removed from the realms of speculation and made a rather painful part of the consciousness of every responsible scientific and political group. During my two days' visit I saw five separate and distinct types of aircraft being studied and handled by our Air Force officials -- with the assistance and permission of the Etherians [Nordics]!

I have no words to express my reactions. It has finally happened. It is now a matter of history. President Eisenhower, as you may already know, was spirited over to Muroc one night during his visit to Palm Springs recently. And it is my conviction that he will ignore the terrific conflict between the various 'authorities' and go directly to the people via radio and television – if the impasse continues much longer. From what I could gather, an official statement to the country is being prepared for delivery about the middle of May.[263]

In his letter, Gerald Light pointed to a "terrific conflict between the various 'authorities'". What is being suggested here is that proponents of a continued nuclear policy involving thermonuclear weapons, led by USAF Chief of Staff Nathan Twining, directly clashed with Eisenhower who had been willing to negotiate away America's nuclear dominance with his 1953 "Atoms for Peace" proposal. Putting America's nuclear technology under the auspices of the United Nations was firmly opposed by the US Air Force, RAND policy strategists, and the national security establishment. This is where Light's reference to Eisenhower going directly "to the people via radio and television" becomes significant. Apparently, Eisenhower was ready to disclose the truth about the extraterrestrial visitors and their concerns over the nation's nuclear weapons policy.

Of course, no such formal announcement was made, and Light's description of the Eisenhower meeting with Nordics has either been the best-kept secret of the twentieth century or the fabrication of an elderly mystic known for his out-of-body experiences. The events Light described in his meeting regarding the panic and confusion of many of those present, the emotional impact of the alleged landing, the tremendous difference of opinion on what to do in terms of telling the public, and the disputed question of how to respond to the extraterrestrial visitors, are all plausible descriptions of what may have occurred. Indeed, the psychological and emotional impact Light described surrounding the senior national security leaders at the meeting is consistent with what could be expected in the face of such a 'life-changing event' involving a policy debate over nuclear weapons.[264]

The main take away from the failed diplomatic negotiations between the Nordics and the Eisenhower administration was the insurmountable precondition: in order for any technological assistance to be given, the United States had to abandon its nuclear weapons development program. While President Eisenhower himself was prepared to take such a step, it was not acceptable to the USAF, the national security establishment or the Deep State. These entities not only wanted to pursue nuclear weapons development, they also wanted to receive technological assistance from the Nordics which could be used to develop improved weapon delivery systems. Presumably, the USAF justified their position based upon the long-term threat posed by the Antarctic Germans and their Draconian allies. While the Air Force had strategic bomber fleets at its disposal and was in the process of developing Intercontinental Ballistic Missile systems (ICBMs), their ultimate goal was to develop their own fleets of spacecraft that could be used for multiple purposes. Included in this category was the ability to deliver thermonuclear weapons in any future confrontation with extraterrestrials, especially the German-Draco alliance. While the Nordics fully understood the threats posed by the German-Draco alliance, they nevertheless were not prepared to hand over advanced technologies that would only fuel the USAF's plans to build more powerful nuclear weapons and delivery systems. The Nordics did not want to repeat those mistakes which had led to the destruction of the planet that once occupied the asteroid belt (Maldek/Lucifer/Tiamat), and caused the destruction of the civilization on Mars.

As shown in General Twinings' "White Hot Report", the Truman administration understood that an agreement had to be reached with extraterrestrials to reap the benefits of the retrieved flying saucer technologies that the USAF was studying at its Wright Patterson facilities. If the Nordics weren't willing to reach a technology exchange agreement, then the Eisenhower administration and the USAF would have to go elsewhere. The Edwards AFB meeting was soon followed by others, including one at Holloman Air Force Base in February 1955. The Holloman meeting laid the foundation for agreements to be reached finally, but with the German-Draco alliance which, unlike the Nordics, did not demand that the Eisenhower administration abandon its thermonuclear weapons development. Instead, this German-Draco group offered technology exchange agreements which were deemed vital by US leaders for unlocking the secrets of their prized retrieved craft, which would make possible a future USAF secret space program.

# CHAPTER 10

## Extraterrestrial & German Delegations at Holloman AFB

*Beware that you do not lose the substance by grasping at the shadow.*

—Aesop

On February 10, 1955, President Dwight Eisenhower flew on Air Force One from Washington DC to Thomasville, Georgia, for a "hunting vacation". He was accompanied by a chartered plane filled with the press. Later that afternoon after landing, Eisenhower disappeared from press view for the next 36 hours. James Hagerty, his press secretary, told reporters that 'Ike' and his valet were "treating a case of the sniffles..."[265] In reality, according to several eyewitnesses, the president had secretly traveled to Holloman Air Force Base, New Mexico, on February 11 to meet with the occupants of a landed flying saucer involved in the 1952 Washington DC flyover.

The Holloman meeting occurred almost one year after the February 20, 1954, meeting at Edwards, during which human-looking Nordic extraterrestrials attempted to dissuade the Eisenhower administration from developing thermonuclear weapons. The Nordics had also warned Eisenhower about another group of extraterrestrials who would come seeking to make an agreement.[266] This group could have been either the Gray extraterrestrials associated with the Roswell Crash or the Draco Reptilians who had helped Nazi Germany establish a breakaway colony in Antarctica during World War II.

The Nordic outreach was rejected by the Eisenhower administration because of the visitors' precondition that thermonuclear weapons development needed to cease before any technology exchange discussions could begin. Eisenhower was open to such an arrangement as he had proved with his "Atoms for Peace" proposal. Sadly, however, the Soviets had declined. The Nordic's precondition was wholly unacceptable to Eisenhower's senior national security officials who now had gained the upper hand with the Soviet rejection. The Air Force was especially opposed because it had ambitious plans to expand its strategic nuclear bombing capacity through new aerospace technologies made possible by rocket developments and captured extraterrestrial technologies. The USAF was intent on reaching an agreement with any group, extraterrestrial or not, that could help it unlock the secrets of the retrieved flying saucers that were being studied at Wright Patterson AFB. This opened the door for agreements to be reached with both the Gray aliens involved in the Roswell Crash and the Antarctic Germans allied with the Draco Reptilians.

There are compelling eyewitness and whistleblower testimonies affirming that President Eisenhower did secretly travel to Holloman AFB to meet with the occupants of a landed flying saucer, and that at least one previous meeting had occurred at Holloman with members of his administration and senior USAF officers. According to information from different sources, the meetings at Holloman respectively involved "Tall Gray" extraterrestrials and Antarctic-based Germans. The USAF and the Eisenhower administration were exploring possible partners for a technology assistance deal that could help develop viable antigravity spacecraft and other advanced technologies. In determining who exactly Eisenhower met with, when, and what was agreed upon, it's worth beginning with eyewitness reports of the events that occurred at Holloman on February 11, 1955.

## Eyewitness Accounts at Holloman

UFO researcher Art Campbell has investigated the secret Eisenhower trip to Holloman and the meeting that occurred there in great detail.[267] He was able to track down several eyewitnesses and interviewed them or family members. One of the eyewitnesses was a former USAF medic named Bill Kirklin, who from March 1, 1954, until August 5, 1955, was stationed at the Holloman Air Force Base hospital. He wrote a paper describing what happened, which Campbell passed on to me in December 2005.[268] After subsequently interviewing Kirklin, I received his DD214 and military service records directly from the US National Archives confirming that he was stationed at Holloman just as he claimed. The documents (see Figure 30) prove that he was an eyewitness to the extraordinary events surrounding President Eisenhower's secret visit there on February 11, 1955. It is worth examining Kirklin's testimony closely to see what actually occurred there and what he heard from others at the base.

SECTION 1—ARTICLES OF WAR—READ AS REQUIRED

| | | | | | | |
|---|---|---|---|---|---|---|
| 27 MAR 53 | | | | | | |
| 20Apr54 | | | | | | |

SECTION 2—APPOINTMENTS, PROMOTIONS, OR REDUCTIONS

| GRADE | DATE | AUTHORITY | DATE ISSUED | DATE OF RANK |
|---|---|---|---|---|
| A/B E-1 | 24 MAR 53 | AFR 39-9 | | 24 MAR 53 |
| ~~A/3C E-2(Perm)~~ | ~~25 MAY53~~ | ~~SO 22, 3277BMTS, 3275BMTG, PAFB, Calif 25 MAY53~~ | [illegible] | ~~25 MAY53~~ |
| A/3C E-2(Perm) | 20 MAY53 | SO 22, 3277BMTS, 3275BMTG, PAFB, Calif 20 MAY53 | | 20 MAY53 |
| A/2C E-3 (Perm | 19Feb54 | SO 9 George AFB Calif | | 1 Feb 54 |

SECTION 3—ORGANIZATIONS TO WHICH ASSIGNED AND ATTACHED

| M/R DESIGNATION OF UNIT | STATION | ASGD OR ATCHD | FROM | TO (Inc.) |
|---|---|---|---|---|
| 3277 BMTS 3275 BMTG | PARKS AFB CALIF | ASGD | 24 MAR 53 | 11Jun53 |
| Hq 3882d Sch Gp GB-USAF SAM P/L | Gunter AFB, Ala | Asgd | 12 Jun 53 | 31 Jul 53 |
| Hq 479th Air Base Group | George AFB, Calif. | Asgd | 1 Aug 53 | 1 Nov 53 |
| 4452nd USAF Hosp | George AFB, Calif, | Asgd | 2 Nov 53 | [illegible] |
| Hq. 6580th Test Gp. | Holloman AFB N Mex | Asgd | 1 Mar 54 | 24 Oct 54 |
| 6580th USAF Hosp | Holloman AFB, N Mex | Asgd | 25 Oct 54 | 5. AUG 55 |
| 36th Air Rescue Squadron | APO 994 | Asgd | 23Aug55 | |

**Figure 30. Bill Kirklin's Service Record showing he was stationed at Holloman AFB from March 1, 1954 to August 5, 1955**

Kirklin says he received prior notification of an impending visit by Eisenhower in February 1955. He wrote in his paper:

> ... we heard that the president was coming to Holloman. I knew there was going to be an honor parade for him. Captain Reiner asked me if I wanted to participate in the parade. I said, "No." He said, "Fine. You will be on duty." The Parade was scheduled for early in the morning. The day before it was to take place it was called off.[269]

At the end of the day of Eisenhower's visit, Kirklin reports that he saw Air Force One leave the base and fly over a restricted area:

> After work I was in my barracks room when I was called out to see Air Force One fly overhead. It flew over the residential area of the base. This is a NO FLYING zone for all military aircraft. Only the President could get away with it.[270]

The above is compelling eyewitness testimony that Eisenhower was not recuperating in Georgia as his press secretary falsely claimed to the press. Instead, Eisenhower was secretly over 2000 miles away at Holloman Air Force Base. Various aspects of Eisenhower's actual encounter with UFOs and their occupants are also revealed by Kirklin and other eyewitnesses.

Kirklin claims that he heard several people commenting on the flying saucers that had arrived at Holloman AFB during Eisenhower's visit. He also says he had an exchange about the craft with a colleague named Dorsey:

> "Kirklin, did you see the disc hovering over the flight line?"
>
> "No." I am thinking something small you hold in your hand like a discus as the only craft I knew capable of hovering were the choppers and the Navy's

hovercraft. There weren't that many helicopters around Holloman. "What's it made of?" I am thinking of a wooden disc with a steel edge. "Looks like polished stainless steel or aluminum. You know just bright metallic and shiny."

I asked, "How big is it?"

"Twenty to Thirty feet in diameter. Do you want to see it?"

"Sure. But with my luck it wouldn't be there."

Dorsey replied. "It was there when I took my wife to the Commissary and it was there when we got out thirty minutes later. Go out to the front of the hospital and take a look."[271]

If Kirklin's account of what his colleague saw is correct, then at least one flying saucer was hovering over the flight line of the base for at least 30 minutes during Eisenhower's visit.

Later Kirklin heard more when he went to the mess hall:

On the way back I followed two pilots. The one on the left was in Khakis, the one on the right in winter Blues. I followed them and listened to their conversation.

Left: "Why the Blues?"

Right: "I'm the Officer of the Day, I was at Base Ops when Air Force 1 came in. Did you see it?"

L. "Yes. It's a big bird isn't it?"

R "Yes. They landed and turned around and stayed on the active runway. We turned off the RADAR and waited."

L. "Why did you turn off the RADAR?"

R. "Because we were told to. I think the one at Roswell that came down was hit by Doppler Radar. It was one of the first installations to have it in the U.S. Anyway, they came in low over the mountains, across the Proving Grounds.

Interrupted by L. " I heard there were three and one landed at the Monument."

R "One might have stayed at the Monument. I didn't see it. I only saw two. One hovered overhead like it was protecting the other one. The other one landed on the active [runway] in front of his plane. He got out of his plane and went towards it. A door opened and he went inside for forty or forty-five minutes."

L. "Could you see? Were they Grays?"

R. "I don't know. They might have been. I couldn't see them. I didn't have binoculars." ...

L. "Do you think these were the same ones that were in Palmdale last year?"

R. "They might have been." ...

R. "It might have been. I just don't know."

L. "Did you see them when he came out?"

R. "No. They stayed inside. He shook hands with them and went back to his plane."

Importantly, these two pilots reveal that Eisenhower disembarked from Air Force One and met with the occupants of the flying saucer that had landed at the end of the flight line for at least 45 minutes. It's interesting that one of the pilots speculated about the occupants being Gray extraterrestrials but wasn't sure. It's also significant that the pilots referred to the Palmdale (Edwards AFB) meeting in February 1954 as being connected, indicating that knowledge of what had happened there was widespread in the Air

Force officer ranks. Perhaps most noteworthy is the handshakes at the end of the meeting. These indicate that an understanding or an agreement had been reached with the saucer's occupants. As I will reveal, this is indeed what happened on that day.

The family of a base electrician who had worked at Holloman and witnessed a flying saucer approach the area where Air Force One was positioned later contacted UFO researcher Art Campbell. He was given a letter by the family from the electrician explaining what had happened:

> So the day the President came we went out in the truck to a job where we were replacing some wire down the flight line.... So we heard the President's plane in the morning lining up for an approach and watched it land on the far runway. So we waited for it to taxi over to the flight line so we could see him, but we didn't hear it anymore. It had shut down somewhere out there ... one of the men ... said he can see out there from that pole over there, so why don't one of us go up the pole and see where the plane is? Well I had one of my climbers on and ... started up with my back to the sun, a safety measure, which also put my back to the runway where we thought his Connie was. Connie was a nickname for the big Constellation the President flew.... A few minutes later ... I could not believe what I saw. There was this pie tin like thing coming at me about 150 feet away. I thought it was remote controlled or something. 25 to 30 feet across and I started down the pole as fast as I could go.... While I was running towards the big hangar I looked back and it had stopped and it was just sitting there.[272]

The electrician's story is very revealing since it is additional eyewitness testimony that Eisenhower's Air Force One plane had landed at the end of the flight line and was waiting to be met by the arriving flying saucer.

A third eyewitness is an airman whose plane was delayed at Holloman AFB on the morning of Eisenhower's arrival to the base. The airman, Staff Sergeant Wykoff, related what happened in an interview with Campbell:

> We had to haul a load of stuff down there. Parts that they needed, and the runway is like this. I [had] never seen anything like that before. And anyway as we were there we saw Air Force One come in, and we didn't know who it was. And then an officer comes around and said you can't leave. The pilot said we have to leave. And he said well President Eisenhower is here and you can't leave the field until he's gone. And I said, you can hear it over loud speakers, but it didn't do any good. I would have liked to have seen him... We didn't have the clearance to go into the mess hall and one of the other officers, a higher ranking officer came [over] to us and said would you like to go in and eat, and listen to his speech. And most of us said yes, because I'd like to see him. I didn't get to shake his hand or anything, because I didn't have the right badge, clearance to go in, but they did let us go in at the very end and we ate and listened to his speech.[273]

SSgt Wykoff is another eyewitness who actually observed Eisenhower's secret arrival at Holloman for an undisclosed purpose. The classified nature of Eisenhower's activities is revealed by Bill Kirklin who recalled base personnel being taken into a large hangar and debriefed in groups of 225.[274] In addition to the three eyewitness accounts, there are also whistleblower testimonies detailing what they learned and read in briefing documents about the secret meetings at Holloman.

## Agreement with Gray Extraterrestrials

William Cooper has supplied documents proving that he worked on the intelligence briefing team of the commander of the US Pacific Fleet from 1970 to 1973. During this period he claims he "saw and read" classified material marked "Top Secret/MAJIC", which he subsequently discussed in papers, lectures and an underground best-selling book, *Behold a Pale Horse*.[275] Cooper says that he learned from the briefing documents that an agreement had been reached at Holloman after the failure of an earlier meeting. Cooper gave a date prior to Eisenhower's February 1955 visit, thereby suggesting that there may have been two different groups meeting with President Eisenhower and/or officials from his administration to work out an agreement.[276]

This makes sense since in addition to the Nordics and the German-Reptilian alliance in Antarctica there were other alien visitors. For example, the occupants of the Roswell crash, the Grays, appeared to be an independent alien group with their own distinct agenda. Cooper, in fact, identified the Grays as the group that Eisenhower met first after the failed Nordic negotiations:

> Later in 1954 the race of large nosed Gray Aliens which had been orbiting the Earth landed at Holloman Air Force Base. A basic agreement was reached. This race identified themselves as originating from a Planet around a red star in the Constellation of Orion which we called Betelgeuse. They stated that their planet was dying and that at some unknown future time they would no longer be able to survive there.[277]

Cooper's reference to a landing at Holloman in 1954 immediately raises the question of whether he had merely mistaken the year and was really referring to the February 1955 encounter. However, Cooper doesn't refer to Eisenhower being present at the 1954 Holloman meeting; he only states that a "basic agreement" had been reached with Tall Gray aliens.

It's helpful to recall from Bill Kirklin's account of the two USAF pilots' conversation that they weren't sure who Eisenhower had met with during his February 1955 visit to Holloman, only that he went alone into the flying saucer craft and emerged after shaking hands with occupants inside the craft. It's very unlikely that Eisenhower would have entered alone if the meeting had involved the signing of a formal agreement. Eisenhower's behavior was more suggestive of a face-to-face diplomatic meeting to establish an understanding, which in turn would lead to an agreement down the road once aides had worked out the details. The earlier 1954 meeting that Cooper had referred to may have involved the "large-nosed Gray Aliens" in a parallel set of negotiations during which a basic agreement had been reached.

Cooper explained the terms of the agreement and subsequent treaty reached with the Gray extraterrestrials at Holloman sometime later in 1954 as follows:

> The treaty stated that the aliens would not interfere in our affairs and we would not interfere in theirs. We would keep their presence on earth a secret. They would furnish us with advanced technology and would help us in our technological development. They would not make any treaty with any other Earth nation. They could abduct humans on a limited and periodic basis for the purpose of medical examination and monitoring of our development, with the stipulation that the humans would not be harmed, would be returned to their point of abduction, would have no memory of the event, and that the alien nation would furnish Majesty [sic] Twelve with a list of all human contacts and abductees on a regularly scheduled basis.[278]

Cooper's account of the agreement's content shows that it involved the highly sought technology transfer program at a price the administration was this time willing to pay. in exchange for this technology, the Grays would be given access to humans for biological experiments. Interestingly, the agreement made no demand for the dismantling of the thermonuclear weapons program that the Eisenhower administration was pursuing, which the Nordics had made a precondition for preceding with the pursuit of a further agreement on technological assistance. If Cooper is correct, then the USAF got a commitment from the Gray extraterrestrials for transfers of technology that could directly help the research and development efforts underway at Wright Patterson AFB involving retrieved extraterrestrial craft.

Another whistleblower source concerning an agreement having been made in the mid-1950's with 'Tall Gray' aliens is Phil Schneider, a former geological engineer who was employed by the US military and major corporations to build underground bases including one at Dulce, New Mexico.[279] He worked extensively on highly classified construction projects, some of which involved extraterrestrials. Schneider revealed his own knowledge of the extraterrestrial agreement as follows:

> [U]nder the Eisenhower administration, the federal government decided to circumvent the Constitution of the United States and form a treaty with alien entities. It was called the ... Greada Treaty, which basically made the agreement that the aliens involved could take a few cows and test their implanting techniques on a few human beings, but that they had to give details about the people involved.[280]

Schneider's knowledge of the alleged "Greada Treaty" would have come from his familiarity with a range of compartmentalized black projects and his interactions with other personnel working with extraterrestrials. Like Cooper, Schneider said the treaty involved biological experiments on abducted humans who would be returned unharmed, only this time the extraterrestrials would provide details about the people they abducted to US authorities.

Schneider went on to describe the extraterrestrials as Tall Grays which he personally had encountered in a firefight at the underground Dulce base in 1979.

Yet another whistleblower source for an agreement being signed is Dr. Michael Wolf, who claims he worked for twenty-five years as a scientist as well as serving on various policy-making committees responsible for extraterrestrial affairs.[281] Wolf's testimony has been investigated by a number of veteran UFO researchers who found him to be credible.[282] He claims that the Eisenhower administration entered into a treaty with an extraterrestrial race from Zeta Reticulum, "the Grays"', and that this treaty was never ratified as constitutionally required:

> In the 1950s-1960s, the U.S. administration entered into classified agreement discussions with the Zetas (so-called Greys) from the fourth planet of the star system Zeta Reticuli, and other star peoples, but these agreements were never ratified as Constitutionally required. The Zetas shared certain of their technological advances with government scientists, apparently often while prisoner "guests" within secure underground military installations in Nevada, New Mexico and elsewhere.[283]

Wolf's information points to a series of meetings and negotiations between Gray extraterrestrials and the Eisenhower and subsequent presidential administrations.

Finally, there is the testimony of Lt Col Phillip Corso, who served on a key intelligence committee that reported to President Eisenhower's National Security Council. Corso was a liaison for then Major General Arthur Trudeau who had just been appointed as Chief of Army Intelligence (G-2) in 1954. Corso had previously served in Army Intelligence under Trudeau and had impressed the general so much that Trudeau brought Corso with him to Washington DC in 1954 to fulfill an important assignment. Corso's service record (DA-66) confirms that he served as the Army Intelligence staff officer for the Operations Coordinating Board from February 1954 to June 1956. The Operations Coordinating Board was an inter-services board that had been created in 1953 by an Executive Order from President Eisenhower.[284]

The Operations Coordinating Board dealt with covert operations from all US departments and agencies and was, therefore, a key committee used by the Majestic 12 Group to implement its policy decisions concerning extraterrestrial life and the Antarctic German space program. During his two years on the Operations Coordinating Board, Corso would have learned about the secret meetings and agreements that occurred from 1954 to 1955 and reported on them to Trudeau and US Army Intelligence.

With his direct access to classified information about extraterrestrial intrusions into US air space, the deliberations of the Majestic 12 Group, and the operations of the Army's Interplanetary Phenomenon Unit, Corso was able to learn the truth about the secret meetings and agreements. In his book, *The Day After Roswell*, Corso referred to a secret agreement reached in the 1950's with the Grays:

> These creatures weren't benevolent alien beings who had come to enlighten human beings. They were genetically altered humanoid automatons, cloned biological entities actually, who were harvesting biological specimens on Earth for their own experimentation. As long as we were incapable of defending ourselves, we had to allow them to intrude as they wished. And that was part of what the working group [MJ-12] had to deal with. We had negotiated a kind of surrender with them as long as we couldn't fight them. They dictated the terms because they knew what we most feared was disclosure.[285]

Corso's acknowledgment that "we couldn't fight them" suggests that the advanced technologies of the extraterrestrial vehicles were too much for the US military, particularly the USAF which had unsuccessfully implemented an intercept and shoot down policy with disastrous consequences ever since the May 1947 White Sands missile tests.

Based on the whistleblower accounts of Cooper, Schneider, Wolf, and Corso, agreements had first been reached with Tall Gray extraterrestrials after the failed meeting with the Nordics in February 1954. Holloman Air Force base appears to have been adopted as the preferred site for these secret meetings, no doubt due to its close proximity to White Sands Proving Ground where ballistic missiles were being developed with the aid of German scientists from Operation Paperclip. But it was not just Gray aliens involved in secret negotiations with the Eisenhower administration in the mid-1950's. According to other whistleblower accounts, a meeting on February 11, 1955, at Holloman Air Force Base involved Germans from Antarctica.

| 18. RECORD OF ASSIGNMENTS | | | | | |
|---|---|---|---|---|---|
| EFFECTIVE DATE | MOS | DUTIES PERFORMED | ORGANIZATION AND STATION OR THEATER | NON-DU-TY DAYS | TYPE G. BEFORE |
| 23Feb42–9Dec42 | | Officer Candidate School, AA School Cp | Davis, NC (Enl Svc) | | |
| 10Dec42 | 1179 | AAA Staff O | Cp Edwards, Mass | | |
| 22Jan43 | 2700 | Student Officer | Mil Intel Tng Center, Cp Ritchie, Md | 40 | |
| 13Apr43 | 9301 | Prisoner of War Interrogation | G-2 Sec, Hq ETOUSA | 2 | |
| 15Sep43 | 9332 | Interpreter (Italian) | 2680 Hq Co. MIS NATOUSA | C | MOP |
| 21Feb44 | 9301 | Liaison G-1 Interpreter | Hq Rome Area Allied Coml, MTOUSA | 069 | 67 |
| 1Jul44 | 9301 | Asst G-2 | Rome Area Allied Coml, MTO | | |
| 28Nov45 | 9301 | G-2 (Tdy CONUS for R&R) | Rome Area, MTO | 90 | |
| 7Jun46 | 9301 | G-2 | Rome Area, MTO | 45 | |
| 23Mar47 | 9301 | G-2 (TDY) | OACofS, G-2, Wash DC | | |
| 21Apr47 | 0001 | Dy Unasg | The Ground Gen Sch, Ft Riley, Kansas | | |
| 28Apr47 | 2520 | Prep&Grader O, SpecProjects Br | The Ground Gen Sch, Ft Riley, Kansas | | |
| 22Sep47 | 2520 | Exec O, Extension Crs Div | The Ground Gen Sch, Ft Riley, Kansas | 12 | 67-1 |
| 1May48 | 2525 | Exec O, Ext Crs Div | S&F The Ground Gen Sch, Ft Riley, Kansas | 15 | 67-1 |
| 1Nov48 | 2525 | Exec O, Dept of Non-Resident Instr | Hq TOGS, TOGSC ASU 5021, Ft Riley, Kansas | 31 | |
| 1Nov49 | 2525 | Sr. Instr Dept of Non-ResidentInstr | Hq TOGS, TOGSC ASU 5021, Ft Riley, Kansas | | 67-1 |
| 1Apr50 | 2120 | Records O,Dept of Non-ResidentInstr | Hq Army Gen Sch, Ft Riley, Kansas | 0 | |
| 12May50 | 2700 | Student O, Strategic Intel Sch | OAC of S, G-2, 8533rd AAU, WashDC | 33 | Acad |
| 11Sep50 | 9301 | Intel Staff O, Plans&EstimatesBr, Theater Intel Div, G2 Sec | GHQ, FECOM | 14 | |
| 1Jan53 | 9301 | Chief Spec Projects Br, G-2 Sec | Hq AFFE, 8000th AU | 30 | |
| 15Jul53 | 9300 | Intel O, Western Br | OACofS G-2, 8533rd AAU, Wash DC | 0 | Abrv |
| 26Aug53 | 9300 | TDY StateDept dy/w Pay Strat Ad | OACof S G-2, 8533rd AAU, Wash DC | 0 | 67-3 |
| 24Feb54 | 9301 | Intelligence Staff O | 8720 DU,ArmySec,OprCoordinatingBd,WashDC | 15 | Itr |
| 1Jun54 | 9301 | Intelligence Staff O | 8720 DU,ArmySec,OprCoordinatingBd,WashDC | 25 | 67-3 |
| 1Jun55 | 9301 | Intelligence Staff O | 8720 DU,ArmySec,OprCoordinatingBd,WashDC | 39 | 67-3 |
| 1Jun56 | 9301 | Intelligence Staff O | Army Sec,OprCoordinatingBd(8720), WashDC | 20 | 67-3 |
| 20Oct56 | 0001 | Casual (P27DASO190dtd4hSep56) | Enroute to Ft Bliss, Texas | 1 | None |
| 21Oct56 | 4400 | StuO Assoc SAMOA(44-0-40)#1-57(TDY BtryD,2dBn(StuOff)SchBrig,AAA&GM Sch(4054)FtBliss,Tex) | 1st GM Brig(4055)AAA and GMCen,FtBliss Tex | 0 | Acad |
| 7Feb57 | 2520 | Insp of Tng | Hq1stGMOp(SAM)FtBlissTexas | 29 | None |
| 8Mar57 | 1180 | Umpire TDYw/ExRingColeFtPolkLa. | Hq1stGMOpFtBlissTexas | 42 | None |
| 19Apr57 | 2520 | Insp of Tng | Hq1stGMOpFtBlissTexas | 12 | Adm |
| 1May57 | 2520 | Insp of Tng | Hq1stGMOpFtBlissTexas | 18 | None |
| 20Jun57 | 1180 | BnComdr,O/S#2TDYw/BtryAlstOMBn (SAM)1stGMOpFtBlissTexas | USAREUR (Ger) | | |
| 20Jun57 | 1180 | (1)BnComd, (2)Op Exec O | Hq1stGMOpTDYw/Hq2dGMOpFtBlissTexas | 9 | 67-4 |
| 27Aug57 | 1180 | BnComdr,O/S#2TDYw/BtryAlstOMBn(SAM) 1stGMOpFtBlissTex | USAREUR (German) | 0 | 67-4 |
| 13Sep57 | 0001 | Casual | Enroute to USAREUR (Germany) | 22 | None |
| 31Oct57 | 1180 | Battalion Comdr | Hq Btry 5524 AAA Msl Bn (NIKE) | 67 | 67-4 |
| | | | | 0 | 67-4 |

Figure 31. Col Philip Corso's Service Record shows he served on the Operations Coordinating Board

## Agreement Reached with the Antarctic Germans

Former spacecraft operator, Clark McClelland extensively interacted with German scientists such as Dr. Werner von Braun, Dr. Kurt Debus, and many others while working at NASA facilities in Florida. McClelland describes what Dr. Ernst Steinhoff, another German scientist brought to the U.S. under Operation Paperclip, told him about the Holloman Air Force Base visit by Eisenhower. At the time of the secret meeting, Steinhoff was visiting Holloman because of his pending transfer there.

> Dr. Steinhoff ... did say, [he] was there during what was called a surprise visit by USA President Eisenhower who flew in with no early notice of those I spoke with that worked there.... The base United States Air Force Officer that managed Holloman, Colonel Sharp, did have prior knowledge of his arrival but not all who worked there.... It was a big surprise to all others who saw his large plane land... [286]

This account is consistent with what Kirklin and others have said about Eisenhower arriving at Holloman in a surprise visit.

McClelland went on to say that Steinhoff told him that the meeting at Holloman involved Germans from a secret space program which had been established in South America and Antarctica:

> It was a German Flying Saucer that he [Steinhoff] and others saw at this base. President Eisenhower, being from German heritage realized that when he was met by a German officer as he boarded that Saucer. The President then realized why none came forth to greet him as he entered that German advanced flying machine.[287]

What McClelland learned here is a stunning admission – it was

Antarctic German space program personnel that had landed one of their vehicles at Holloman on February 11, 1955, not Gray extraterrestrials as previously believed.[288]

McClelland also points out the connection between the German craft that landed at Holloman and those that overflew Washington DC in July 1952:

> I recall something I heard Dr. Kurt Debus say to Dr. Knoth, the Senior Scientist at KSC [Kennedy Space Center], as I entered his office one day. He was speaking of a V-7 craft and my entrance startled both of them. I apologized for walking in on them. Dr. Debus said it was "OK, Clark" to me. Later, I discovered through another German Scientist that the V-7 was the code name for a German Saucer shaped craft that was developed below the South Pole Ice Cap. The same type that overflew Washington, DC and startled President Truman and the Pentagon Chiefs in 1952.[289]

It's worth emphasizing that the former Nazi scientists working at NASA had knowledge of the German space program operating out of Antarctica and knew that it was applying political pressure to US leaders through overflights of major cities such as Washington DC. According to McClelland, this appeared to be widespread information among German scientists, thereby raising the question of the political allegiance of the 1500 or more former Nazis who had been brought to the U.S. under Operation Paperclip.[290] It's highly likely that some, if not the majority of the German scientists, were *fifth columnists* for the Antarctica-based colony that was actively infiltrating the US Military-Industrial Complex.

William Tompkins has also confirmed that the Holloman meeting involved President Eisenhower conducting negotiations with the breakaway German Antarctica colony. In an April 30, 2017, interview, I asked Tompkins directly about the meeting:

Salla: [#26] Did Antarctica-based Nazi spacecraft land at Holloman Air Force base in February 1955 to meet with President Eisenhower?

Tompkins: And I have to say on #26 that's a YES also. Essentially, Eisenhower accepted like a defeat in the war without it actually being that. There was really nothing he could do, like it's really [a] one-sided situation.

Salla: So, basically, it was a negotiated surrender.

Tompkins: Yeah. Like he lost that war. He surrendered.[291]

Significantly, other whistleblowers claim that the agreement secretly reached during the Eisenhower administration was a negotiated surrender. This includes Philip Corso who stated that we "had negotiated a kind of surrender with them as long as we couldn't fight them".[292] While Corso's book focused on a cover-up of extraterrestrial life, making it logical to assume that the negotiated surrender was with off-world visitors, Corso made clear that there was a connection between the visitors and the flying saucer program that had been created by Nazi Germany. Consequently, when Corso spoke of a negotiated surrender with extraterrestrials, he was aware of a link connecting the Germans as well. Based on the similarities of the craft that crashed at Roswell in comparison with the prototypes developed in Nazi Germany, Corso wrote:

> Twining had suggested the crescent shaped craft look so uncomfortably like the German Horten wings our flyers had seen at the end of the war that he had to suspect the Germans had bumped into something we didn't know about. And his conversation with Werner von Braun and Willy Ley at Alamogordo in the days after the crash confirmed this. They didn't want to be thought of as verruckt

[insane] but intimated that there was a deep story about what the Germans had engineered. No, the similarity between the Horten wing and the craft that had pulled out of the arroyo was no accident. We always wondered how the Germans were able to incorporate such advanced technology into their weapons develop[ed] in so short a time during the Great Depression. Did they have help? Maybe we were now as lucky as the Germans and broke off a piece of their technology for ourselves.[293]

Consequently, the negotiated surrender that Tompkins and Corso have referred to involved different extraterrestrial groups, along with the Antarctic Germans, imposing unfavorable terms on the Eisenhower administration in a series of agreements reached at secret USAF locations such as Holloman beginning in 1954/1955. Corey Goode also claims that an unfavorable agreement had been reached with the German Antarctic colony in the mid-1950's, enabling them to accelerate their infiltration efforts of the US Military-Industrial Complex through Operation Paperclip.[294]

The USAF was at the epicenter of the agreements reached with the Gray extraterrestrials and German-Draco alliance. The USAF was hoping to develop advanced aerospace technologies derived from the retrieved alien spacecraft from the Los Angeles Air Raid, Roswell, and other incidents. The ultimate goal for the Air Force was to gain the necessary advanced technology assistance to understand and reverse engineer the captured extraterrestrial craft so that major US aerospace corporations could build future fleets of antigravity spacecraft. As far as the Germans were concerned, their real goal in the agreements was to gain US resources, hobble the Air Force's efforts to build effective antigravity craft, and to persuade Air Force leaders to channel their resources into rocket propulsion technologies for a more conventional space program that would pose no threat to the Germans' Antarctic operations. Operation Paperclip provided the critical means for such a deception since there were many German scientists placed in key positions within the US Military-Industrial Complex who were part of a fifth column for the Antarctic German colony. In the next chapter, I will focus on the agreements reached with Gray extraterrestrials, and how they assisted the USAF in its ambitious space program goals.

# CHAPTER 11

# USAF Begins Technology Assistance Programs with Gray Extraterrestrials

> Beware the bearers of FALSE gifts & their
> BROKEN PROMISES. Much PAIN but still time.
>
> –The Chilbolton "Arecibo Answer" Crop Circle Message

The 1942 Los Angeles Air Raid led to the crash retrieval of two remotely controlled saucer-shaped drones according to William Tompkins. At the time, Tompkins was being groomed by the Office of Naval Intelligence for a covert espionage program designed to infiltrate the German flying saucer program and gather data on the extraterrestrial help the Germans were receiving in developing their own advanced aerospace technologies using antigravity and other types of exotic propulsion systems. Documents from the Los Angeles incident confirm that no bodies were recovered, which corroborates Tompkins' claim that these were unmanned drones.[295] In contrast, the 1947 Roswell Crash provided the Army Air Force/Air Force with the first opportunity to begin working directly with the remnants of a piloted spacecraft and to study the bodies to fully understand how the pilots interacted with the spacecraft's advanced technologies.

Several Majestic documents refer to the occupants of the retrieved Roswell flying saucer as all being dead at the site. One of these documents was specifically prepared for President-elect Eisenhower for a November 18, 1952, briefing on the topic. The "Eisenhower Briefing Document" was given a "High Level of Authenticity" rating by Dr. Robert Wood and Ryan Wood.[296] It says:

> On 07 July 1947, a secret operation was begun to assure recovery of the wreckage of this object for scientific study. During the course of this operation, aerial reconnaissance discovered that four small human-like beings had apparently ejected from the craft at some point before it exploded.... All four were dead and badly decomposed due to action by predators and exposure to the elements during the approximately one week time period which had elapsed before their discovery.[297]

A guide for crash retrieval operations, called the "Special Operations Manual", was written in April 1954 and is also a leaked Majestic document that was classified Top Secret MAJIC. After extensively investigating the validity of the document, the Woods gave it a "High Level of Authenticity" rating.[298] The "Special Operations Manual" was written for covert UFO retrieval teams such as the Army's Interplanetary Phenomenon Unit and its USAF successors, Project Moon Dust and Operation Blue Fly.

Project Moon Dust and Operation Blue Fly were associated with the 4602d Air Intelligence Service Squadron (AISS) which was created in 1953, and one year later, officially assigned under Air Force Regulation 200-2. The AISS had "a direct interest in the facts pertaining to UFOBs [aka UFOs]" and for conducting field investigations "to determine the identity of any UFOB."[299] Regulation 200-2 further stated that the 4602d AISS would have highly mobile units of "specialists trained for field collection and investigation of matters of air intelligence interest."[300] A Freedom of Information Act document called the "Betz Memo", which was dated November 13, 1961, and released in 1979, described the peacetime functions of the 4602d:[301]

- UNIDENTIFIED FLYING OBJECTS (UFO) - A program for investigation of reliably reported unidentified flying objects within the United States.
- PROJECT MOONDUST [sic] - A specialized aspect of the U.S. Air Force's over-all material of the exploitation program to locate, recover, and deliver descended foreign space vehicles.
- OPERATION BLUE FLY – [A unit] to facilitate expeditious delivery to the Foreign Technological Division (FTD) of Moon Dust and other items of great technical intelligence interest. [302]

Put simply, while Project Moon Dust would find, secure and retrieve extraterrestrial artifacts found anywhere around the world, Operation Blue Fly would ensure the delivery of these artifacts to proper technical facilities at Wright Patterson AFB. The "Special Operations Manual" was used by personnel in Project Moon Dust and Operation Blue Fly to conduct their operations from their headquarters at Fort Belvoir, Virginia.[303]

The "Special Operations Manual" says that, at the time of its writing, no live aliens had been retrieved from any of the UFO crash sites that had been investigated by covert UFO retrieval teams.

> Several dead entities have been recovered along with a substantial amount of wreckage and devices from downed craft all of which are now under study at various locations. No attempt has been made by extraterrestrial entities, either to contact authorities or to recover their dead counterparts or the downed craft, even though one of the crashes was the result of direct military action.[304]

The recovery of alien bodies from the crash sites gave Air Force,

Navy and government scientists the opportunity to begin studying how the pilots interacted with the advanced craft which was controlled by a pilot's mind directly interfacing with the crafts advanced propulsion and navigation systems.

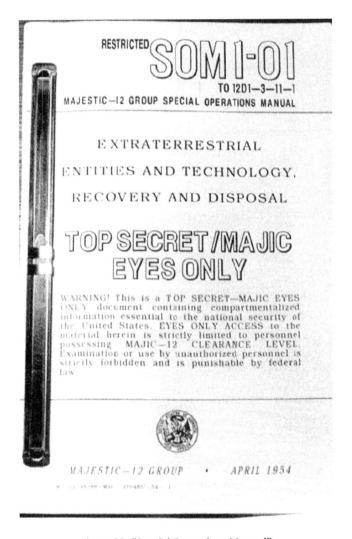

Figure 32. "Special Operations Manual"

It would take Air Force and Navy scientists many years to understand how the incredibly advanced mind-interface technology systems could be replicated for future reverse engineered spacecraft. According to Philip Corso, the mind-interface technology was a problem that had earlier even stumped the Germans who were far ahead of the U.S. in understanding and reverse engineering extraterrestrial technologies.

Furthermore, the fiber optic cables discovered in the Gray spacecraft were a completely new form of communications technology in 1947. It would take more than a decade for scientists to understand and develop fiber optics for classified USAF programs and eventually release them to the public. The term "fiber optics" came into widespread usage through a 1960 article published in *Scientific American* which introduced the new topic of 'fiber optical communications' to the world.[305] This corroborates Corso's claims that fiber optics were among the technologies from the Roswell craft that were seeded to US corporations from 1961 to 1963 for development.[306]

Despite the Gray pilots not surviving the crashes, their bodies and the interior components of the craft provided valuable clues into the vehicle's operations and performance. Multiple sources refer to the "Extraterrestrial Biological Entities" (EBEs) from the Roswell crash as "Short Grays", which appear to be a cloned synthetic life form used by different extraterrestrial groups. The "Special Operations Manual" (SOM1-01) referred to the Grays as the second of two types of EBEs found at UFO crash sites up to 1954. Section 10 from the "Special Operations Manual" says:

> Examination of remains recovered from wreckage of UFOBs indicates that Extraterrestrial Biological Entities may be classified into two distinct categories as follows:
>
> a. *EBE Type I.* These entities are humanoid and might be mistaken for human beings of the Oriental race if seen from a distance. They are bi-pedal, 5-5 feet 4 [sic] inches in height and weigh 80-100 pounds.

Proportionally they are similar to humans, although the cranium is somewhat larger and more rounded. The skin is a pale, chalk-yellow in color, thick, and slightly pebbled in appearance. The eyes are small, wide-set, almond-shaped, with brownish black irises with very large gray cast. The ears are small and set low on the skull....

b. *EBE Type II*. These entities are humanoid but differ from Type 1 in many respects. They are bi-pedal, 3 feet 5 inches – 4 feet 2 inches in height and weigh 25-50 pounds. Proportionally, the head is much larger than humans or Type 1 EBEs, the cranium being much larger and elongated. The eyes are very large, slanted, and nearly wrap around the side of the skull. They are black with no whites showing. There is no noticeable brow ridge, and the skull has a slight peak that runs over the crown. The nose consists of two small slits which sit high about the slit-like mouth. There are no external ears. The skin is a pale bluish-gray color, begins somewhat darker on the back of the creature, and is very smooth and fine-celled. There is no hair on either the face or the body, and these creatures do not appear to be mammalian. The arms are long in proportion to the legs, and the hands have three long, tapering fingers and a thumb which is nearly as long as the fingers.[307]

Corso and others have described the Short Grays as a synthetic life form, or clones, developed and used by other extraterrestrial races for a variety of purposes, which included reconnaissance missions on dangerous planets such as Earth. One of the groups using the Short Grays were the "Tall Grays" who negotiated agreements with the Eisenhower administration. According to Dr. Arthur Horn, author of *Humanity's Extraterrestrial Origins*, "Tall

Grays" play an overseeing role vis-à-vis Short Grays that piloted the Roswell craft:

> The short greys are overseen within their own ranks by the taller seven to eight foot tall greys. These greys are the ones that actually carry out "diplomatic" missions, such as secretly negotiating treaties with heads of human governments. As mentioned the greys in general, and the small three to five foot greys in particular, have been likened to mercenaries.[308]

The Tall Grays first became known to the Eisenhower administration in 1953 when their large motherships began appearing in Earth orbit, according to William Cooper.

As previously discussed, Cooper stated that an agreement was reached at Holloman AFB in 1954 with "large nosed Grays" who were permitted to conduct biological experiments on abducted humans in exchange for some technological assistance. Similarly, Phil Schneider said that a Treaty was signed with the Tall Grays approving such biological experimentation in exchange for technological assistance. The 3-4 foot Short Grays have most commonly been sighted performing the abductions permitted under these "treaties" and have been described by multiple witnesses as being controlled by taller beings, often referred to as Tall Grays.

Dr. David Jacobs concluded from an exhaustive analysis of sixty case studies in *Alien Encounters* that the Tall Grays played an overseeing role in abduction cases.[309] The Tall Grays were found to be present in many of the abductions that Jacobs examined through hypnotic regression of abductees and were in command of the Short Grays who performed most of the abductions. According to Jacobs, the Tall Grays would assist in regaining control in abduction situations where humans fought against the Short Grays while performing invasive procedures:

If the abductee gets out of control, the Small Beings usually back off and let the Taller Being deal with the situation, and the proper procedures for regaining control are instituted. Yet some abductees have learned the areas where defiance and self-assertion are possible.[310]

Figure 33. Alleged photo of Gray Alien. Credit: Robert Dean

Consequently, when technology assistance agreements were reached with extraterrestrials at Holloman AFB, these were negotiated with Tall Gray extraterrestrials who subsequently allowed the Short Gray clones to directly assist the USAF in its research and development efforts dealing with crashed flying saucers.

## Short Grays and USAF Technology

Bill Uhouse served as a fighter pilot for the US Marine Corps during World War II and the Korean War. He attained the rank of Captain before he was honorably discharged after ten years of service. In 1954, he joined the USAF as a test pilot for experimental aircraft at Wright Patterson AFB. While serving here, he was asked if he wanted to participate in a classified program involving a flying saucer simulator:

> While I was at Wright Patterson, I was approached by an individual who [wanted] to determine if I wanted to work in an area on new creative devices. Okay? And, that was a flying disc simulator. What they had done: they had selected several of us, and they reassigned me to A-Link Aviation, which was a simulator manufacturer. At the time they were building what they called the C-11B, and F-102 simulator, B-47 simulator, and so forth. They wanted us to get experienced before we actually start work on the flying disc simulator, which I spent 30-some years working on.

> I don't think any flying disc simulator went into operation until the early 1960s – around 1962 or 1963. The reason why I am saying this is because

the simulator wasn't actually functional until around 1958. The simulator that they used was for the extraterrestrial craft they had, which is a 30-meter one that crashed in Kingman Arizona, back in 1953 or 1952. That's the first one that they took out to test flight.[311]

According to Ryan Wood's book, *MAJIC Eyes Only*, the Kingman crash happened on May 21, 1953, and involved a flying saucer that was largely intact.[312] Wood analyzed a number of witness testimonies that described various aspects of the crash retrieval of the flying saucer and its occupant(s). Particularly noteworthy is the testimony of an Air Force serviceman who served on the crash retrieval team, who described the craft as being "without any sign of structural damage."[313]

This makes Uhouse's claim plausible that the craft recovered at Kingman in 1953 was sufficiently undamaged to be test flown and used to develop a flight simulator for future reverse engineered USAF flying saucers, and other USAF held extraterrestrial vehicles. Uhouse described the workings of the flight simulator as follows:

> The disc simulator didn't have a reactor, [but] we had a space in it that looked like the reactor that wasn't the device we operated the simulator with. We operated it with six large capacitors that were charged with a million volts each, so there were six million volts in those capacitors. They were the largest capacitors ever built. These particular capacitors, they'd last for 30 minutes, so you could get in these and actually work the controls and do what you had to – to get the simulator, the disc to operate.
>
> So, it wasn't that simple, because we only had 30

minutes. Okay? But, in the simulator you'll notice that there are no seat belts. Right? It was the same thing with the actual craft – no seat belts. You don't need seat belts, because when you fly one of these things upside down, there is no upside down like in a regular aircraft – you just don't feel it. There's a simple explanation for that: you have your own gravitation field right inside the craft, so if you are flying upside down – to you – you are right side up....

Because the disc has its own gravitational field, you would be sick or disoriented for about two minutes after getting in, after it was cranked up. It takes a lot of time to become used to it. Because of the area and the smallness of it, just to raise your hand becomes complicated. You have to be trained – trained with your mind, to accept what you are going to actually feel and experience. [314]

Uhouse explained the difficulty in conducting flight tests and adapting the retrieved flying saucer for human pilots since the interior of the craft had been originally designed for the small 3-4 foot Short Gray aliens. He also pointed out that the disks couldn't be fitted with conventional weapons systems – a similar problem to the one Nazi Germany had contended with over twenty years earlier, but had solved by the time of Operation Highjump in 1947:

Each engineer that had anything to do with the design was part of the startup crew. We would have to verify all the equipment that we put in – to be sure it [worked] like it was supposed to, etc.... The whole problem with the disc is that it is so exacting in its design and so forth. It can't be used like we use aircraft today, with dropping bombs and having machine guns in the wings.

The design is so exacting that you can't add anything – it's got to be just right. There's a big problem in the design of where things are put. Say, where the center of the aircraft is, and that type of thing. Even the fact that we raised it [the entrance] three feet so the taller guys could get in....[315]

Uhouse and others working on the disc simulation project were given the opportunity to talk with a Gray alien to get answers to engineering and technical questions about the flying saucers. Significantly, Uhouse mentioned that Dr. Edward Teller had accompanied the Gray alien to such meetings and was directly involved in the flying saucer simulation project:

We had meetings and I ended up in a meeting with an alien. I called him J-Rod – of course, that's what they called him.... The alien used to come in with [Dr. Edward] Teller and some of the other guys, occasionally to handle questions that maybe we'd have. You know? But you have to understand that everything was specific to the group. If it wasn't specific to the group, you couldn't talk about it. It was on need-to-know basis. And [the ET] he'd talk. He would talk, but he'd sound just like as if you spoke – he'd sound just like as if you spoke... The preparation we had before meeting this alien was, basically, going through all of the different nationalities in the world.... So basically, the alien was only giving engineering advice and science advice... Sometimes you'd get into a spot where you [would] try and try and try, and it wouldn't work. And that's where he'd [the alien] come in. They would tell him to look at this and see what we did wrong.[316]

Nearly three decades later, in 1988, Teller would recruit Bob

Lazar, whom he had met back in 1982 when Lazar was working for Los Alamos National Laboratory to understand better the propulsion system in one of the flying saucers used by the Gray aliens.[317] The Lazar story illustrates how slowly the USAF progressed in its research and development of flying saucer technologies, despite the help that the Gray aliens were providing as a result of the agreements reached with the Tall Grays in 1954/1955. If the Short Grays technological assistance was piecemeal and minimally offered, perhaps working directly with the Tall Grays would provide the USAF with the kind of technological assistance they required to expedite achieving their dream of future fleets of antigravity flying saucer craft.

## Charles Hall on Extraterrestrial Assistance for USAF Atomic Powered Spacecraft

Arguably, the most detailed account of the agreements reached with the Tall Grays and how the USAF has operationalized these for the reverse engineering of extraterrestrial spacecraft comes from Charles Hall.[318] Hall was a duty weather observer for the Air Force from 1963 to 1967 and was stationed at the Indian Springs facility of Nellis Air Force Base, Nevada. He claims that beginning in 1965 until he completed his military service, he frequently interacted with extraterrestrials which he called the "Tall Whites" (aka Tall Grays), who regularly met with senior Air Force leaders at the remote location where he was stationed. Hall further claims that a secret underground base was built at Indian Springs to house the extraterrestrials and their advanced interstellar ships. Rumors of the extraterrestrials at Nellis AFB date back to the mid-1950's amongst personnel serving there and were also acknowledged by Hall, creating a direct link to the alleged agreements reached during the Eisenhower administration at Holloman AFB. Confirmation has been found for some of Hall's claims concerning anomalous events he witnessed at Indian

Springs by a former airline pilot-turned-investigator, David Coote, who tracked down and interviewed three Airmen who served with Hall.[319]

**Figure 34. Artists impression of the "Tall Whites". Credit: Teresa Barbatelli**

Insights have been provided by Hall into the agreements reached between the USAF and the Tall Whites in terms of the legalistic way in which the extraterrestrials interpreted their provisions. In an interview, Hall explained how he was chosen to liaise with the Tall Whites and why he was safeguarded in this role:

... the decision to send me, and no one else, out to

the ranges, was made by a committee of individuals that included the Tall Whites as well as high ranking USAF Generals and other high ranking members of the U.S. Government. The Tall Whites are very meticulous about keeping their agreements and expect the U.S. Government to be equally meticulous about keeping its agreements as well. If I were victimized or threatened by anyone, The Tall Whites would interpret that to mean that the U.S. Government could not be trusted to keep its agreements. The consequences would be enormous.[320]

Hall has described how the Tall Whites strictly limited the advanced aerospace technologies they shared with the USAF so that the latter would be limited to developing vehicles capable of only short interplanetary flights. The more advanced technologies the extraterrestrials used for their own deep space or interstellar flights were kept away from USAF officials, according to Hall. Most significantly, in exchange for this limited technological assistance, Charles Hall revealed that the USAF supplied many resources including basing rights to the Tall Whites.

In a December 3, 2004 email interview with Hall, I asked him to clarify a claim that he had made in his book series, *Millennial Hospitality,* that the Air Force "Generals were desperate for technology exchanges with Tall Whites" that could be used for space travel.[321] In response, Hall wrote:

> Tall Whites exchanged technologies such as radio and communications systems, but not faster than light speed technology. Technology exchange was done on the basis of only those technologies that would benefit Tall Whites, such as good radios and communications that they could use as well if necessary....

Tall Whites would help with a nuclear powered craft but not propulsion systems for deep space travel. Anti-gravity technologies were not shared with the generals which were deep space capable. American generals were sometimes in the scout craft of the Tall Whites so the technology for the scout ships was shared to an extent since the scout ships were made on Earth using materials here with the assistance of the US military. Tall Whites sometimes participated in classified meetings and helped with technology development.... Basically, the Tall Whites would participate in classified meetings by sitting in and helping with some well-placed questions.

As Hall explained from his direct interactions with the Tall Whites, high ranking Air Force generals, as well as senior Pentagon officials, were hoping to gain any technology they could from their agreement, but the technologies acquired were not nearly as advanced as they had hoped. In a subsequent interview on December 16, 2004, I asked Hall to explain how the Tall Whites built spacecraft at Nellis AFB using materials supplied by the USAF and corporate contractors, yet were able to strictly limit access to them by Air Force personnel:

[Key: M.S.–Michael Salla; C.H.–Charles Hall]

**M.S. You describe the tall whites as building scout crafts using materials found on Earth. Can you elaborate on how you came to know this information?**

C.H. This information is based on my personal observations. In book two [Millennium Hospitality II] I describe the afternoon when The Teacher and Range Four Harry [Two of the Tall Whites] were showing me the inside of one of the scout craft. Many of the items, such as the seats and the overhead compartments still carried the mold markings placed on them by various American industries such as Boeing aircraft and Lockheed Corporation. The overhead compartments were obviously "off-the-shelf" items from companies such as Airstream corporation. Many of the clothing items that the Tall Whites were wearing were obviously purchased straight out of the Sears and Montgomery Ward Catalogs....

**M.S. Did the U.S. play any role in the actual construction of the Tall White's scout craft?**

C.H. I'm not sure. From what I saw, the Tall Whites performed all of the construction activities themselves. I'm certain that the Tall Whites performed all work on the propulsion systems and on the fiber optics windings themselves.

**M.S. Did the military have personnel observe the Tall Whites in the construction process or observe their construction facilities?**

C.H. I'm certain that the Tall Whites did not allow any USAF personnel to observe the construction process. In book two, in the chapter entitled "Two Games on one Board" I describe the day that I was able to observe their repair activities from a short distance. To the best of my knowledge, I am the only human that was allowed to view those activities from as close as I was.

**M.S. You described the hangar that was used as a**

base for the larger interstellar spacecraft used by the Tall Whites. Was this built solely by the Tall Whites or did the US military play a role in this?

C.H. The hanger appeared to have been entirely constructed by the USAF for use by the Tall Whites. For example, the inside of the hanger looked just like any other ordinary aircraft hanger. It included ordinary fire extinguishers, arrows marking exits, etc. In addition to writing and signs on the walls in English, it also included hieroglyphics and icons used by the aliens. The alien writing was done in pink paint against a white background.

**M.S. When in your view was the hangar supplied by the USAF for the Tall Whites built?**

C.H. I have no idea. However, I'm certain that the legends of Range Four Harry go back at least as far as 1954. The hanger construction (i.e. its steel supports, its other materials, it's lighting designs, its concrete doors that raised up in narrow sections by being lifted up from the top, etc.) were consistent with the construction techniques used in the late 1940's and early 1950's....

**M.S. Are all the bases used by Tall Whites in the US supplied and constructed by the USAF?**

C.H. I have no idea. My guess is that the answer is yes because I only saw USAF generals and personnel out on the Nellis Ranges.

**M.S. You contrast nuclear-powered propulsion as elementary for the Tall Whites which they supplied**

to the US military. You contrast this with anti-gravity propulsion which they did not supply. Are these the only two propulsion systems that the Tall Whites openly discussed or make known to the US authorities or to you?

C.H. Yes, as far as I know.

**M.S. You mention that the Tall Whites supplied the know-how for nuclear powered scout craft. Did the Tall Whites actually fly nuclear powered scout craft when on Earth or did they use anti-gravity propulsion in their scout craft?**

C.H. The Tall Whites always used the anti-gravity powered spacecraft and scout craft. The nuclear powered craft were only for use by the USAF.

**M.S. What kind of speed were the nuclear powered scout craft capable of, in contrast to the anti-gravity propelled scout craft?**

C.H. Vastly slower. The anti-gravity craft were capable of speeds in excess of the speed of light where as the nuclear powered were capable of just ordinary rocket powered speeds.

This exchange is highly significant since it indicates that in the mid-1960's, the Air Force was limited by the Tall Whites in developing anything faster than the speeds that could be attained through nuclear powered propulsion systems. In chapter 13, I will discuss how the USAF had to move forward with the slower nuclear powered rocket propulsion system when creating its network of Earth orbiting space stations to establish a permanent foothold in space.

While the USAF hoped that technological assistance from the Grays would lead to rapid breakthroughs in reverse engineering captured extraterrestrial spacecraft, they soon realized that in reality, the progress would be slow and difficult. In fact, it would take decades – a stark contrast to Nazi Germany where rapid progress occurred in a relatively few short years after agreements had been reached with the Draco Reptilians. In addition to receiving technological assistance from the Grays, the USAF also loftily hoped that the 1955 agreements with the Antarctic Germans would lead to significant technological gains from them and their Draco Reptilian allies.

# CHAPTER 12

# NASA Created as a Front

*What a thing was this, too, which that mighty man wrought and endured in the carven horse, wherein all we chiefs of the Argives were sitting, bearing to the Trojans death and fate!*

–Homer, *The Odyssey*

As a result of the agreements reached between the Eisenhower Administration and the Antarctic German space program dating back to the first meeting at Holloman AFB in February 1955, both sides eagerly looked forward to the benefits the deals would bring, but for very different reasons. The Antarctic Germans (aka Fourth Reich) was to be given access to the enormous industrial resources and population base possessed by the United States. America's aerospace industry could quickly mass produce the necessary components for the Antarctic Germans to expand their polar operations rapidly, and also possessed a trained workforce that could be contracted to fulfill vital production and logistic tasks. Critically, the Antarctic Germans could select from a pool of hundreds of thousands of highly trained servicemen from the different branches of the US military who would be given temporary assignments to perform key combat-related assignments in deep space and on planets such as Mars.[322]

For the Eisenhower administration and the USAF in particular, the Antarctic Germans had the necessary technical knowledge and experience to help them reverse engineer captured extraterrestrial technologies. The Antarctic Germans were widely acknowledged to be as much as two decades ahead of the U.S. in understanding advanced aerospace technologies, such as rocket propulsion and jet engines.[323] The technological gap when it came to flying saucers was even more pronounced since the operation of such craft involved very different applied and theoretical physics principles, and US scientists had to be thoroughly retrained to understand and apply these principles.

In order to facilitate the sharing of resources and technology, and to hide the secret agreements from the American public, the Eisenhower Administration and the Antarctic Germans came up with a cunning solution. President Eisenhower created a civilian space agency which would act as a façade for the sharing of resources and technology through multiple cover programs. USAF historians generally accept that NASA was created as a smokescreen for military space activities. According to Lt Col Mark Erikson: "Eisenhower's space-for-peace policy was designed to overtly emphasize civilian scientific exploration and divert attention away from covert reconnaissance satellites". [324] However, the deception went much further than Erikson and other USAF historians suspected.

Created on July 29, 1958, the National Aeronautics and Space Administration (NASA) had former Nazi Scientists filling its senior leadership positions. Placed in these key positions with responsibility over numerous NASA programs and vast resources, the former Nazi Scientists secretly sent the German Antarctic Space Program whatever was needed for its rapid expansion towards becoming a colonial space power. Furthermore, NASA had the responsibility of hiding the truth about Antarctic German space operations by setting up scientific missions which could be used to explain away any Antarctic German activities observed in deep space and on planets or moons. Finally, NASA provided the necessary cover allowing US military servicemen to be recruited into covert service for the Antarctic German space program, which would develop by the 1970's into an expansionist colonial power in deep space and other solar systems.

## The Antarctic German Space Program and NASA

In 1915, a naval appropriations bill funded the creation of the National Advisory Committee for Aeronautics (NACA) whose primary goal was to help develop military aviation in response to the outbreak of World War I. According to USAF historian Lt Col Mark Erickson, "the history of NACA and its R&D was closely tied to national security and the fortunes of the military services."[325] Erikson described in detail the extent to which NACA continued to be dependent on the US military for funding and resources throughout its history.[326]

According to conventional historians, the course and destiny of NACA was dramatically changed by the Soviet Union's launch of Sputnik 1 on October 4, 1957, which allegedly sent shockwaves through the United States. Three months later, on January 14, 1958, Hugh Dryden, the director of NACA published a paper titled: "A National Research Program for Space Technology" in which he stated:

> It is of great urgency and importance to our country both from consideration of our prestige as a nation as well as military necessity that this challenge [Sputnik] be met by an energetic program of research and development for the conquest of space.... It is accordingly proposed that the scientific research be the responsibility of a national civilian agency working in close cooperation with the applied research and development groups required for weapon systems development by the military. The pattern to be followed is that already developed by the NACA and the military services.... The NACA is capable, by rapid extension and expansion of its

effort, of providing leadership in space technology.
[327]

Dryden's reference to the challenge posed by Sputnik is most often given as the reason for the creation of NACA's successor, NASA. According to President Eisenhower, NASA's "whole program was based on psychological values.... The furor produced by Sputnik was really the reason for the creation of NASA."[328]

Given the military history of NASA's predecessor, and the continuing military interest in space affairs at a time when there was no commercial space industry, it is very strange that US policymakers felt the need to create a civilian-run space agency to conduct, as Dryden put it, "an energetic program of research and development for the conquest of space." Wouldn't a military space agency have made more sense if the goal was to maintain the national security interests of the U.S. in space through the "conquest of space"? Indeed, some NASA researchers have claimed that NASA has never really been a civilian agency at all, but instead, a military-run institution from its inception. Richard Hoagland and Mike Bara, co-authors of *Dark Mission: The Secret History of NASA* wrote:

> NASA ostensibly is a *"civilian* agency exercising control over aeronautical and space activities sponsored by the United States".
>
> But contrary to common public and media perception that NASA is an open, strictly civilian scientific institution, is the legal fact that the Space Agency was quietly founded as a direct adjunct to the Department of Defense."[329]

To prove their point, they cite from the original NASA Charter passed by Congress in 1958, which states:

> The Administration shall be considered a defense agency of the United States for the purpose of chapter 17 of title 35 of the United States Code.[330]

There is, however, another more compelling explanation for NASA's true purpose other than it simply being a front for the Pentagon's interests in space conquest. In reality, NASA was not created as a genuine response to the launch of Sputnik to get Americans into space in an effort to restore lost US pride – the "psychological values" as Eisenhower put it. National security officials briefed about visiting extraterrestrial life and the German Antarctic program viewed Sputnik in comparison to these dire national security threats as little more than a joke. NASA was created as a front organization, though not for the Pentagon as its founding charter suggested, but for the Antarctic German space program. NASA was to be secretly run behind-the-scenes by a consortium of groups tied to the US military and the Antarctic Germans who saw NASA as the ideal institutional façade for implementing their secret agreements.

Hoagland and Bara, to their credit, got very close to the truth in *Dark Mission* where they recognized the Nazi element as one of three "fringe elements" really running NASA programs:

> These literal "fringe elements," then, are divided into three main groups inside the Agency, as best as we can tell at present. For the purposes of this volume, we shall them the "Magicians," the "Masons" and the "Nazis" – and deal with each group separately.

> Each "sect" is led by prominent individuals, and supported by lesser-known players. Each has stamped their own agenda on our space program, in indelible but traceable ways. And each, remarkably, is dominated by a secret or "occult" doctrine, that is far more closely aligned with "ancient religion and mysticism" than it is with the rational science and cool empiricism these men promote to the general public as NASA's overriding mantra. [331]

There are three good reasons for believing NASA is an

institutional façade for implementing the secret U.S.-German agreements. First is the timing of NASA's creation, which came relatively soon after the 1955 Holloman meeting that set the stage for further negotiations and the subsequent agreements reached in the months and years that followed. The agreements required that significant resources be transferred to the Antarctic Germans, while the Germans in return would provide the scientific know-how for helping America's space program move forward. In order to implement the agreements without the US Congress or the American public being informed about the vast resource transfers that would be taking place, an institutional façade had to be created. NASA fit the bill since it could be presented as a civilian-run organization to gain widespread public support for the "conquest of space" while really being run, behind the scenes, by the Pentagon, USAF and the Antarctic Germans.

The second and most compelling reason to view NASA as an institutional facade for secretly implementing agreements with the Antarctic Germans is the number of Operation Paperclip scientists appointed to key leadership positions within the organization. It is well known that Nazis who worked on the V-2 rocket program were brought to the U.S. under Operation Paperclip to start up America's rocket program. Rather than being rehabilitated Nazis who achieved public prominence in America due to their scientific acumen, they were *de facto* representatives of the Antarctic Germans /Fourth Reich. Their real mission was to run the NASA programs as effective covers for the space activities by the Fourth Reich out of Antarctica, and to implement the secret agreements requiring vast resources and manpower to pour into the Antarctic German space program without the American public or Congress learning the truth.

The number of Operation Paperclip scientists that headed up key NASA installations and projects from their inception is astounding. The leadership positions they attained at the apex of NASA space operations support the conclusion that they were *de facto* representatives of the Antarctic Germans. Indeed, the Apollo program (aka Project Apollo), which was originally conceived in mid-1950 during the Eisenhower Administration,[332] and only later associated with President Kennedy's May 25, 1961 pledge to get man to the Moon by the end of the decade, was the primary program for achieving a deeper purpose arising out of the secret agreements.[333] The Apollo program would covertly provide the Antarctic Germans with all of the human and technical resources they needed to expand their operations on the Moon and Mars, and help the Germans extend their presence into deep space even to reach other solar systems. Put simply, while Apollo got Americans to the Moon, it got the Antarctic Germans to Mars, Alpha Centauri and beyond.

This astounding conclusion is supported by the 'who's who' of German scientists in the leadership positions of the key installations and facilities associated with the Apollo, Gemini and Skylab missions. Soon after the creation of NASA, Dr. Wernher von Braun, the famed former "Nazi SS officer", transferred his entire German team of former V-2 rocket scientists who had been working with him at the Army Ballistic Missile Agency to the newly created George C. Marshal Space Flight Center (MSFC) in Huntsville, Alabama. It's important to review a little of von Braun's wartime activities to get an idea of how closely he was aligned with Nazi SS ideology, despite claims that he was merely a scientist following orders to implement the V-2 rocket program. Richard Hoagland and Mike Bara had the following to say about von Braun and his Nazi past:

> Von Braun during World War II was nothing less than a Major in the SS.... Linda Hunt found survivors of the Nazi missile factories at Mittlework and Peenemunde who told her that Von Braun not only "witnessed executions and abuse of prisoners at those facilities," but on at least one occasion *ordered executions.*[334]

Figure 35. Heinrich Himmler, visits Peenemünde in April 1943. Wernher von Braun is standing behind Himmler and wearing a black Nazi SS uniform.

Despite his dubious Nazi SS past, von Braun became Marshall Space Flight Center's first director from its inception on July 1, 1960, heading it until February 1970. NASA describes the resources transferred to the MSFC with its newly won autonomy:

> The Marshall Space Flight Center was activated on July 1, 1960, with the transfer of buildings, land, space projects, property, and personnel from the Development Operations Division of the U.S. Army Ballistic Missile Agency. It was dedicated on September 8, 1960, by President Dwight David Eisenhower. The Center was named in honor of General George C. Marshall, the Army Chief of Staff during World War II, Secretary of State, and Nobel Prize Winner for his world-renowned "Marshall Plan."

> Shortly before activating its new field Center in 1960, NASA described the Marshall Center as "the only self-contained organization in the nation which was capable of conducting the development of a

space vehicle from the conception of the idea, through production of hardware, testing and launching operations."[335]

It's worth highlighting the phrase "the only self-contained organization in the nation" to understand that there was something strange about the autonomy granted to the MSFC and the vast Army resources that had been transferred to it. Once placed in charge of this key NASA facility, von Braun and his team of former Nazi rocket scientists were responsible for all aspects of the Saturn program that provided the launch vehicles for Apollo space missions.

Von Braun's deputy at MSFC, Eberhard Friedrich Michael Rees, who on March 1, 1970, was elevated to the position of director, had previously served under von Braun on the Nazi SS V-2 program. As director of the MSFC, Rees oversaw the development and construction of the Skylab space station until his retirement on January 19, 1973.[336] This meant that the MSFC, during the bulk of the Apollo program, was headed by two former Nazis occupying both the director and deputy director positions, along with other members of von Braun's team of German Paperclip scientists taking other senior positions. This glaring fact alone should immediately raise suspicions about what the German Paperclip scientists were really up to at Marshall. Why transfer resources and personnel from US Army control to a civilian installation, and incredulously, have it run by former Nazis with virtual independence and responsibility for designing and contracting out the building of the highly sensitive Saturn launch vehicles for NASA's space program?

William Tompkins has provided an important clue to what was really happening. He claims that when he arrived at the MSFC on July 1, 1962, with a model design he had built of the launch facility for the Apollo program (the future John F. Kennedy Space Center), he witnessed a US flag with a translucent Swastika over it prominently displayed at the former US Army facility. Tompkins' Nordic extraterrestrial secretary, Jessica, had given him instructions before his departure detailing what he could expect to see when he arrived at the Marshall facility. According to Tompkins, Jessica's description was entirely accurate:

> "You will turn off the highway, onto the entrance road leading to the base entry gates. Do not slow down, because the gates will open and you will just drive right through, past four military guards carrying automatic rifles.

> "They will not challenge you to stop. You won't even be challenged for carrying a large, unmarked gray crate that could be holding, an A-Bomb to blow up the entire base. After driving through very heavy base traffic, stopping and starting at the stop signs, passing Army tank convoys security vehicles, and with everyone looking at you, no-one will be pointing rifles at your head, demanding that you halt.

> "Getting into von Braun's tower will not be trouble either. But finding it will be something else; trees cover everything. You will not be able to see the tower. Don't worry, you will be amused as you drive out into a clearing. von Braun's white tower will be right in front of you. Continue driving part way around the circular drive, around the American flag that has a translucent swastika in it. You pull right up and stop in front of the steps of the twelve story building. That's von Braun's tower.[337]

The flag was a brazen display of the Antarctic German/Fourth Reich's triumph of having established a major foothold within the US Military-Industrial Complex, skillfully achieved through its acquired control of the key NASA facility responsible for designing and organizing the construction of the rockets powering the Apollo program.

Furthermore, Tompkins said that as he proceeded into the MSFC, he was introduced to Dr. von Braun by his secretary, who said. "I am Connie and this is Dr. von Braun, the conceiver and implementer of the world's first penetration into the galaxy."[338] Now, this is a highly significant and revealing statement. A trip to the Moon can hardly be considered "penetration into the galaxy". However, if von Braun and the MSFC were involved in facilitating the construction of spacecraft that were taking German astronauts outside of our solar system, her statement makes sense. This information supports the conclusion that von Braun and his team at Marshall had two distinct functions: coordinating, designing and planning for the construction of spacecraft for the Apollo program, and also, doing the same for the spacecraft for the Antarctic German space program.

It is understandable why such an arrangement would have been acceptable to US national security authorities and the USAF. They knew that the MSFC was secretly facilitating the construction of craft for the Antarctic Germans using leading US aerospace companies as contractors, but believed the expertise acquired by Lockheed, Boeing, General Dynamics, etc., could in the future be applied to building similar craft for the USAF. Indeed, the collaboration between U.S. and German companies in fulfilling contracts for NASA and the Antarctic Germans grew quickly – especially in deep underground Antarctic bases where slave labor could be used.[339]

Tompkins has provided several documents supporting his claim that he did travel to Marshall Space Flight Center in 1962 to deliver his design model, where he witnessed what was really happening at Marshall.[340] One of the documents proves that he was placed on a design team for the planned launch operations facility for the Apollo program (see Figure 36). At Marshall, Tompkins met with von Braun and another former Nazi SS officer who had only recently been appointed as the first director of the yet to be built Launch Operations Center – Dr. Kurt Debus.

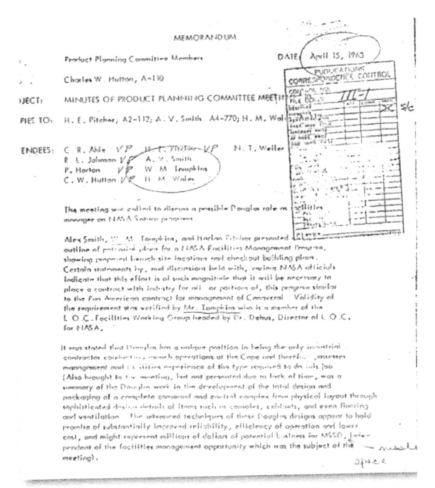

MEMORANDUM

DATE April 15, 1963

Product Planning Committee Members

Charles W. Hutton, A-110

JECT: MINUTES OF PRODUCT PLANNING COMMITTEE MEETING

PIES TO: H. E. Pitcher, A2-112; A. V. Smith A4-770; H. M. Wolf

ENDEES: C. R. Able VP H. E. Pitcher VP N. T. Weller
R. L. Johnson VP A. V. Smith
P. Horton VP W. M. Tompkins
C. W. Hutton VP H. M. Wales

The meeting was called to discuss a possible Douglas role as manager on NASA Saturn programs.

**Douglas Memo Confirming Tompkins Appointment to Kurt Debus
Working Group for Designing Launch Operations Center**
Source: William Tompkins, *Selected by Extraterrestrials*, p. 351

Figure 36. Documentary evidence from W. M. Tompkins

As director of the Launch Operations Center (renamed John F. Kennedy Space Center in late 1963 after President Kennedy's assassination), Debus was in charge of NASA's only launch facility for the Apollo program, as well as the Gemini and Skylab programs. Debus oversaw the facilities' design and construction, making the choices required for launching the huge Saturn V rockets that would be used for the Apollo and other space-related NASA programs. Debus also worked with von Braun in choosing the multiple aerospace companies that would build the different rocket stages and vehicle components that had been designed by von Braun's team at Marshall. Debus remained in charge of the Kennedy Space Center until his retirement in November 1974. Consequently, with von Braun and Rees running the Marshall Space Flight Center, and Debus running the Kennedy Space Center, former Nazi SS officers were effectively running the NASA programs with all their enormous resources right up to 1974/1975.

This takes me to the third reason for supporting the view that NASA, from its inception, was an institutional front for implementing the secret U.S.-German agreement. From 1964 to 1970, the Apollo program consumed between 54%-70% of the entire NASA budget of approximately four billion dollars annually.[341] With former Nazis in charge of all aspects of the design, construction, and launches of the Apollo program space vehicles, and having complete autonomy at Marshall, there was clearly a means available for siphoning off a significant portion of these funds and resources to the Antarctic German/Fourth Reich operations. However, siphoning off NASA funds intended for Apollo would not have been sufficient for the Fourth Reich Germans in Antarctica to build their own fleets of ships for deep space and interstellar operations.

## CIA Funds Antarctic German Space Program Through Apollo Program

With the agreements in place, the stage was set for a major intelligence game to begin. The Eisenhower administration had the weaker hand in this match and would allow significant resources and manpower to be given to the Antarctic Germans with the hope of eventually learning enough to catch up in understanding the precious alien technology they held. First, however, they would have to get their hands dirty. The US agency responsible for raising the necessary funds for the Antarctic Germans would become the Central Intelligence Agency (CIA) which was headed at the time by Allen Dulles, a key figure in the US-Fourth Reich negotiations.[342] Indeed, Dulles was later discovered to be the head (MJ-1) of the MJ-12 Committee that oversaw the negotiations between the Eisenhower administration and the Germans from Antarctica.[343]

Soon after its formation in September 1947, the CIA was given the statutory power to send or receive money without regard to statutory law as the following clause in the 1949 CIA Act makes clear:

> ... any other Government agency is authorized to transfer to or receive from the Agency such sums **without regard to any provisions of law** limiting or prohibiting transfers between appropriations. Sums transferred to the Agency in accordance with this paragraph may be expended for the purposes and under the authority of sections 403a to 403s of this title without regard to limitations of appropriations from which transferred. [emphasis added] [344]

Essentially, back in 1949, the CIA had been given a Congressional stamp of approval for the creation of a 'black budget', which was to be used for the funding of covert programs without any Congressional oversight – an arrangement that maintained plausible deniability. The most highly classified programs were those authorized by the Majestic 12 Group that dealt with extraterrestrial life and the Antarctic German problem. After agreements had been reached with the Eisenhower administration after the 1955 Holloman AFB meeting, a significant portion of the CIA's black budget funds ended up being sent to the Antarctic Germans.

The CIA's "deep black budget" was estimated to be as high as $1.7 trillion per year by the end of the Clinton administration in January 2001.[345] To fully appreciate the significance of such a vast sum, consider that the proposed Pentagon budget was $686 billion for 2019.[346] This means that the deep black budget, back in the year 2000, was *more than double the entire Pentagon budget today*! In 1966, the Apollo Program received 66% of NASA's $4.5 billion budget appropriation.[347] The Pentagon budget in 1966 was $356 billion, dwarfing the funds available to the Apollo Program.[348] If the CIA's black budget at the time was, at the very least, comparable to the Pentagon budget, that would mean that a significant portion of $356 billion or so that the CIA had raised through a variety of illicit fund raising operations could have gone to the Fourth Reich German's Antarctic operation.

There's no way of knowing exactly how much the CIA was channeling through the MSFC to the Antarctic Germans. However, given that funding for the Antarctic space program would have been a high priority for the Fourth Reich, it can be guessed that they would have demanded and received a significant share of what the CIA was raising through covert operations. Speculating on what the Germans would have demanded, something in the order of 30% or so of the agency's black budget does not seem an unreasonable estimate. This figure would equal approximately $100 billion, using the projected size of the CIA's black budget back in 1966.

If the estimate of $100 billion is anywhere near what the Antarctic Germans were secretly getting from the CIA on an annual basis, this would have dwarfed the $3 billion received by the Apollo program in 1966.[349] The Apollo program officially ended in 1972, with the remaining Saturn rockets being used for Skylab missions up to November 16, 1973, and a joint Soviet-U.S. mission on July 15, 1975. By 1975, US born scientists had replaced the former Nazi SS officers running the Marshall Space Flight Center and the Kennedy Space Center. The end of Apollo and Saturn rocket production marked the end of an era that was effective on two counts. NASA had successfully built the launch vehicles for manned spaceflights to the Moon and low Earth orbit, and secretly, it had supplied the resources, manpower and spacecraft components for the Antarctic German space program to travel to Mars, deep space and neighboring solar systems.

NASA's role in helping establish the Antarctic Germans as a colonial space power was something that the USAF wished to leverage in its own effort to develop a space program. The Gray/Tall White extraterrestrials were not as helpful as the Air Force had hoped in developing functional space travel technologies using electrogravitic, antimatter and other exotic propulsion systems. Similarly, the Germans were not as helpful as promised when it came to reverse engineering captured alien spacecraft. However, when it came to rocket propulsion technologies, the German scientists at NASA had been eager to help and were, in fact, enthusiastic about the potential of rocket propulsion for the Air Force's ambitious plans for space. Lt Col Erikson described the thinking among USAF officers about projecting a presence into space:

> In a 28 January 1958 speech, Brig Gen Homer A. Boushey posited, "the moon provides a retaliation base of unequaled advantage.... It has been said that 'He who controls the moon, controls the earth.' Our planners must carefully evaluate this statement, for, if true (and I for one think it is), then the United States must control the moon." The Air Force's deputy chief of staff for development, Lt Gen Donald L. Putt, supported a military base on the moon while testifying to Congress in March 1958 and declared this was "only a first step toward stations on planets far more distant from which control over the moon might be exercised.[350]

Von Braun, in particular, was very optimistic that rocket propulsion could be used not only for manned lunar missions but also for establishing an Earth-orbiting space station[351] and manned missions to Mars, as outlined in his 1952 book, *The Mars Project*.[352] Was von Braun's enthusiasm genuine or was he under orders from his Antarctic brethren who closely monitored NASA activities, and even sent senior officials to NASA during the Apollo program?

Former NASA spacecraft operator Clark McClelland reported seeing Nazi SS Lieutenant General Hans Kammler at NASA in the mid-1960's. McClelland says that he met Kammler in the director's office at Kennedy Space Center while Kurt Debus was the director (1962-1974):

> I opened his office door and saw two people I had not seen at KSC. He introduced me to both men. He only gave me their first names during the introduction.... One was introduced to me as Siegfried and the other was introduced as Hans.... Both had the look of Nazi High Command Officers.... Today I am certain of who these two men were. I eventually learned from other German scientists that one of them was Siegfried Knemeyer. He was a very high ranking Nazi Oberst Officer in the Luftwaffe... The other man was difficult to recognize until I saw an older photo of him after he had later entered the USA. He was in my opinion Heinz (Hans) Kammler.... There were rumors after WWII that Kammler had made a deal with General George Patton to turn over German Top Secret technology for his support in getting Kammler into the USA. That may have actually happened. I personally believe it did happen.[353]

As a result of these informal visits from senior German officials from the Antarctica program, von Braun and other NASA-affiliated Germans steered the USAF towards the use of conventional rockets as a known and reliable propulsion system for their future space program.[354] With the more exotic propulsion technologies still in development and years away from deployment, the USAF worked with the Antarctic Germans, through NASA, in moving forward with a rocket-powered secret space program during the Apollo era.

# CHAPTER 13

# Manned Orbiting Laboratory and Secret Space Programs

Lunch with 5 top space scientists. It was fascinating. Space truly is the last frontier and some of the developments there in astronomy etc. are like science fiction, except they are real. I learned that our shuttle capacity is such that we could orbit 300 people.

– Ronald Reagan, *Presidential Diaries*, June 11, 1985

Air Force interest in developing a manned Earth-orbiting space station began in 1960 with a proposal for a two-phase program: a "Military Test Space Station" to be launched before 1965, followed by a more advanced version within five years.[355] The two-phase process was necessary first to evaluate what would be needed to construct a more complex model of the manned station which was intended to stay in Earth orbit for a prolonged period. The publicly stated goal of this endeavor was to create a stable space platform for reconnaissance missions, and possible use against enemy satellites. The hidden purpose, known only to those officials with "MAJIC" level security clearance, was to monitor space traffic from different visiting extraterrestrial groups and the Antarctic Germans, all of whom continued to present a potential national security threat despite the agreements that had been reached.

Determination drove the USAF to develop and deploy a manned space station. This is reflected in a statement by General James Ferguson, deputy chief of staff for Research and Development, who said on February 12, 1962, "We are convinced that a manned, military test space station should be undertaken as early as possible".[356] The "Military Test Space Station" was given the more innocuous name of "Manned Orbiting Laboratory" (MOL), and on August 25, 1962, the Secretary of the Air Force approved the program.[357]

President Kennedy signed off on the proposal to begin feasibility studies, and later President Johnson would approve its phase-one construction. It was decided that the USAF was to collaborate with NASA in building its military space station, which could operate in orbit from between 70 to 400 nautical miles (130-740 km).[358] On December 10, 1963, a press release by the Office of the Secretary of Defense provided details about the proposed MOL:

> Secretary of Defense Robert S. McNamara today assigned to the Air Force a new program for the development of a near earth Manned Orbiting Laboratory (MOL).
>
> The MOL program, which will consist of an orbiting pressurized cylinder approximately the size of a small house trailer, will increase the Defense Department effort to determine military usefulness of man in space....
>
> MOL will be designed so that astronauts can move about freely in it without a space suit and conduct observations and experiments in the laboratory over a period of up to a month. The first manned flight of the MOL is expected late in 1967 or early in 1968...[359]

The Pentagon press release went on to give more details about the launch vehicles to be used for the MOL program:

226

The MOL will be attached to a modified GEMINI capsule and lifted into orbit by a TITAN III booster. The GEMINI capsule is being developed by NSA [National Security Agency] for use in the APOLLO moon shot program. The TITAN III is being developed as a standardized space booster by the Air Force.

Astronauts will be seated in the modified GEMINI capsule during launch, and will move to the laboratory after injection into orbit. After completion of their tasks in space, the astronauts will return to the capsule, which will then be detached from the laboratory to return to earth.

The design of the MOL vehicle will permit rendezvous in space between the orbiting laboratory and a second GEMINI capsule, so that relieved crews could replace original crews in the laboratory. Such an operation would be undertaken if man's utility in space environment were demonstrated and long operations in the space laboratory were needed.

The MOL program will make use of the existing NASA control facilities. These include the tracking facilities which have been set up for the GEMINI and other space flight programs of NASA and of the Department of Defense throughout the world.[360]

The MOL had a number of experimental and living areas together with a "transtage" (a re-startable engine} that could be used for maneuvering in space (see Figure 37).[361] The laboratory module was essentially a third stage added to a modified Titan II rocket. The Titan II had been successfully used both for the NASA Gemini program, which had performed 12 launches from 1964 to 1966 (10 of which were manned), as well as an Intercontinental

Ballistic Missile launch for USAF Strategic Air Command.[362]

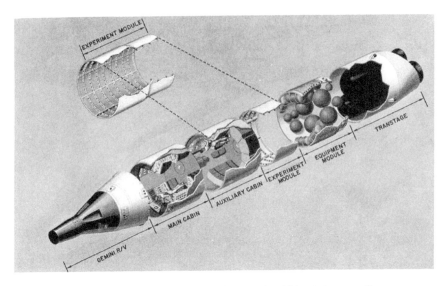

**Figure 37. Proposed design of "Manned Orbiting Laboratory"**

In order to launch both a Gemini capsule and the MOL, the Titan IIIC was developed by Martin Marietta (later incorporated into Lockheed Martin in 1995). The Titan IIIC would have two booster rockets attached to the first stage to provide the necessary thrust to get the MOL into space.

Nearly two years after the Secretary of Defense gave the go-ahead for MOL feasibility studies to begin, President Johnson appeared at an August 25, 1965, press conference to publicly authorize the construction and testing of MOL. It was given a budget of $1.5 billion beginning in fiscal year 1965/66. Douglas Aircraft Company was authorized to build the cylindrical laboratory and the modified Gemini capsule for re-entry, while General Electric would build the experimental equipment.[363]

What neither the 1963 Pentagon press release nor Johnson's 1965 press statement revealed is that USAF astronauts would be conducting surveillance operations over designated targets, as a declassified top secret document to Major General Ben Frank, commander of the USAF Space Systems Division, dated March 10, 1964, confirmed.[364] Another declassified top secret USAF document dated May 25, 1965, declared: "The Initial objective is to develop and demonstrate at the earliest time an operationally useful high resolution manned optical reconnaissance system".[365] A PBS documentary presented interviews with former astronaut participants – dubbed "astrospies" – who pointed out the likely primary targets to be in the former Soviet Union:

> The plan was to launch a two-man Gemini capsule atop a 56-foot-long laboratory module.... the astrospies could look through a view-port and observe and photograph high priority targets in Russia and elsewhere. When the 30-day mission was completed, the astrospies would return to Earth in the Gemini capsule, leaving the laboratory module to de-orbit and burn up in the atmosphere.[366]

## Polar Orbit Missions Designed to Spy on Antarctic Germans, Not Russians

Declassified National Reconnaissance Office (NRO) documents give the false impression that the Soviet Union/Russia was the primary target of MOL.[367] However, as stated in a previous chapter, former CIA agent Kewper/Stein revealed that the space activities of the Germans launching out of Antarctica and visiting extraterrestrials were the real targets of advanced aerial surveillance technologies. Kewper/Stein explained that the Pentagon was trying to distinguish between the extraterrestrial and German craft, which differed significantly in terms of their flight performance.[368]

Other declassified NRO documents reveal that satellites intended for polar orbits required their launch to take place from Vandenberg Air Force Base, California, since a take-off from Kennedy Space Center required a difficult dog-leg maneuver which consumed more fuel, thereby requiring heavier payloads and a danger to the Florida coast population.[369] Secretary of the Air Force Harold Brown wrote the following letter explaining why a polar orbit was required for MOL missions:

> Because polar orbits are required, the TITAN IIIC MOL launches can be easily performed by flying south from Vandenberg Air Force Base. Flying due south from Cape Kennedy would result in the trajectory passing directly over southern Florida, and this would be totally unacceptable from range safety considerations.[370]

However, if the primary goal of MOL was to spy on the Soviet Union, then polar orbits were a very inefficient way of doing this. A geosynchronous orbit over the USSR would have provided the optimal surveillance options, as explained in scholarly research by Major A. Andronov titled "American Geosynchronous SIGINT Satellites".[371] A major disadvantage of a satellite in geosynchronous orbit is that it has trouble monitoring activities near the poles. In contrast, a polar orbiting satellite traveling at a near 90 degrees inclination would provide a clear view of the poles for monitoring the activity of the German colony and their extraterrestrial allies operating from Antarctica. In addition, a declassified document titled "Application of MOL to Astronomical Observations" reveals MOL was intended to be used for planetary observation also, which is best achieved by a polar orbit which allows observation of virtually every part of the Earth. This supports the idea that among MOL's real goals as a manned space station was to monitor space traffic entering and leaving the Earth – both from the Antarctic Germans and extraterrestrials.[372]

## MOL, NASA, and the Secret Astronaut Corps

Of particular significance is the arrangement reached between the Department of Defense and NASA when it came to funding and running MOL. In May 1964, Secretary of Defense Robert McNamara had reached an agreement with NASA Administrator James Webb on MOL, and on September 25 they came up with the following management plan:[373]

1. DOD and NASA will agree that the MOL is the flight forerunner to the definition of a scientific or military operational space craft.

2. NASA will accept the responsibility of manager of the scientific program to be carried out using the MOL. All such scientific experiments (except those that may be classified for security purposes) will be selected, developed, and analyzed by NASA. The funding for this work will be provided by NASA. Our planning indicates that substantial weight and volume can be made available in the MOL after accommodation of particular classified experiments. In addition, we would be prepared to consider variations in MOL configuration for specific scientific purposes on particular flights of the program.[374]

NASA, through its Antarctic German *de facto* representatives, would effectively be monitoring what the USAF was planning to do with its MOL program and pass on whatever intelligence data was gained to the Fourth Reich leaders in Antarctica. In January 1964, the USAF arranged for military pilots to start training for the MOL program, using the cover of NASA's civilian astronaut program. A declassified "Proposed Release" revealed that "Astronaut candidates will be military test pilots and graduates of the Aerospace Research Pilot School at Edwards AFB, California."[375] A transcript from a PBS documentary described the strict secrecy surrounding the parallel military astronaut training program:

> NARRATOR: Run by Chuck Yeager, the first man to fly faster than the speed of sound, ARPS [Aerospace Research Pilot School] was a school where some military pilots, with the right stuff, were groomed to become astronauts in NASA's civilian space program. This year it was different.
>
> RICHARD TRULY: As we went through our student year, and got toward the fall, we realized that something funny was going on. And the thing that was funny going on was, they were actually conducting a secret, I guess you'd say crew selection.
>
> JAMES BAMFORD: Without them knowing it, they were actually competing with each other for this program. They were being watched and being evaluated by these people to see who would make the best astronauts. The program was so secret, it was even kept from the potential astronauts themselves.[376]

The military pilots being secretly trained in 1964 for classified space missions worked alongside colleagues who would later serve as NASA astronauts. The USAF believed that their pilots/MOL astronauts would learn what was necessary to be able to successfully work in space from the Operation Paperclip scientists running NASA's Apollo and Gemini programs. After all, USAF leaders were aware that Antarctic German astronauts were flying into deep space and other solar systems while the Air Force was still trying to establish a presence in near Earth orbit. Whatever the Germans had learned from their space operations, assurances had been made that some of this would be passed on to Apollo and Gemini program astronauts through von Braun, Debus, and other

Operation Paperclip scientists in touch with their German brethren in Antarctica. In turn, the USAF hoped its pilots enrolled with NASA's astronaut corps would learn what it took to operate in space for extended periods. This kind of mutual sharing of resources and knowledge was at the core of the secret agreements reached at Holloman AFB in 1955. MOL, as publicly announced, was only the preliminary phase of something much more ambitious than the USAF was planning.

## Advanced Phase of MOL Links Multiple Modules Together

The Manned Orbiting Laboratory was conceived as only the first phase of a more ambitious Air Force military space station plan. Phase II involved expanding the modular system to allow more diverse operations to be performed in space over longer periods in orbit, as explained by Lt Col Mark Erikson:

> The system itself consisted of a permanently orbiting station module, an earth-based spacecraft comprised of a modified-Gemini capsule for ferry purposes, and a new launch vehicle, probably the Titan III. The crew of four could remain in the 1,700-cubic-foot-station module for 30 days without resupply, while the station itself would remain in orbit for at least a year. The USAF fully expected MODS [Military Orbital Development System] to grow: "Ultimately, as MODS is expanded through modular extension, it will serve as a base from which experimental military space vehicles can be developed, tested and employed.[377]

Correspondence on October 1963 between the Department of Defense (DOD) and NASA officials, as part of a "DOD-NASA Coordination Agreement", described in more detail what Phase II would entail:

> Phase II (a) is aimed at the selection of major subsystems such as docking system, artificial g system, data handling system, etc., and will incorporate results of the modified Gemini and Apollo Ferry Studies ... A broad look at subsystem integration is planned utilizing inputs from such current studies as Nuclear Isotope Power ... and Integrated Life Support System... [378]

What this correspondence shows is that the completed space station would have an artificial gravity system and use nuclear power. The presence of an artificial gravity system divulges that Phase II would involve a design in which the space station would rotate around a central axis. This suggests a design similar to what Werner von Braun had proposed back in 1946, which has been dubbed the "Von Braun Station". It incorporated 20-cylindrical sections connected end-to-end:

> The 1946 version used 20 cylindrical sections, each about 3 m in diameter and 8 m long, to make up the toroid. The whole station was about 50 m in diameter and guy wires connecting and positioning the toroid to the 8 m-diameter central power module. This was equipped with a sun-following solar collector dish to heat fluid in a ball-shaped device. The heated fluid would run an electrical generator. Presumably visiting spacecraft would dock or transfer crew at the base of the power module. Two narrow transfer tubes allowed the crew to move between the living and work quarters in the toroid and the power module.[379]

In the completed Phase II space station, each section would be made up of MOL cylinders sent up separately on a Titan III rocket, and then assembled together in space. Given that each MOL cylinder was anticipated to be 56 feet (17 meters) in length (double what von Braun envisaged), the final diameter of the space station would be over 300 feet (91 meters) in diameter.[380]

A declassified document titled "MOL Program Advanced Planning", issued on March 17, 1967, shows that the funding and planning for Phase II were completely separated from the basic MOL program.[381] This made it possible for the basic MOL program to be publicly canceled down the track while the advanced portion (Phase II) involving more complex projects and objectives (i.e., assembling a type of 'Von Braun Station' in space) could continue uninterrupted as a covert program.

Figure 38. Von Braun Space Station

## How the Public Was Misled: Cancellation of MOL

On July 1, 2015, the National Reconnaissance Office approved the declassification and release of 825 official documents describing the official history of MOL and its different phases.[382] What the documents reveal is that an "unmanned version" of MOL was covertly developed using the same laboratory modules developed by Douglas Aircraft Company for the manned version. Cancelation of Phase I of the MOL program became the cover for implementing the advanced Phase II of MOL; construction of a Von Braun Station using the modules designed and developed by Douglas but built by different contractors such as Martin Marietta.

In a declassified April 17, 1969, top secret memorandum by the Secretary of Defense to President Nixon, it was outlined out how the manned version of the MOL program could be canceled, but an unmanned version would secretly continue (out of Vandenberg AFB using Titan IIID rockets).[383]    The manned MOL was officially canceled on June 9, 1969, in a memorandum by the Deputy Secretary of Defense to the Secretary of the Air Force which gave instructions for the continuance of the unmanned version with a new contractor competition:

> The Air Force is hereby directed to terminate the MOL Program except for those camera system elements useful for incorporation into an unmanned satellite system optimized to use the TITAN III D. Directions to MOL contractors should be issued on Tuesday morning, June 10, at which time we will also notify the Congress and make a public statement that MOL is canceled...
>
> All future work on the camera and an unmanned system will be part of the NRP [National Reconnaissance Project]. As a security measure, appropriate elements of the MOL Project Offices and the camera system contracts should be transferred to the Air Force NRP Special Projects

Offices at an early date. Overt MOL activities should be phased out in conjunction with closeout of MOL Program activities.[384]

This meant that Air Force Systems Command, which had been initially in control of building the manned version of MOL, had to transfer primary responsibility over to the National Reconnaissance Office, through the rubric of the National Reconnaissance Project (NRP). This arrangement would make it easier to hide USAF involvement in the advanced MOL program.

Another declassified top secret memorandum, dated June 7, 1969, two days before the official public cancelation, made clear that the "unmanned MOL program" being covertly continued had been shown in studies to be significantly less efficient than the manned version. Strikingly, it would use the entire "MOL Mission module" originally intended for use in the more costly manned version:

ALL CONCERNED RECOGNIZE THAT AN UNMANNED SYSTEM USING THE MOL CAMERA WILL HAVE POORER AVERAGE RESOLUTION, BE LESS CAPABLE AND FLEXIBLE, BE LESS RELIABLE, ETC., THAN THE MANNED MOL WOULD HAVE BEEN. ...

THE GE [General Electric] AND EK [Eastman Kodak] CONTRACTS ARE TO BE REDUCED IMMEDIATELY TO ONLY THOSE CAMERA SYSTEM EFFORTS APPLICABLE TO A NOT-YET-DEFINED UNMANNED SATELLITE SYSTEM THAT WILL USE THE COMPLETE MOL MISSION MODULE.[385]

This document is critical since it shows that despite the cancellation of the manned MOL program, the "complete mission module" that had been designed and developed by Douglas Aircraft Company would continue to be used for the unmanned version.

It's worth pointing out that the unmanned version of MOL was NOT the "Hexagon" unmanned camera satellite that was being simultaneously planned as part of the KeyHole 9 series of spy satellites that the USAF would launch in conjunction with the NRO out of Vandenberg AFB on Titan IIID rockets from 1971 to 1986.[386] While both the unmanned MOL and Hexagon would be launched on Titan IIID rockets, they were entirely separate programs. In fact, the Keyhole satellites were a very effective cover program for the more highly classified advanced MOL program.

The June 7, 1969, memorandum went on to describe that the different contractors originally involved in the design and construction of the manned MOL would be allowed to compete for contracts related to future unmanned MOL missions:

> EARLY IN FY 1970, IT IS PLANNED TO HOLD A FUNDED COVERT SPACECRAFT COMPETITION BETWEEN LMSC/GE/EK AND MAC-D /GE/EK. MAC-D WILL BE ENCOURAGED TO PROPOSE A REPACKAGING OF APPROPRIATE MOL COMPONENTS AND SUBSYSTEMS...[387]

It was later decided that the McDonnell Douglas Company (MAC-D), which was created out of a merger between Douglas Aircraft Company and MacDonnell Aircraft in 1967, would not be awarded the contract for building the modules to be used in the advanced unmanned MOL program, thereby leading to mass layoffs for the company. It was a major McDonnell Douglas competitor, Lockheed Missiles and Space Company (LMSC) that was given the covert contract.

Lockheed had been granted the resupply contract for the original manned MOL program and was able to adapt their resupply efforts for the more complex advanced unmanned MOL program prior to the cancellation of the basic MOL program. A declassified 1967 document, titled "Lockheed DORIAN Resupply Study" confirms that Lockheed was in charge of resupply missions using Titan IIID rockets from Vandenberg. The study showed a modular design that could be interlinked, as in a Van Braun Station. This study is evidence that Lockheed got the covert contract for the advanced phase of MOL.[388]

Another declassified NRO document, titled "Advanced MOL Planning: Missions and Systems" shows General Electric's plan for reconfiguring MOL so it could hold a space crew of either 12 or 40, depending on how many sections of modules were linked together in the advanced MOL program.[389] The document describes the command module characteristics of MOL and states that by using "three mated cylinders" (each cylinder section is composed of modules), a crew of up to 40 could be accommodated. Interestingly, the proposed space station would have self-defense capabilities and be powered either by solar arrays or another power supply which is blacked out in the document's text but is presumably a nuclear generator (see Figure 39).

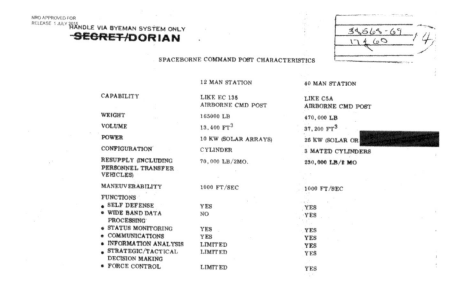

Figure 39. General Electric Proposal for Spaceborne Command Posts

The "Advanced MOL Planning: Missions and Systems" document also illustrates both the configuration of the advanced station and some of its various Earth surveillance activities that General Electric envisaged for the program (see Figure 40). Significantly, the proposed space station was designed to monitor threats posed by Intercontinental Ballistic Missiles (ICBMs), submarine-launched ballistic missiles (SLBMs) and aircraft by relaying real-time surveillance of these to Strategic Air Command. It's possible that the self-defense capability of the space station could eventually enable it to intercept and neutralize such threats.

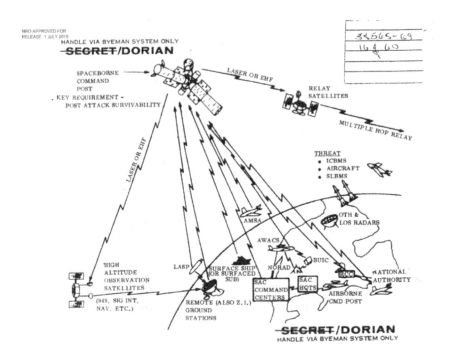

**Figure 40. General Electric's illustration of a three section station built with MOL modules that could accommodate 40 astronauts.**

The declassified "Advanced MOL Planning: Missions and Systems" proposal by General Electric answers a perplexing question raised by the June 11, 1985, entry in the daily presidential diary of Ronald Reagan. In it, he wrote that he had been briefed about the 300 astronauts the U.S. could simultaneously place into orbit.[390] Reagan was not writing about NASA's space shuttle fleet which could accommodate a crew of 10 on each of the five shuttles, making a maximum of 50 people put into space if all the shuttles took off simultaneously. However, if the USAF/NRO/NASA had created two or more secret space stations using MOL modules assembled in space into a configuration similar to the Von Braun

Station, then it is easy to fathom the truth behind what Reagan was told in a briefing which he documented in the presidential diary. The training of Air Force and Navy pilots for the advanced MOL program very likely duplicated the same process used in the canceled basic MOL program, in which pilots were covertly recruited for the classified space program although they were led to believe they were being trained to be NASA astronauts.

## Von Braun and the Real Purpose Behind the USAF/NRO/NASA Space Station

Of particular interest when examining the developments leading to the different phases of the space station is the role of Fairchild Industries in the canceled manned MOL program and its likely continuation in the same role for the advanced MOL program. Fairchild Industries' role included life support functions for the military astronauts:

> NASA MAY REPEAT MAY PICK UP THE FAIRCHILD-HILLER EFFORT, PART OF THE IBM EFFORT (PRINTERS), THE MOLECULAR SIEVE, AND SOME OF THE LIFE SUPPORT ITEMS, BUT THOSE APPEAR TO BE ABOUT THE ONLY ONES.[391]

The involvement of Fairchild Industries (formerly Fairchild Hiller up until 1971) in running the life support functions for the canceled basic manned MOL program, which in turn could be used in the advanced "unmanned program", is highly significant. After Wernher von Braun officially retired from NASA and the Marshall Space Flight Center in 1972, he immediately moved over to Fairchild Industries located in Germantown, Maryland, and became their vice president.[392] It's worth emphasizing that von Braun had done the most work in conceptualizing a viable space station that began with his 1946 von Braun Station proposal which was to be built in a modular fashion.

At Fairchild Industries, von Braun led the secret effort to link together in space the MOL modules originally designed by McDonnell Douglas for the manned missions into the required modular sections, just as the Von Braun Station conceptualization had proposed. Lockheed Martin, General Electric and Fairchild Industries were all deeply involved in building the stealth space station(s) for the NRO/USAF/NASA. Fairchild Industries' connection to the Antarctic German program was firmly established with von Braun's tenure as vice president. While at Fairchild, von Braun met Carol Rosin, who was employed there as a corporate manager from 1974 through 1977. She became von Braun's spokesperson after he was stricken with cancer, and in 2000 she publicly disclosed von Braun's concerns over secret corporate plans for a false alien invasion in order to weaponize space.[393] The ultimate agenda was to use such a staged attack to justify major countries such as the U.S. giving up their national sovereignty to a one-world government, which would be secretly controlled by the Antarctic Germans and their Draconian allies.

Later, in 1996, Fairchild Industries took over Dornier which had assembled the Haunebu flying saucers for Nazi Germany from key components provided by other German aviation companies during World War II.[394] Dornier scientists were among those that continued to secretly work with the Antarctic Germans in constructing new spacecraft out of their hidden bases using the latest technological advances by the U.S. and other nations. Rosin's testimony suggests that the Antarctic Germans planned to use the space stations secretly assembled under the advanced MOL program in their plan to stage a false alien invasion.

Advanced holographic technologies would be beamed to Earth from either the manned stations or Earth orbiting satellite networks connected with a NASA project called "Blue Beam". Project Blue Beam was first publicly exposed in 1994 by the journalist Serge Monast, who mysteriously died of a heart attack in 1996 even though he had no history of heart disease.[395] Monast wrote about Project Blue Beam as an elaborate multistage attempt to set up a one-world religion using the holographic projections of religious figures. He confirmed that Project Blue Beam also involved staging a false alien invasion, as described by Rosin and others, to get public support for radically reforming political institutions to pave the way for a one-world government.

There are many conclusions to be drawn from the space station programs. The public had been deceived into believing that MOL had been canceled in 1969. In fact, only the basic MOL program was terminated, not the more ambitious and advanced "unmanned" MOL program using complete MOL modules which were secretly launched into space from Vandenberg AFB. Throughout the 1970's and 1980's, these unmanned MOL modules were initially assembled into the three-section configuration proposed by General Electric, and eventually, into larger configurations such as the 20-section Von Braun Station. It's also important to keep in mind that the modules offered life support systems and were fully capable of supporting a crew.

According to Corey Goode, he saw electronically archived classified documents of the assembled USAF/NRO space station during his "20 years and back" secret space program service from 1987 to 2007. Goode described the space station as having nine modular cylindrical sections joined together into a circular arrangement – a nonagon – similar in overall shape to the Von Braun Station, but with less than half the sections. Using General Electric's crew estimates for a three-section configuration capable of housing 40 crew, a nine-section configuration would hold approximately 120 crew. Goode said that there are currently two or three of these space stations in operation which have been outfitted with advanced stealth technologies.[396] Goode's covert service began in 1987, so it's possible that by then two or three nonagon stations had already been built, based on Reagan's diary entry about his briefing where he identified 300 crew members orbiting in space.

While Goode believes the stations were created using "space junk" from the disused stages of the many rockets used to launch satellites, it makes more sense that these were in fact the MOL modules sent up on Titan IIID rockets from Vandenberg AFB. Goode provided a graphic illustration of the nonagon-shaped space station he witnessed during his covert space service (see Figure 41). The declassified NRO files corroborate Goode's claim that the USAF/Military-industrial Complex secret space program had assembled multiple nine-sided versions of the Von Braun station which are currently in operation.

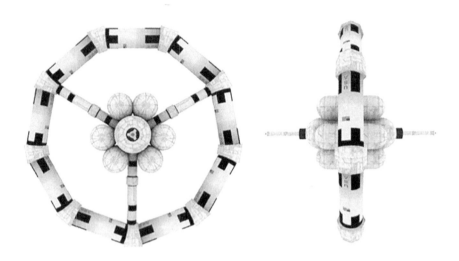

Figure 41. Space Station comprised of nine cylindrical sections.
Credit: Sphere Being Alliance

It is certain that the USAF, NRO, and NASA had competing agendas for the manned space stations that had been covertly set up. For the USAF, the covert space stations offered the high ground for dealing with future national security threats from Earth or in space using advanced weapons systems that would be added over time, such as "Rods of God" (a kinetic energy projectile) technology. For the NRO, the space stations offered an unprecedented vantage point for global and space surveillance of both the Antarctic Germans and visiting extraterrestrial civilizations. For NASA, which had been deeply infiltrated by the Antarctic Germans, the space stations offered a means of monitoring USAF/NRO activities, learning about the different extraterrestrial groups visiting Earth, and using such information to stage a Project Blue Beam global religious event or alien invasion.

# CHAPTER 14

# USAF Antigravity Alien Reproduction Vehicles

> I can tell you about a contract we recently received ... The Skunk Works has been assigned the task of getting E.T. back home.

> – Ben Rich, Director of Lockheed's Skunkworks

From the early 1970's, Vandenberg Air Force Base was used to launch Titan III rockets carrying Manned Orbiting Laboratory (MOL) modules and, when necessary, military astronauts in Gemini B capsules who could assemble together the modules in space to create the different modular configurations before returning to Earth. One configuration revealed by declassified National Reconnaissance Office (NRO) documents was a station design proposed by General Electric using "three mated cylinders" (cylindrical sections) that could accommodate 40 astronauts.[397] As more modules were delivered into Earth orbit throughout the 1970's and into the early 1980's, the three-section station configuration could be expanded into completely different and larger configurations. This would make it possible to assemble the nine-sided nonagon that Corey Goode had read about, whose design was based on von Braun's original 1946 idea of building a space station by assembling cylindrical modular sections in a circular formation.

As mentioned in chapter 13, a declassified NRO document

titled "Advanced MOL Planning: Missions and Systems" showed in great detail General Electric's plan for configuring MOL modules so they could hold a space crew of either 12 in a "one-cylinder station" or 40 in a "three-cylinder station".[398] A close up of the proposed space station "Command Post" shows modules for crew living quarters, general quarters and housekeeping, a resupply area and combat information center, along with docking ports and solar array energy panels (see Figure 42).[399] Another source, including nuclear generators could replace the large solar array panels designed for supplying energy to the three-section station.

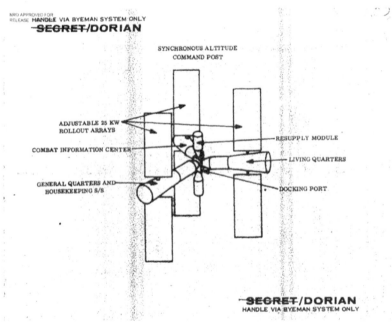

**Figure 42. Declassified NRO document showing General Electric Proposal for assembling a "Command Post" Space Station from MOL modules**

The larger configurations undoubtedly existed by the time of President Reagan's 1985 diary entry referring to the U.S. having the capability of placing 300 astronauts in orbit. If 40 astronauts could fit into General Electric's three-section configuration of MOL modules (with roughly 13 astronauts per section), then it would take approximately 22 sections comprising various MOL modules to accommodate permanently 300 astronauts in Earth orbit. This estimate is close to the number of sections originally envisaged by von Braun for his 1946 space station design. Alternatively, there could have been up to three of the nine-sided configurations that Goode learned of during his secret space program service.

Assembling space stations using different configurations starting from the basic three-section modular system, to nine (as Goode disclosed) or 20 (as von Braun conceived) would take multiple trips by the Saturn III rockets ferrying the modules, military astronauts and additional equipment up on particular missions. The "Advanced MOL Planning: Missions and Systems" document also contains a flow chart showing how manned operations would begin in 1972, and by 1979, the three-section Command Post would have been completed.

However, accommodating 300 astronauts in orbit required more than the three-section design presented by General Electric. Many more sections made up of the MOL modules were needed, thereby requiring significantly more time. Just how much more time is not clear from the declassified NRO documents, but we can get an idea from how long it took a consortium of nations comprising the U.S. (NASA), Japan (JAXA), Canada (CSA) and Russia (Roscosmos) to assemble the smaller International Space Station (ISS) that also used a modular construction process. It took the international consortium over 40 space missions and 13 years (from 1998 to 2011) to build the ISS, which can accommodate a permanent crew of up to "six astronauts"[400] –half of what only one section of the advanced MOL program could accommodate according to the General Electric proposal. Consequently, it is highly likely that after the 1969 cancellation of the basic MOL program and the start of the advanced program, the period spanning the 1970's into the early 1980's was spent assembling the larger USAF/NASA/NRO space station(s).

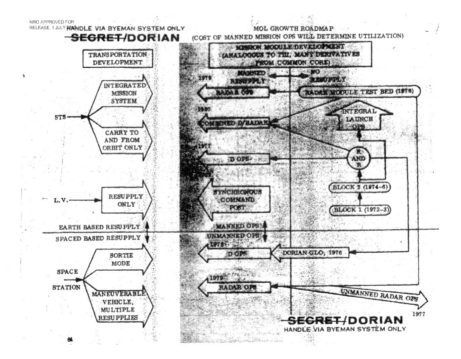

**Figure 43 Declassified NRO document showing General Electric Proposal for Assembling Space Station over 1970's**

To have already placed 300 astronauts in space by 1985, the USAF/NASA/NRO would have had to quietly develop a delivery system far more advanced than the Titan III rockets with their Gemini B capsules that could carry only two astronauts at a time. A more sophisticated transportation system was indeed developed to ferry astronauts and supplies to the space stations. This ambitious feat was achieved by using antigravity technologies acquired through the successful reverse engineering of captured alien spacecraft.

## Reverse Engineering a Better Delivery System

In chapter 3, it was mentioned that the USAF had retrieved four saucer craft from Nazi Germany under Operation Paperclip. The craft had been taken first to Wright Field/Wright Patterson AFB for study and reverse engineering purposes. Sometime before 1958, the four craft were then taken to Area 51's S4 facility for further study and storage alongside three extraterrestrial craft that had been retrieved from New Mexico. However, German scientists brought to the U.S. under Operation Paperclip provided little help to the USAF in understanding such advanced technologies. Aside from a few like Professor Otto Schumann, most of the Paperclip scientists' area of expertise were rockets rather than exotic propulsion systems using high voltage electrostatics and highly pressurized plasma circuits.

Only after agreements were reached between the Eisenhower Administration and the Antarctic Germans in the 1950's did the USAF begin to receive reliable help in their study and reverse engineering efforts of the captured German and extraterrestrial craft. German Paperclip scientists placed in senior NASA positions were in regular touch with their Antarctic brethren, as evidenced by Hans Kammler being witnessed at NASA's Kennedy Space Center in the mid-1960's.[401] Since the Apollo, Gemini and Skylab programs were a means for NASA to covertly move enormous resources and manpower into NASA programs that directly helped the Antarctic Germans develop large space battle groups capable of interstellar travel. The Germans in return assisted the USAF to reverse engineer the comparatively small antigravity flying saucers of both German and extraterrestrial origin stored at Wright Patterson and Area 51. The USAF was helped in developing smaller surveillance craft using advanced antigravity technologies because they posed no strategic threat to the Antarctic German's far larger space battle groups capable of interstellar travel.

In addition to their precarious German partners, the USAF also worked tenuously with Gray aliens under the agreements reached in the mid-1950's. Engineers such as Bill Uhouse have revealed how the Gray aliens helped the USAF understand propulsion and navigation systems on the flying saucers, yet progress remained very slow in modifying the Gray's spacecraft for the significantly larger human pilots. According to Charles Hall, by the 1960's the USAF was also receiving help from the Tall Whites in developing prototype nuclear powered saucer-shaped craft capable of reaching the Moon. Hall described how parsimonious the Tall Whites were in allowing senior USAF officers or scientists to witness their construction process, denying access to the more advanced aerospace technologies needed for interstellar travel. Whatever spacecraft the USAF developed with, or gained from, the Tall Whites was very limited in range and capacity. Nevertheless, such craft could be used to ferry military astronauts and supplies into low Earth orbit for constructing the USAF/NRO/NASA manned space stations. This means that in addition to the Titan III rockets fitted with Gemini B capsules holding only two member crews, the USAF could also ferry military astronauts and equipment into Earth orbit using the prototype flying saucers produced from their various reverse engineering projects. We can get an idea of the size and holding capacity of these saucers from what whistleblowers have revealed about these "Alien Reproduction Vehicles".

## Three Kinds of Alien Reproduction Vehicles

The most detailed description of what these reverse engineered craft looked like comes from an aviation patent illustrator, Brad Sorensen, who witnessed three flying saucer craft at Edwards AFB on November 12, 1988. After attending a USAF airshow at Norton AFB in San Bernardino, California, Sorensen was then taken over to Kern County in Southern California to a restricted area within Edwards AFB. There he saw the three flying saucer craft which were hovering with other adjacent exhibits, along with a video demonstration of the flight performance and specifications of the smallest of the saucer-shaped craft. Mark McCandlish, a professional aviation illustrator who was a friend and colleague of Sorenson, had declined an invitation to attend the Norton airshow but later explained what Sorensen had told him:

> He said that the smallest was somewhat bell-shaped. They were all identical in shape and proportion, except that there were three different sizes. The smallest, at its widest part, was flat on the bottom, somewhat bell-shaped, and had a dome or a half of a sphere on top...

> On the inside of the crew compartment was a big column that ran down through the middle, and there were four ejection seats mounted back-to-back on the upper half of this column. Then, right in the middle of the column, was a large flywheel of some kind.

> Well, this craft was what they called the Alien Reproduction Vehicle; it was also nicknamed the Flux Liner. This antigravity propulsion system – this flying saucer – was one of three that were in this hangar at Norton Air Force Base....[402]

Sorensen sketched the smallest craft and gave it to McCandlish. From Sorensen's detailed sketch, McCandlish was able to render a more comprehensive illustration of the "Alien Reproduction Vehicle" (ARV) which Sorenson later saw and confirmed as being highly accurate.[403] McCandlish's illustration showed how the flying saucer was divided into a top cupola section seating a crew of four, and a bottom section filled with equipment designed for power generation, electrical energy storage, propulsion, and navigation (see Figure 44). Sorensen identified 48 large capacitor stacks capable of holding large electrostatic charges which could be used to store electrical energy for powering the craft. The 48 capacitor stacks also had another very important purpose. They were designed to generate antigravity propulsion using the Biefeld-Brown Effect, which occurs when a large electrostatic charge is generated on the anode or positive plate of a capacitor.[404]

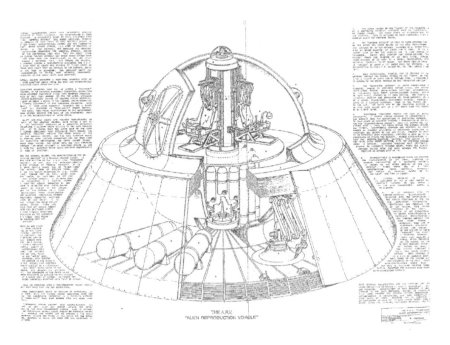

Figure 44. Cutaway of ARV showing 48 large capacitor stacks of eight plates each and oxygen cyclinders. Copyright (March, 1989): Mark McCandlish. All rights reserved.[405]

The Biefeld-Brown Effect dates back to 1923 when Thomas Townsend Brown first learned that high voltage electrostatic charges could provide a previously unknown thrust to a capacitor. He cooperated with a professor of physics at Ohio's Denison University, Dr. Alfred Biefeld, in improving his experiments and measurements. Brown filed for a British patent in August 1927, which was granted to him in November 1928.[406] The title for his successful patent was: "A Method of and an Apparatus or Machine for Producing Force or Motion."

In the accompanying text of the patent, Brown described his invention:

> This invention relates to a method of controlling gravitation and for deriving power therefrom, and to a method of producing linear force or motion. The method is fundamentally electrical. The invention also relates to machines or apparatus requiring electrical energy that control or influence the gravitational field or the energy of gravitation; also to machines or apparatus requiring electrical energy that exhibit a linear force or motion which is believed to be independent of all frames of reference save that which is at rest relative to the universe taken as a whole, and said linear force or motion is furthermore believed to have no equal and opposite reaction that can be observed by any method commonly known and accepted by the physical science to date.[407]

The application of the Biefeld-Brown effect is succinctly presented in McCandlish's explanation of how navigation was achieved by manipulating the electric charge in each capacitor/plate so that it would send the Alien Reproduction Vehicle in one direction or another:

> Why so many different capacitor sections? If you just have one big disc ... in that case, you have levitation, but you don't get any control. You have this thing floating around, and it's just sort of floating on whatever this field is that it's producing, but you don't have any control.

So, what happens? You break that disc up into 48 different sections, and then you can decide how much electricity you want to put on this side or over there on that side, so you can control the amount of electricity and the amount of thrust and vectoring that you get. You can make it go straight up, you can make it bank and turn and pitch – whatever you want to, by virtue of the fact that you can control where the electricity goes in those 48 different sections. If you ever take a circle and divide it up into 48 equal sections, you'll find that those are really thin little slices. So, you have these 48 individual capacitors, and you have one big Tesla coil. You've got to have some kind of a rotating spark gap, just like the distributor in your car, that sends the electricity out to each of those sections. Then, you have to have some way of controlling how much electricity goes to each one.[408]

Put simply, McCandlish has pointed out that the craft could be steered by sending a high voltage electrostatic charge to a specific capacitor plate among the 48, thereby generating thrust in the direction that the positive plate/anode was pointing.

As to the question of what powered the ARV, McCandlish said that it could tap into the zero-point energy field of the vacuum. This would make the ARV capable of generating sufficient electrical power for all its energy needs:

There is a scientist in Utah by the name of Moray B. King – he wrote a book called *Tapping the Zero Point Energy*. What he maintains is that this energy is embedded in space-time all around us: it's in everything we see. I think it was James Clerk Maxwell who speculated that there's enough of this flux, this electrical charge, in the nothingness of space, that if you could capture all the energy that was embedded in just a cubic yard of space, you'd have enough energy to boil the oceans of the entire world.... This Alien Reproduction Vehicle, this Flux Liner, has a way of doing this somehow electronically. Now, Brad had described the fact that this central column has a kind of vacuum chamber in it. The vacuum chamber is one of the things that all of these scientists describe in these over-unity or free-energy devices they build. They all have some kind of vacuum tubes, vacuum technology. [409]

McCandlish went on to describe the mass cancellation effect that would be created as energy was drawn from the vacuum flux, thereby allowing the ARV to defy the laws of inertia. This would enable the ARV to accelerate to incredible speeds and travel great distances without its occupants experiencing undue stress due to changes in inertia.

## Sources Corroborating Sorensen's 1988 ARV Sighting

Mark McCandlish says that in 1992 at an Edwards AFB airshow, he spoke with Kent Sellen, a former Air Force non-commissioned officer (NCO) who had worked as a crew chief for an aircraft being test flown at Edwards in 1973. Sellen told McCandlish that in 1973, he saw a flying saucer hovering in a Quonset-style hangar which was under heavy military security. McCandlish recounts Sellen's story:

He [told me] one night [his] shift supervisor [had told him], "Go out to North Base — they've got a ground power unit for an aircraft that's leaking or failed or something, so we need you to take a tow vehicle out there. Go out, pick it up, bring it back, drop it off at the repair depot; then you can go for the night, because we've finished all our other work." Well, instead of going around the big perimeter road that goes up to the main entrance of North Base, Kent Sellen drove straight across the dry lake bed at Edwards to the North Base facility. He [came] up off the dry lake bed, [rolled] right up on the tarmac, and [was] going down these rows of hangars - they [were] all Quonset-style hangars back then. He [stopped] in front of the first one with the doors cracked, expecting to find this defective ground power unit, and what [did] he see? He [saw] this flying saucer sitting in the hangar, hovering off the ground.

... I [asked], what happened? He [said], "This thing was flat on the bottom, [with] sloping sides, little cameras in these little plastic domes all over, [and] there was a door on the side. I wasn't there for 15 seconds, [when] I heard footsteps running up to me, and before I could even turn around and look, there was a machine gun barrel at my throat." A gruff voice [said], "Close your eyes and get on the ground, or we're going to blow your head off."

They put a hood over his head, blindfolded him, hauled him off, and they spent 18 hours debriefing him. While they did, they told him things about this vehicle that my buddy Brad [Sorensen] didn't even know.[410]

The saucer Sellen accidentally saw in 1973 matched the one that Sorensen would view and sketch in 1988. In addition, McCandlish says that each of the ARVs witnessed by Sellen and Sorensen had a similar design to a craft that was photographed in Provo, Utah, back in 1966 – a photo which was discussed in the Condon Committee. Based on Sellen's testimony and the 1966 Provo UFO photo, McCandlish felt that he had found independent corroboration for Sorensen's story.

Figure 45. Provo, Utah, UFO – July 1966

## Comparison of ARVs with Recovered Nazi Flying Saucers

As a former Air Force NCO and crew chief working on test aircraft, Sellen's corroboration of Sorensen's ARV is compelling and begs further exploration related to all three of the saucer-shaped craft the aviation illustrator witnessed at Edwards, and how these compare to the Nazi flying saucers brought into the United States after World War II. One important observation which will be shown is the remarkable similarity in the sizes of the Air Force ARVs compared to the Nazi flying saucers. First, here is what McCandlish said about the different sizes of the vehicles that Sorensen viewed:

> Now, he [Sorensen] said there were three vehicles.
> The first one - the smallest, the one that was
> partially taken apart, the one that was shown in the
> video that was running in this hangar November 12,
> 1988 at Norton Air Force Base – was about 24 feet
> in diameter at its widest part, right at the base. The
> next biggest one was 60 feet in diameter at the base
> ... the largest of these vehicles was about 130 feet in
> diameter.[411]

According to McCandlish, the 24 ft vehicle could carry a crew of
four in the top cupola section, but details were not provided for
how many crew the larger ARVs could hold. However, a rough
estimate can be made by a comparison with the Vril and Haunebu
series of flying saucers built by Nazi Germany.

In chapter 3, Information was given about four Nazi flying
saucer craft which had been secretly brought to the U.S. under
Operation Paperclip to be closely studied at Wright Patterson AFB
and then at the S4 facility at Area 51.[412] Let's examine what is
known of the Vril and Haunebu flying saucer craft that had been
successfully developed by Nazi Germany, the most advanced of
which were taken to Antarctica before the end of World War II. Of
these, at least four craft (two Vril and two Haunebu) were captured
and brought into the United States.

In my book, *Antarctica's Hidden History*, a detailed analysis is
given reviewing the Nazi SS documents that were released at the
end of the Cold War by a former member of the Bulgarian Academy
of Sciences, Vladimir Terziski.[413] In 1991, Terziski claims he came
into the possession of a leaked documentary film from the Nazi SS
archives that had been shared among Warsaw Pact countries,
which revealed different types of flying saucer craft built in Nazi
Germany.[414] The film displayed documents that date from late
1944 and early 1945, with detailed data on four different sized
flying saucers that were being flight tested and developed by the
Nazi SS in different underground research facilities. The first three

of the four Nazi flying saucers are comparable in size to those Sorensen saw at Edwards AFB. The similarities are such that it can be assumed the Air Force designs for the ARVs were at least partially a result of them having successfully reverse engineered the Vril and Haunebu series saucers that had been brought into the U.S. after the war.

The first Nazi SS document from the film details the production statistics of four flying saucer models in terms of how many had been built and the number of times each was test flown [see Figure 46]. The document shows that 17 Vril I craft were built, and flight tested 84 times. This model used a "Schuman Levitator" propulsion system, based on the work of Professor Otto Schumann who was among the German scientists brought into the U.S. under Operation Paperclip. The Schuman Levitator was an electrogravitic propulsion system that used high voltage electrostatics to generate thrust similar to the Biefeld-Brown effect. Additional Nazi documents gave the crew size of the Vril I as two, with a flight duration of 5.5 hours and a top speed of 7,200 mph (12, 000 km/hr).[415]

The 38 ft (11.5 meters) diameter Vril I is comparable to the 24 ft (7.3 meters) Air Force ARV witnessed by Sorensen. This is significant given that we know that two of the Vril I craft came into the U.S. for study and reverse engineering. The Vri-1 was propelled by a "Schumann SM-Levitator" which incorporated a zero-point/free energy device that had been created by German inventor Hans Coler, with a propulsion system developed by Otto Schumann.[416] The charged capacitors in the Vril I could only power the craft for 5.5 hours, suggesting that the zero-point energy device was still quite limited in its power generation capacity. In addition to miniaturization and developments in modern battery/capacitor technologies, we know from McCandlish's analysis of Sorensen's design sketch that the reverse engineered USAF craft could sufficiently extract energy from the vacuum field for its power needs. Consequently, the smaller ARV had a flight time that was significantly longer than its Nazi forerunner.

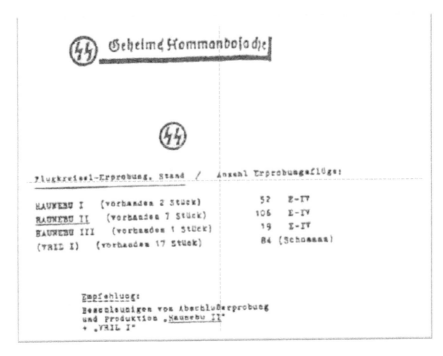

**Figure 46. Production statistics for Nazi SS Flying Saucers**

The Nazi SS production statistics document stated further that two Haunebu I craft were built and flight tested 52 times. At least one of these two Haunebu I craft was found and brought to the U.S. under Operation Paperclip. The Haunebu I was fitted with E-IV propulsion systems called a "Tachyonator-7 drive" which was:

> ... a revolutionary electro-magnetic-gravitic engine which improved Hans Coler's free energy machine into an energy Konverter coupled to a Van De Graaf band generator and Marconi vortex dynamo [a spherical tank of mercury] to create powerful rotating electromagnetic fields that affected gravity and reduced mass. It was designated the Thule Triebwerk [Thrustwork, aka Tachyonator-7 drive].[417]

Additional Nazi documents put the Haunebu I crew size at eight with a top speed of 10,200 mph (17,000 km/hr).[418]

The 82 ft (25 meters) Hanuebu I is comparable in size to the 60 ft (18.3 meters) Air Force ARV. Given that the smaller 24 ft ARV held twice the number of crew than the somewhat larger Vril I craft, it can be assumed that the 60 ft ARV could hold more crew than its somewhat larger Haunebu I counterpart, perhaps as much as double the eight crew size of the latter, therefore as many as 16 astronauts. Most importantly, the 60 ft ARV possessed an advanced engine capable of generating an independent power supply from the vacuum rather than relying on stored electrical energy. The Haunebu I had a flight endurance of 18 hours, so it can be assumed that the Air Force would have significantly improved the vacuum energy device that powered its ARVs.[419] This meant that the 60 ft ARV was capable of longer-term space missions than its Haunebu I forerunner.

Finally, the Nazi SS document showed that seven Haunebu II saucer craft were built and flight tested 106 times. These were also fitted with E-IV propulsion systems. Additional Nazi documents put its crew size at a maximum of 20 with a top speed of 12,620 mph (21,000 km/hr) and a flight endurance of 55 hours.[420] At 105 ft (32 meters), the Haunebu II is comparable in size to the 130 ft (39 meters) Air Force ARV Sorensen witnessed. Again, this is not accidental given that the Air Force had exhaustively studied the Haunebu II craft that had been taken to Wright Field after World War II before being taken to Area 51 where it was eventually witnessed in 1958.

The fact that the smaller 24 ft ARV held twice the number of crew than its slightly larger Vril I craft counterpart allows us to estimate that the 130 ft diameter ARV would have been able to accommodate roughly 40 astronauts. The largest of the Air Force's ARVs would have possessed an advanced engine capable of generating power like the Haunebu II, yet with modern advancements it likely greatly exceeded the 55 flight hour endurance of the Haunebu II. This also would have made the 130 ft

ARV capable of even longer-term deep space missions.

While the speeds of the Vril and Haunebu craft cited by Terziski sound impressive, even by today's standards of publicly known aerospace vehicles, they are far less than what Sorensen heard from a four-star Air Force general giving a lecture about the saucers on display at Edwards AFB in 1988. The General said that the three ARVs "were capable of doing light speed or better".[421] This suggests that by 1988, the Air Force had successfully developed zero-point energy devices for generating sufficient power for superluminal speeds.

### Table: Comparison of Nazi Germany Saucers and US Alien Reproduction Vehicles

| Nazi Germany 1945 | US Air Force 1988 (Norton AFB) |
|---|---|
| **Vril I:** Diameter 38 ft (11.5 m)<br>Crew: 2<br>Flight Duration: 5.5 hours<br>Speed: 7,200 mph (12,000 kmh) | **Small ARV:** Diameter 24 ft (7.3 m)<br>Crew: 4<br>Flight Duration: Unknown<br>Speed: Superluminal |
| **Haunebu I:** Diameter 82 ft (25 m)<br>Crew: Unknown<br>Flight Duration: 18 hours<br>Speed: 10,200 mph (17,000 kmh) | **Medium ARV:** Diameter 60 ft (18 m)<br>Crew: 8-16 (estimated)<br>Flight Duration: Unknown<br>Speed: Superluminal |
| **Haunebu II:** Diameter 105 ft (32 m)<br>Crew 20<br>Flight Duration: 55 hours<br>Speed: 12,620 mph (21,000 kmh) | **Large ARV:** Diameter 130 ft (39 m)<br>Crew 20-40 (estimated)<br>Flight Duration: Unknown<br>Speed: Superluminal |
| **Haunebu III:** Diameter 233 ft (71 m)<br>Crew: 32<br>Flight Duration: 7-8 weeks<br>Speed: 24,855 mph (40,000 kmh) | No comparably sized ARV known |

## USAF Flying Saucers and Superluminal Travel

While the idea of superluminal travel may at first sound far beyond the reach of USAF capabilities in 1988, it's worth keeping in mind what the former director of Lockheed's Skunkworks, Ben Rich, said when ending public lectures. Beginning on September 20, 1983, he would show a slide of a flying saucer and tell the audience members that Lockheed Martin had gotten a new contract; they had "been assigned the task of getting E.T. back home."[422] At the time, most took this as the punchline of a joke. However, Rich was likely alluding to a contract that had been awarded to Lockheed by the USAF to build spacecraft capable of interstellar travel at superluminal speeds. Rich didn't say when the contract had been awarded, but it's significant that five years after his speech, the USAF was showing off three brand new saucer-shaped craft at Edwards AFB.

On March 23, 1993, two years after his retirement and five years after the airshow at Norton AFB, Rich changed his punchline. At an engineering conference in Los Angeles, he again displayed the flying saucer on screen, but instead said: "We now have the technology to take E.T. home."[423] The audience laughed as before, but privately Rich shared with colleagues what was really going on in the classified programs that Skunkworks was involved in:

> We already have the means to travel among the stars, but these technologies are locked up in black projects and it would take an act of God to ever get them out to benefit humanity…. Anything you can imagine, we already know how to do.[424]

Rich's comment here is consistent with Lockheed being given the contract for building the Air Force's own flying saucers capable of superluminal travel sometime before 1983 and completing them by at least 1988. The saucers were thereafter deployed in the then newly formed USAF Space Command established September 1, 1982.

Figure 47. Slide Ben Rich showed when discussing taking ET home

For further clues into the propulsion and navigation technologies used in the USAF flying saucers that were capable of "getting E.T. back home", two Defense Intelligence Reference Documents written in 2009 discuss how superluminal travel can be achieved through the development of "Warp Drives" and "Traversable Wormholes". The two documents are titled, "Traversable Wormholes, Stargates, and Negative Energy" and "Warp Drive, Dark Energy, and the Manipulation of Extra Dimensions".[425] The introductions to the two documents state that they are part of "a series of advanced technology reports produced in Fiscal Year 2009 under the Defense Advanced Aerospace Weapon System Applications (AAWSA) Program." Both documents were confirmed to be genuine by one of the authors, Dr. Eric Davis, who was surprised to see them publicly circulated since they carried national security markings of "Unclassified: For Official Use Only".[426]

Based on the evidence considered so far, the USAF and contractors such as Lockheed's Skunkworks had succeeded in greatly improving upon the flight performance of the Vril and Haunebu models which had been secretly studied at Wright Patterson and later at the S-4 facility. The craft were capable of superluminal travel, thus placing USAF astronauts in space and getting them to nearby planetary bodies such as the Moon and Mars. What we now know of the USAF saucer-shaped ARVs provides a clear answer to what President Reagan's diary entry on June 11, 1985, was referring to when he said, "our shuttle capacity is such that we could orbit 300 people."[427] Reagan was alluding to one or more squadrons of the different sized ARVs witnessed at Edwards AFB, which could ferry astronauts into low Earth orbit to work on the secret USAF/NRO/NASA space stations using MOL modules sent up on Titan IIID rockets, at least one of which was completed by 1985. Lockheed Martin's Skunkworks not only played a key role in building the Air Force's ARVs but also in assembling the space station(s) in low Earth orbit.

What Brad Sorensen saw in 1988 were operational flying saucers being displayed by their proud owners, the USAF. The Air Force and especially its Space Command was not content, however, to only have squadrons of flying saucers compliments of Lockheed Martin and other aerospace contractors. The Air Force was already working on significantly larger vehicles than the 130 ft ARV witnessed by Sorensen. Now came a radical new aviation concept – flying triangles!

# CHAPTER 15

# USAF Flying Triangles and the TR-3B

I do not think there is any thrill that can go through the human heart like that felt by the inventor as he sees some creation of the brain unfolding to success.

— Nikola Tesla

E dgar Fouche served with the US Air Force from 1967 to 1987 before spending another eight years with defense contractors working on a number of classified aviation programs at Area 51 in the Nellis Air Force Range, Nevada. He has supplied documents corroborating his USAF military service and work with defense contractors on various aviation projects. One document showed Fouche was assigned to Nellis Air Force Base during the period from 1976 to 1979, which is when he says he was first assigned to Area 51. Fouche claims he met people there working on what were believed to be the Air Force's most classified aerospace programs.[428] During his nearly 30 year career in military aviation, Fouche spent a total of 12 years working at Area 51 in various capacities for different employers, including the Defense Advanced Research Projects Agency (DARPA).

When Fouche decided to write a 'fiction based on fact'

book revealing information about the most advanced aerospace technologies currently under development, he consulted with five close friends who had also worked on classified aviation projects.[429] In confidential discussions, the group pooled their information about the classified projects and technologies they had worked on at a number of military facilities across the country.[430] Based on these firsthand accounts, Fouche later claimed that the replacement for the Blackbird SR-71 was a project called "Aurora" involving a group of exotic aircraft. Fouche further claimed that in addition to the development of hypersonic aircraft, Aurora included an even more exotic type of craft using antigravity technology.

Aviation industry discussion of the Aurora program began in March 1990, due to a story by *Aviation Week & Space Technology* magazine that revealed:

> Aurora was inadvertently released in the 1985 US budget, as an allocation of $455 Million for Black aircraft PRODUCTION in FY 1987. Note that this was for building aircraft, not R&D.[431]

Prominent Aviation writers Nick Cook, author of *The Hunt for Zero Point*, and Bill Sweetman, subsequently discussed the Aurora program and its likely existence.[432] After interviewing Area 51 employees, Sweetman concluded in 2006:

> Does Aurora exist? Years of pursuit have led me to believe that, yes, Aurora is most likely in active development, spurred on by recent advances that have allowed technology to catch up with the ambition that launched the program a generation ago.[433]

Regarding the Aurora Project's two hypersonic craft, Fouche said: "The Aurora comprises the SR-75 capable of speeds above Mach 5, and acts as a mother ship for the SR-74 that can travel at speeds of Mach 18 or more into space to deliver satellites."[434] According to Fouche, the SR-74 and SR-75 incorporated advanced stealth capabilities. Such capabilities raise the possibility that these two craft also incorporated electrogravitics as part of their propulsion systems since stealth programs were known to be used as 'cover programs' to hide more sensitive electrogravitic projects.[435] This would help explain how the SR-75 could carry the SR-74 piggyback-style to an altitude of 100,000 feet and reach the minimum hypersonic speed of Mach 5.1 to launch the SR-74, as theoretically required for it to fly at even faster hypersonic speeds into low Earth orbit. Therefore, the SR-74 could act as a delivery vehicle for the USAF/NRO space stations, while its relatively small size would limit how much it could carry in cargo and personnel.

According to Fouche, there is a third vehicle belonging to the Aurora Program called the "TR-3B" which was the most highly classified program he and other workers at the Groom Lake Area 51 facility were aware of:

> The TR-3B is Code named Astra. The tactical reconnaissance TR-3B first operational flight was in the early 90's. The triangular shaped nuclear powered aerospace platform was developed under the Top Secret, Aurora Program with SDI and black budget monies. At least three of the billion dollar plus TR-3Bs were flying by 1994. The Aurora is the most classified aerospace development program in existence. The TR-3B is the most exotic vehicle created by the Aurora Program. It is funded and operationally tasked by the National Reconnaissance Office, the NSA, and the CIA. The TR-3B flying triangle is not fiction and was built with technology available in the mid-80's. [436]

In addition to nuclear power, Fouche said the TR-3B incorporated three rocket engines using conventional fuel sources like hydrogen, oxygen or methane.[437]

( —                              ( —        AUG 20 1979

| NAME L. RATE (LAST, FIRST, MIDDLE I., IAL) | | SSAN | ACTIVE DUTY GRADE |
|---|---|---|---|
| Fouche, Edgar A | | FR4 ▮▮▮▮▮ | TSGT |

*(CHECK APPROPRIATE BLOCK AND COMPLETE AS APPLICABLE)*

| ☐ SUPPLEMENTAL SHEET TO RATING FORM WHICH COVERS THE FOLLOWING PERIOD OF REPORT | ☒ LETTER OF EVALUATION COVERING THE FOLLOWING PERIOD OF OBSERVATION |
|---|---|

| FROM | THRU | FROM | THRU |
|---|---|---|---|
| | | 10 June 79 | 14 AUG 79 |

Precede comments by appropriate data, i.e. section continuation, indorsement continuation, additional indorsement, additional reviewer comments, etc.

FACTS AND SPECIFIC ACHIEVEMENTS:  TSgt Fouche is an outstanding
NCO and technician.  His exceptional performance of all duties,
even under adverse conditions, indicates a dedicated interest in
his work, and a high degree of professionalism.  His wide-
ranging knowledge and diverse expertise in diagnostics, and
mechanical engineering, coupled with his training in advanced
electronics, has helped solve critical mission support problems
consistently.  His awareness of the big picture for future
avionics development and maintenance is an attribute that makes
him a valuable asset to TAC.  TSgt Fouche is considered one of
TACs best R&D team builders in areas of ECM, ATS, and
cryptological support.  He has proven himself recently in the
implementation of the TEWS-TITE bed-down, which was lauded by
TAC Hq LGM.  STRENGTHS:  TSgt Fouche displays excellent
capabilities when given greater responsibilities, which was
demonstrated in his MAJCOM involvement in cryptological asset
training and provisioning.  He has high endorsements from his
chain of command and is considered an excellent candidate for a
command level position.  OTHER COMMENTS:  This out of cycle
report is generated because TSgt Fouche was assigned TDY to the
AFFTC-DET 3, Nellis AF Range from 1 June 79 to 14 August 79.  His
duties and responsibilities for this period have been verified
via a separate report.  RECOMMENDATION:  Promote at the earliest
opportunity.

| NAME OF EVALUATOR, GRADE, ORGANIZATION, AND LOCATION | DUTY TITLE NCOIC F-15 AIS | | DATE 24 Aug 79 |
|---|---|---|---|
| ▮▮▮Brewer, MSgt, USAF | SSAN (INCLUDE SUFFIX) | SIGNATURE | |
| 57 CRS, Nellis AFB Nevada | FR465-▮▮▮▮▮ | *Brewer* | |

AF FORM 770a PREVIOUS EDITION WILL BE USED
NOV 74
☉ U.S. GOVERNMENT PRINTING OFFICE: 1976-211-291/1105

SUPPLEMENTAL SHEET TO AF FORMS
707, 909, 910, 911 AND 475

1968 - OJT Training (On the Job Training)

This is Ed Fouche's OJT Training (On the Job Training) For his "five level upgrade"

**Figure 48. Document confirming Edgar Fouche worked at Nellis AFB adjacent to Area 51**

What makes the TR-3B stand out from other Aurora craft, and
the saucer-shaped Alien Reproduction Vehicles (ARV) built by
Lockheed for the USAF, is that it used a different type of antigravity

effect compared to the electrogravitics systems developed by Townsend Brown. Fouche referred to this antigravity system as a "Magnetic Field Disrupter", which rotates highly pressurized mercury-based plasma around a circular accelerator ring:

> A circular, plasma filled accelerator ring called the Magnetic Field Disrupter, surrounds the rotatable crew compartment and is far ahead of any imaginable technology. Sandia and Livermore laboratories developed the reverse engineered MFD technology. The government will go to any lengths to protect this technology. The plasma, mercury based, is pressurized at 250,000 atmospheres at a temperature of 150 degrees Kelvin and accelerated to 50,000 rpm to create a super-conductive plasma with the resulting gravity disruption.[438]

Fouche explained that the Magnetic Field Disrupter (MFD) technology differs from electrogravitics insofar as while the latter provides a thrust, MFD technology reduces weight by a staggering factor of 89%.

> The MFD generates a magnetic vortex field, which disrupts or neutralizes the effects of gravity on mass within proximity, by 89 percent. Do not misunderstand. This is not antigravity. Anti-gravity provides a repulsive force that can be used for propulsion.[439]

With the weight of the TR-3B reduced by the MFD technology, this means that other propulsion systems like nuclear and conventional jet engines could provide the necessary thrust to make the TR-3B outperform conventional vehicles possessed by the USAF:

> The MFD creates a disruption of the Earth's gravitational field upon the mass within the circular

accelerator. The mass of the circular accelerator and all mass within the accelerator, such as the crew capsule, avionics, MFD systems, fuels, crew environmental systems, and the nuclear reactor, are reduced by 89%. This causes the effect of making the vehicle extremely light and able to outperform and outmaneuver any craft yet constructed – except, of course, those UFOs we did not build. The TR-3B is a high altitude, stealth, reconnaissance platform with an indefinite loiter time. Once you get it up there at speed, it doesn't take much propulsion to maintain altitude.[440]

# USAF Top Secret Nuclear Powered Flying Triangle - The TR-3B

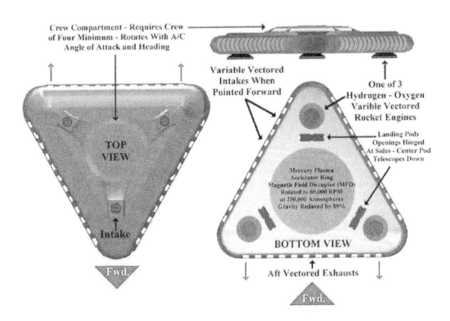

**Figure 49. Illustration of TR-3B. Adapted Edgar Fouche Original Version**

Fouche claimed that the TR-3B was able to silently hover for at least 10 minutes and gave off "a corona of silver blue light" that glowed around it while hovering. [441]

Fouche described the locations where the operational 600 ft versions of the TR-3B were stored:

> The TR-3Bs were stored in a hanger built into aside of a mountain near DARC [Defense Advanced Research Center]. The Nellis Range covers more than 3.5 million acres. One operational TR-3B is now stationed in Scotland, another in Diego Garcia, and the other TR-3B is at Papoose. To my knowledge, there are only three of the 600-feet operational models. I believe there were only 2 or 3 200-foot prototypes built.[442]

In a series of questions and answers, Fouche [EF] provided additional details about the corporate contractors involved in building the TR-3B, numbers built and when test flights began:

> 1. Who is the primary contractor for the TR-3B?
> [EF] Lockheed, Boeing, Northrup, Teledyne Ryan, managed by NRO, NSA and CIA. There were a number of major defense contractors involved to keep different parts of the R&D compartmentalized
>
> 2. What was the date of the first flight of the prototype for the TR-3B?
>
> [EF] They built a lot of DIFFERENT prototypes from the early 70's on. I believe the first gravity warping vehicle, i.e., Triangle was flown in the early 70's.
>
> 3. How many different sizes of the TR-3B were built?
>
> [EF] Approx. 250 feet (prototype) and three were flying before 1994. The operational model was supposed to be 600 feet. Never saw a large

production model. Only saw them over Edwards AFB, AFFTC [Air Force Flight Test Center] and Nevada. [443]

As far as crew capacity is concerned, the 250 ft prototype could carry a crew of four, indicating that the larger 600 ft version could carry significantly more, and likely had a cargo storage area that could accommodate significant equipment and supplies. Furthermore, Fouche identified Ben Rich of Lockheed's "Skunk Works" (pseudonym for "Advanced Development Programs") as the senior project designer in the development of the TR-3B.[444] This suggests that while Rich and Lockheed were developing the Alien Reproduction Vehicle in the early 1970's using the retrieved alien/Nazi flying saucer craft, this company was simultaneously developing a wholly U.S. designed antigravity vehicle: the TR-3B.

Fouche's reference to Scotland as one of the locations where an operational TR-3B was based is significant given flying triangle sightings in both Scotland and nearby Belgium from November 29, 1989, to April 1990. Hundreds of witnesses including police officers saw and photographed these large flying triangles. The sightings were investigated by the Belgium Air Force which resulted in a very well documented incident on March 30, 1990, involving F-16 fighters attempting to intercept the mysterious triangular-shaped craft.[445] Fouche said that based upon his examination of the Belgium photos of the flying triangle and the TR-3B design schematics he had witnessed, they were the one and same craft.[446]

What is especially relevant is that the reported acceleration of the Belgium flying triangle (TR-3B) was 46G according to F-16 fighter instruments.[447] This is far above the 9G forces that most humans can tolerate for an extended period before blacking out.[448] In the case of 16G, this acceleration becomes deadly when experienced for a minute or longer.[449] In short duration G-force circumstances, such as when a fighter jet seat ejects, 32G is considered the safety limit. [450]

The Belgium flying triangle/TR-3B sighting is powerful evidence that antigravity technology along the lines of Fouche's 'Magnetic Field Disruptor' technology was used to significantly reduce the weight and inertia of the craft and its occupants by a factor of 89%. This would explain how the occupants could survive accelerations generating G-forces as high as the 46G estimated by the Belgium F-16's instruments. Essentially, the 46G experienced in a conventionally powered jet or rocket propelled vehicle would convert to approximately 5G being felt by the occupants of a TR-3B when its MFD technology was activated.

According to Fouche's account, the TR-3B became operational in the early 1990's and three were flying by 1994. This suggests that the 1989/1990 Belgium sightings were the result of a TR-3B prototype test program or the larger operational TR-3B was being flight tested before deployment. Corey Goode stated that the TR-3B was a "hand me down" to the USAF's military space program from an even more highly classified space program controlled by the Antarctic Germans and US corporations:

> The TR-3B is considered extremely outdated technology and in many cases has been gifted as "hand me down" craft to "Elites" within the Secret Earth Governments and their Syndicates as something akin to "Company Jets".[451]

Goode's comments suggest that during the Apollo Program when U.S. and German corporations built a vast number of vehicles for the Antarctic space program throughout the 1960's and early 1970's, the TR-3B was among the vehicles that were developed. This might explain how Lockheed was involved in the development of the TR-3B in the early 1970's, but they were only released to the USAF in the early 1990's. Essentially, there was approximately a 20 year lag between the time when US aerospace corporations had developed the TR-3B for the use by the Antarctic German space program and when they were allowed to be transferred to the USAF.

Goode also said the following about the flying triangle phenomenon: "There are so many newer technologies that are of the same general shape as the TR-3B (and models that came afterward) that it would blow people's minds."[452] It is worth remembering this statement for chapter nineteen when photographs of triangular-shaped craft flying over a major USAF base are shown. The craft to be presented appear much smaller than either the 600 ft TR-3B or its 250 ft prototype.

Finally, it's important to bear in mind that the period Fouche gave for the TR-3B becoming operational is well after Ronald Reagan's 1985 presidential diary entry about the U.S. having the capability of maintaining 300 astronauts in Earth orbit. This means that the space station(s) built by the USAF and NRO, under cover of the 'unmanned' MOL project, were initially serviced using Titan IIID rockets, Gemini B capsules atop Titan IIIC rockets, and the Alien Reproduction Vehicles that were already operational.

# CHAPTER 16

## Stargate SG-1 and USAF Sanctioned Soft-Disclosure of Traversable Wormholes

Wormholes were first introduced to the public over a century ago in a book written by an Oxford mathematician. Perhaps realizing that adults might frown on the idea of multiply connected spaces, he wrote the book under a pseudonym and wrote it for children. His name was Charles Dodgson, his pseudonym was Lewis Carroll, and the book was *Through the Looking Glass*.

— Professor Michio Kaku

On July 27, 1997, the television spin-off of the popular 1994 sci-fi movie *Stargate* began to air which featured virtually instantaneous travel using traversable wormholes not only throughout the Milky Way Galaxy but also to nearby galaxies such as Pegasus a staggering three million light years away. What immediately distinguished *Stargate SG-1* was the extraordinary level of support it received from the US Air Force throughout the ten years it appeared on *Showtime* and the *Sci-Fi Channel*. The credits to the series openly acknowledged the cooperation given by the USAF, US Space Command and the Department of Defense.

The USAF granted the series producers access to its secretive underground Cheyenne Mountain complex near Colorado Springs where stock film footage was taken and used throughout the series.[453] At the time of filming, the Cheyenne complex hosted elements from a number of sensitive military commands, including Space Command. USAF jets which included F-15 and F-16's were flown to Vancouver, Canada, for various episodes, and active Air Force personnel were used as extras in multiple episodes.

What really stood out in the series were the cameo appearances made by two serving Air Force Chiefs of Staff, illustrating the extraordinary level of support the show was receiving from the top leadership ranks within the Department of the Air Force. General Michael E. Ryan appeared in the Season 4 episode "Prodigy", and General John P. Jumper appeared during Season 7 in "Lost City, Part 2". The Air Force Office of Public Affairs read every *Stargate SG1* episode script and approved its airing after any "mistakes" were removed. The series producer, Richard Dean Anderson, was given special recognition for the show's positive depiction of the Air Force by the Air Force Association (AFA) at its 57th annual dinner on September 14, 2004. If there was ever a television series designed to show the virtues of a 'USAF secret space program" with its use of exotic stargates to promote American values throughout the Milky Way galaxy and beyond, *Stargate SG-1* was that ideal depiction.

The appearances made by two serving Chiefs of Staff offered a very clear sign that the USAF was doing more than merely signaling its approval of a favorable depiction of the USAF in a sci-fi show. These notable appearances were a sure way of signaling that *Stargate SG-1* was a soft-disclosure initiative officially sanctioned by the highest level within the Department to reveal some of its classified exotic technologies to the public to preempt any possible leaks, and in preparation for a future official disclosure event.

# Screenshot of Stargate SG-1 Credits

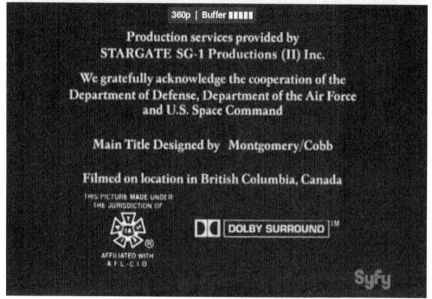

360p | Buffer ▋▋▋▋▋

Production services provided by
STARGATE SG-1 Productions (II) Inc.

We gratefully acknowledge the cooperation of the
Department of Defense, Department of the Air Force
and U.S. Space Command

Main Title Designed by Montgomery/Cobb

Filmed on location in British Columbia, Canada

THIS PICTURE MADE UNDER
THE JURISDICTION OF

DOLBY SURROUND™

AFFILIATED WITH
A F L - C I O

Syfy

**Figure 50. Credits showing USAF & US Space Command
cooperation with Stargate SG-1 production**

The series featured repeated tongue-in-cheek references to its "Stargate Program" and lead mission team "SG-1" being a sanctioned form of soft-disclosure in several episodes using a show-within-a-show formula. For example, two episodes referred to a fictional sci-fi series within the actual series, called "Wormhole X-Treme", which depicted a USAF run program using wormholes to travel throughout the galaxy and the existence of alien 'motherships'. In "Wormhole X-Treme" (Season 5 Episode 12), which aired in September 2001 and commemorated the show's 100th episode, the characters playing USAF leaders in the actual *Stargate SG-1* series approved the fictional series going forward, despite it revealing classified information. The goal was to insulate the truth about the "real" Stargate Program in the *SG-1* series in case future unauthorized leaks occurred. This episode presented an ironic, tongue-in-cheek reference to what the USAF was really doing through its *Stargate SG-1* series, which was disclosing the existence of an extensive galactic-wide system of traversable wormholes – but in a fictional sci-fi setting to insulate the real USAF Stargate Program from unauthorized leaks.

Several whistleblowers have commented that the *Stargate SG-1* series was indeed soft-disclosure of a galactic-wide travel system using traversable wormholes. Among them is Corey Goode who, in a "Cosmic Disclosure" interview with *New York Times* best-selling author David Wilcock that aired on September 15, 2015, included "traversable wormholes" or "portals" among the various kinds of exotic travel technologies used by ancient societies, extraterrestrial civilizations and current day secret space programs.[454] Goode described the difference between various "wormhole" or "portal" technologies that either occurred naturally or were built by a highly advanced ancient civilization unknown even to the current extraterrestrial visitors of Earth.

> There are the natural portal systems that are a part of the known universe. We call it the Cosmic Web. And the ancient portal systems and the current era portal systems use the, or exploit, these natural Cosmic Web portal system to travel from point to point.
>
> The ancient portal systems ... and there are several ancient portal systems that are left behind by several ancient groups that have been found on Earth. They vary in their sophistication.
>
> Some of them do very short point-to-point jumps to reach ... Let's say you want to reach a planet or solar system that is in ... They call them 'hops'.
>
> If there's like let's say 10 solar systems between where you want to go, then you may have to make three or four hops to get to that desired destination. [455]

What Goode described is very similar to the scenario depicted in the *Stargate SG-1* series, where stargates were depicted as being built by a race of beings called "the Ancients". There are significant differences, however. Goode described travel to solar systems within our local stellar cluster as requiring a "series of hops", rather than the "single journey" taken to travel virtually anywhere in our Galaxy as depicted in *Stargate SG-1*.

In his interview with Wilcock (DW), Goode (CG) referred to technologies depicted in *Stargate SG-1* as similar in appearance to how ancient ring-shaped portals or "stargates" actually look:

> CG: The ancient systems . . . these are physical like ancient physical devices. And there's another ancient physical device just like in the show *Stargate* that are spread out across many solar systems.

> DW: And when you say, "Just like *Stargate*", you mean it's shaped like a ring?

> CG: There have been some that look very similar — shaped like a ring ....[456]

Goode explained that another type of portal technology was associated with step pyramids found throughout Central America. Essentially, the four-sided geometry of these pyramids makes it possible for a portal to appear on the platform at their apex which individuals can walk through similarly as with the ring-shaped portals.

This is where the testimony of another secret space program insider, Emery Smith, becomes relevant. Smith claims he participated in several dozen traversable wormhole missions. He says that each mission required three to six months of preparation by teams comprised of 6 or 7 highly trained personnel with multiple specialties. These individuals trained simultaneously for multiple missions and had an average of one mission per month.[457] Like Goode, Smith also stated that there are both natural and artificial portals in existence. Smith described the portals he experienced as looking very similar to the three sides of an upright door frame. In the *Stargate SG-1* series, similar rectangle-shaped portals acted as thresholds bridging different dimensions.

Goode has explained how the navigation system used in the *Stargate SG-1* series can be likened to how the internet operates, where numerical sequences are assigned for individual computers, networks, ISPs, and countries. This is the basis for the IP (Internet Protocol) numbers, which are the backbone for navigating the 'World Wide Web'. Goode claims that a series of numerical coordinates based on hyper-dimensional mathematics designate different galaxies, solar systems, planets, etc., to become the address for a particular portal. The seven-digit vector system used in the *Stargate SG-1* series mirrors a highly advanced mathematical coordinate system based on sacred geometry used in the real wormhole system that is part of the USAF secret space program.

Goode is not alone in describing such a numerical address system being used to navigate the many wormholes or portals. In the following exchange with Goode [CG] on a *Cosmic Disclosure* episode, David Wilcock [DW] mentioned another whistleblower he had interviewed and vetted, referred to only as Daniel, who described to Wilcock a similar system using coordinate points for portal and stargate travel:

> CG: ... They've been buried here. Wars have been fought over them. Finding them and knowing how to use them is a totally different thing. They worked on a – just like in the TV show *Stargate* – they worked on an addressing system. It was much like the Mac address or the IP address on computer systems.
>
> DW: That's exactly what my contact, Daniel, had said. In fact, he gave me the complete numerical address of the Earth as a ... There was a series of single digits – three single digits – that could be anywhere between 1 and 9.

Then there's a series of three digits that are between 1 and 99. And then the last number can be between 1 and 999 depending on what you need it for... And I've never leaked the whole sequence, but I have leaked that Earth's number apparently was 606 and that Mars was 605. That's the last three digits of the address...

Now, Daniel had told me that almost any number you dial in is going to take you somewhere, because there's a great amount of ... Like all the addresses are used and he had said that there's some kind of ancient race that comes along and puts one of these ancient gates on a planet and that the planet has one central gate.

And that when you gate into that particular number address, it will route you to that particular stargate on that planet. And it seems like these ... the beings that do this, will put a stargate on a planet once it starts to have intelligent life so that they can eventually find it and be able to travel. Does that line up with anything that you heard?

CG: I heard that there were two gates and that, like I said, there were several different ages of these ancient gates going all the way back to the ancient builder race ... And there were other gate systems that have been found on the Earth that have been much younger and look different, but pretty much were reverse engineered or use the same technology of addressing as the most ancient gate travel.

"Daniel" is a pseudonym used by the whistleblower who disclosed information to Wilcock, and Daniel has also presented more material in anonymous papers uploaded to the internet.[458]

Among the first to publicly emerge claiming to have witnessed a traversable wormhole system is the whistleblower Dr. Daniel "Dan" Burisch, a microbiologist who says he was employed at the S-4 facility at Area 51. While there is considerable controversy over Burisch's claims since he went public in 2003, researchers have conducted extensive investigations into them.[459] Among the investigators was William Hamilton, a former USAF security service NCO (1961-1965), who wrote a book about Burisch's experiences called *Project Aquarius: The Story of an Aquarian Scientist*. Hamilton's research into the whistleblower's background and claims brought him to the conclusion that Burisch was very credible.[460]

Burisch claims that "artificial portals" are instruments that tap into natural portals in order to operate. He believes there were approximately 50 artificial portals present on the planet coinciding with an undisclosed number of "natural portals".[461] He argued that information for the development of artificial portals came from "cylinder seals" extracted from Iraq during both the first (1991) and second (2003) Gulf War military interventions by the U.S. and its allies. The cylinder seals contained extraterrestrial information given to the ancient Sumerians who faithfully replicated this information on cuneiform texts. Burisch stated that he was part of a covert team that went into Iraq to find its portal technologies and the cylinder seals discussing the use of such technologies.[462]

Corroboration for Burisch's claims comes from another whistleblower, Henry Deacon (aka Arthur Neumann), who began publicly disclosing his information regarding portals and stargates in a September 2006 interview.[463] Deacon confirmed that a portal had been discovered in Iraq and that Burisch was correct in his account of how portal technologies operated.[464] There is much circumstantial evidence supporting Burisch and Deacon's claims that the real reason behind the U.S. led invasion and takeover of Iraq was to find ancient portal technologies and information, as I also proposed and wrote about back in 2003.[465]

A key element of the cosmic portal grid is our Sun, which Corey Goode has described as a natural portal that is used by spacecraft to enter or exit our solar system.[466] Elaborating further in an interview with David Wilcock, Goode explained that solar filaments form torsion fields that create traversable wormholes for portal travel between suns in our galaxy:

> [CG] NASA recently released that our sun has basically a portal or a magnetic filament connection to every planet in our solar system, and anything that has enough mass to cause a gravitational pull or a torsion in our space-time is going to create a magnetic and gravitational relationship with the host sun.
>
> And these filaments that they are just now releasing – these filaments are the portals ... And they are strong electromagnetic filaments.
>
> DW: But if people don't include the torsion component, they're not going to understand how this works. (Right.) So it's an electromagnetic tube that also has a strong torsion field and would act like a traversable wormhole.
>
> CG: Right. And it's happening within the torsion field of each solar system. The galaxy is a giant torsion field. All of the stars are constantly moving around the center of the galaxy and stars closer in are moving at a slightly different speed. And the magnetic relationships are always changing. These filament relationships are always changing between each star...
>
> It's just like electricity. Electricity takes the path of least resistance. And if you want to travel to a star that's on the other side of the center of the galaxy, you have to wait or calculate just the right time to travel, because if you don't, if you begin your travel and this star has just changed its position and the

electrical field, or the electrical connection, changes to another star – the path of least resistance – you're going to find yourself in a different solar system.[467]

Goode, along with Smith, has asserted that the natural and ancient portal systems form a "cosmic web" that spans the universe. By jumping from one portal location to another, one can travel incredible distances between galaxies and across the universe. Goode and Smith's disclosure of a cosmic web of portals, which is backed by Daniel, Burisch and Deacon's earlier claims, closely corresponds to what was depicted in the *Stargate SG-1* series.

Goode says that traversable wormhole travel was first developed by the Antarctica-based Germans, who received assistance from Reptilian extraterrestrials in establishing bases on the Moon and Mars:

> DW: Okay. So did the Germans have a plan to try to use local materials to make a sustainable base when they got there? Was that always the intention?
>
> CG: Yes. Just as on the Moon, they planned to take a certain amount of resources to Mars – lime and all the different things they need[ed] to mix with local resources to make concrete and whatever they needed to build structures that they could then pressurize and use as temporary shelters. They had to make quite a few trips to bring people and materials over in the beginning.
>
> This was in the beginning of when they were using stargate or portal travel…. And in this early era, they were using the portals to transport materials and not people or organics.
>
> DW: Did they have trouble with the organics having damage – like to their life cycle?

STARGATE SG-1 & USAF SANCTIONED DISCLOSURE OF TRAVERSABLE WORMHOLES

CG: Yes, as in killing them in a very gruesome way.... Until they figured out the proper way to do it with help from some of these allied extraterrestrial groups.

DW: So what's the year that you know where portals started to be used to transport materials?

CG: Materials have been ... They've been using portals to transport materials ... They've known how to do that since the '30s and the '40s.

DW: Wow! So that's well before the Philadelphia Experiment.

CG: Right. And you saw how badly that went with the people.

DW: Absolutely.

CG: It wasn't until the '50s that they were able to start transporting people consistently without them suffering – I believe they called it temporal dementia. People that would teleport intact from here to Mars, but they would look fine. But then after a number of days, they would suffer some sort of dementia they would call temporal dementia.

And the Germans did a lot of work in this field that helped us figure out how to do this properly – them working alongside their ET allies.[468]

Support for this early historical use of portal travel comes from Andrew J. Basiago, J.D., who says he was first exposed to it in 1968 as a 6-year old test subject. Basiago's participation began in a top secret project after the technology was first developed by the Ralph M. Parsons Engineering Company on behalf of the Advanced Research Projects Agency (the forerunner to the Defense Advanced Research Projects Agency or DARPA), and the CIA.[469] The project was named "Pegasus" and involved many children participating in the assessment of its suitability. Later, from 1980 to 1984, Basiago says portal technology was used to take him on trips to Mars, something that whistleblowers associated with the "Montauk Project" (involving time travel experiments) similarly have claimed. 470

Goode's testimony, along with several other credible whistleblowers such as Smith, Burisch, Daniel, Deacon, and Basiago, all suggest that portal and wormhole travel has been secretly developed and used for decades by the USAF, DARPA and the CIA. Their collective testimonies raise the question: Is the idea of traversable wormholes scientifically feasible or not?

## Carl Sagan and Scientific Validation for Traversable Wormholes via Singularity Points

A "natural stargate" is similar to an *Einstein-Rosen bridge* or "wormhole" which has been theorized to form when space-time is distorted by the intense gravitational fields generated by collapsing stars. Collapsing stars compress to the extent that they form a point of singularity, wherein time slows down in proportion to proximity to the singularity when viewed by outside observers. Singularity points have been hypothesized to exist at the center of black holes, and it is possible that they exist in other locations where an intense gravitational field significantly distorts space-time.

At a singularity point, time would effectively stop as it dilates down to infinity, as predicted by Einstein's 'General Theory of Relativity'. This would make it possible for anyone surviving, even briefly, the intense gravitational fields accompanying a singularity point to emerge at whatever future date they desired. Physicists also consider it theoretically possible for an Einstein-Rosen bridge to form wherein one could travel backward in time. So a singularity point, at least in theory, offers a means to travel forward or backward in time, and also through space. A singularity point, however, is highly unstable and requires enormous quantities of energy to keep it open – for travel through one to become viable.

The first reference to a traversable wormhole in mainstream scientific literature came through Carl Sagan, famed astronomer, cosmologist, and astrophysicist, who wrote the novel *Contact*, which became a famous movie depicting the feasibility of traversable wormholes as a means of instantaneous travel throughout our galaxy. Sagan asked leading astrophysicist Kip Thorne from the California Institute of Technology (Caltech) about the scientific validity of wormholes as a means of transportation, given what was known about singularity points. In mid-1985, Sagan wrote:

> Sorry to bother you, Kip, but I'm just finishing a novel about the human race's first contact with an extra-terrestrial civilization, and I'm worried. I want the science to be as accurate as possible, and I'm afraid I may have got some of the gravitational physics wrong. Would you look at it and give me advice?"[471]

Thorne began to think about how traversable wormhole travel would work, and then responded to Sagan's request with the necessary scientific equations making it all feasible. In *Contact*, Sagan acknowledged Thorne's contribution: "Professor Thorne took the trouble to consider the galactic transportation system described herein, generating fifty lines of equations in the relevant gravitational physics." [472]

Thorne told Sagan that as long as an exotic form of matter could be deployed to keep the singularity point open, then a "traversable wormhole" was theoretically possible. Consequently, in theory at least, a singularity point that is kept open offers a means for objects or life forms to move through time and space. Thorne's work on traversable wormholes began with Sagan's request. This led to Thorne collaborating with a Ph.D. candidate at Caltech, Michael Morris (Thorne was Morris' thesis advisor), in producing the first published scientific paper on wormholes titled: "Wormholes in spacetime and their use for interstellar travel", which was published in the *American Journal of Physics* in 1988.[473] Thorne went on to write about traversable wormholes in his book, *Black Holes and Time Warps* (1995) and became the science consultant for the movie, *Interstellar* (2014). [474] In the movie *Contact*, a spherical-shaped stargate device was depicted, which was Thorne and Sagan's conception of the technological means for traversable wormhole travel.

Sagan and Thorne's correspondence stimulated other scientists to explore the feasibility of traversable wormholes via singularity points generated by black holes. Among them was Professor Paul Davies, a popular physicist, author, and professor at Arizona State University, who believed it would be possible to create a time machine if a singularity point could be found, and if technology was developed to keep it open for the time traveler. Davies described a process whereby advanced civilizations might control a traversable wormhole: "wormholes may have formed naturally in the big bang, so that an advanced space-faring civilization might expect to discover one in the galaxy and commandeer it..."[475]

It's well known that Sagan's request to Thorne for advice sparked scientific speculation about the viability of traversable wormhole travel. What's not well known is that from 1976 until at least 1989, Sagan was a member of the "Majestic 12" committee that had been set up as a policy making body running classified programs involving extraterrestrial life and technologies. In 1976, Sagan had replaced the ailing Dr. Donald Menzel, who was identified in the Eisenhower Briefing Document as MJ-10, a member of the Majestic 12 Group.[476] Veteran UFO researcher Stanton Friedman investigated Menzel's appearance on the list of MJ-12 members and found his inclusion very credible based on Menzel's security clearances and classified work with various government agencies.[477]

In his capacity as "MJ-10", Sagan was informed of the existence of traversable wormhole technology that was being utilized by extraterrestrials as a means of bridging the vast distances between solar systems in our galaxy. Sagan's book and subsequent movie, *Contact*, can therefore be included among the soft-disclosure initiatives backed by various government and military bodies. Thorne's reply to Sagan on the feasibility of traversable wormholes focused on singularity points connected with black holes. Remarkably, however, a different way of understanding how a traversable wormhole would operate without the need for black holes or singularity points appeared in a scientific paper authored by Dr. Eric Davis that was published in 2010 by the Defense Intelligence Agency (DIA).

## Dr. Eric Davis and Traversable Wormholes via Exotic Matter

On December 17, 2017, SSP whistleblower Corey Goode uploaded two documents to his website that dealt with advanced technologies such as traversable wormholes and warp drives, which had been given to him by a confidential source.[478] The source believed that the documents corroborated some of Goode's startling claims concerning advanced transportation programs used in secret space programs. The two documents were part of a collection of 38 reports commissioned by the DIA and have since been validated as authentic. The two documents are titled "Traversable Wormholes, Stargates, and Negative Energy" and "Warp Drive, Dark Energy, and the Manipulation of Extra Dimensions". The first was authored by Dr. Eric Davis, and the second co-authored by Dr. Richard Obousy and Dr. Davis.

Both documents stated that they were part of "a series of advanced technology reports produced in Fiscal Year 2009 under the "Defense Advanced Aerospace Weapon System Applications (AAWSA) Program" and were marked "Unclassified // For Official Use Only".

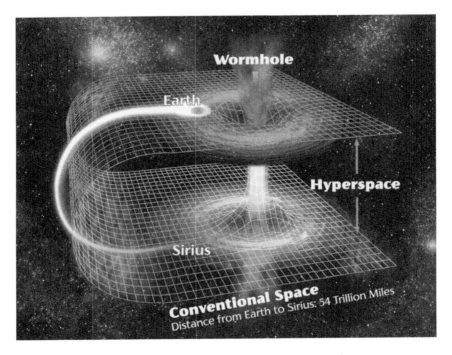

**Figure 51. Intra-Universe Wormhole as a Hyperspace Shortcut through Conventional Space**

Veteran FOIA and UFO investigator John Greenewald confirmed that the two documents' initial public release occurred through Goode. Greenewald wrote on January 12, 2018:

> I saw these documents back in late December and early January, but dismissed them as they are largely sourced/credited to Corey Goode, a very controversial figure to begin with. If they are genuine (and they may be) these documents do not appear that they were released under any official channels. So the biggest question is, "How did Corey get them?" I have not found an 'older' source, but am open if anyone else has. Veteran investigator George Knapp, has been able to find the names of all 38 DIA studies which are called "Defense Intelligence Reference Documents", and where the

above two documents first leaked by Goode can be found.[479]

On June 2, Australian UFO researcher Keith Basterfield similarly acknowledged Goode's role in the initial public release of the two documents through his website.[480] George Knapp discovered that the two documents released by Goode were part of a collection of 38 DIA studies, which were collectively titled "Defense Intelligence Reference Documents" (DIRDs).[481]

When British UFO researcher Isaac Koi contacted Dr. Davis about the DIRDs, Davis expressed his surprise that two of the documents he had authored had been released to the public, thereby officially confirming their authenticity: "I don't know how you got two of my DIRD reports" – "Yes! All of my DIRD reports are in the set of 38 total."[482] Furthermore, in a June 24 appearance on *Coast to Coast* radio, Davis said:

> ... that 2 or 3 of these papers had been "leaked" onto the Internet, by "someone on the beltway." The beltway is a highway that encircles Washington. A reference to "inside the beltway" means matters of importance to US government officials; lobbyists; and government contractors.[483]

Davis, a leading world scientist, had effectively stated that "someone on the beltway" was Goode's source for the two documents. This is an astounding admission! For the first time since Goode's public emergence in late 2014, a notable scientist went on the record confirming that Goode was working with a Washington insider to reveal significant details of advanced space technologies being studied by the DIA.[484] More significantly, one of the papers leaked by Goode presented startling corroboration for his testimony regarding traversable wormholes being used in different SSPs.

In Davis' report "Traversable Wormholes, Stargates, and Negative Energy", the feasibility of traversable wormhole travel was affirmed by laboratory experiments demonstrating that exotic matter, in the form of "negative energy", could be created:

> Implementation of faster-than-light (FTL) interstellar travel via traversable wormholes generally requires the engineering of spacetime into very specialized local geometries. The analysis of these via Einstein's General Theory of Relativity, plus the resultant equations of state, demonstrates that such geometries require the use of "exotic" matter. It has been claimed that since such matter violates the energy conditions, FTL spacetimes are not plausible. However, it has been shown that this is a spurious issue. The identification, magnitude, and production of exotic matter are seen to be a key technical challenge however. These issues are reviewed and summarized, and an assessment on the present state of their resolution is provided.[485]

In his paper, Davis makes the critical point that a black hole's singularity point is not required for the creation of a traversable wormhole, as originally conceived by Thorne and relayed to Sagan. All that is required is sufficient "exotic matter" or "negative energy" for its creation anywhere, including on the Earth's surface. This is something that would be impossible if a black hole were to be used for wormhole travel. Stunningly, it makes traversable wormhole technology something that could be developed as an alternative to conventional space travel.

Davis describes the limitation of conventional propulsion systems that achieve relativistic speeds (near the speed of light) requiring years to travel to nearby star systems, as well as pointing out the significant time dilation effect to be encountered where centuries or millennia would pass on Earth in the meantime. Instead, the use of traversable wormholes would make it virtually instantaneous, allowing travel back and forth, as Davis explains:

The solution to this problem is to dispense entirely with long interstellar voyage times or the undesirable outcome of relativistic time dilation. Explorers could deploy a wormhole-stargate near the Earth' surface, in Earth's orbit, or anywhere in the solar system they like and just pass through the "stargate" and come out the other side in remote spacetime within seconds, moving through the throat at low cruises speeds (30 mph!) and with no time dilation effects. Explorers could travel through the wormhole-stargates in small scout ships or send probes unencumbered by either enormous propellants mass rations or extensive life support provision. Effective travel time through the Cosmic Neighborhood via stargates would become irrelevant but could be estimated to be many times or thousands of times the speed of light. [486]

Davis even suggests that a time travel element could be incorporated into wormhole travel allowing someone to arrive back slightly before their departure, which would be a very effective security provision to avoid unsuccessful missions:

If explorers were to really push the envelope, they would design their stargate so they could return from their voyage in time to wave goodbye to themselves as they see themselves depart on their journey.[487]

Davis gave various design proposals for a stargate, one of which offered a distinct flat-face at the wormhole entrance similar to the ring-shaped version Goode described, which also appeared in the *Stargate SG-1* series:

It is a straightforward exercise to design a real "stargate" from wormhole physics. A stargate is essentially a traversable wormhole with a flat-face shape for the throat as opposed to the spherical-shaped throat of the Morris and Thorne wormhole...[488]

Davis went on to contrast the differences between the spherical-shaped wormhole entrance, or stargate, conceived by Thorne with that of a flat-faced hypersurface design. Their contrast is notable when peering through them into other space-times or dimensions:

> The entrance to the spherically symmetric Morris & Thorne wormhole looks like a sphere that contains the mirror image of a whole other universe or remote region within our own universe, incredibly shrunken and distorted ... A flat-faced wormhole, or stargate, which is also a hypersurface, would not distort the mirror image of the remote space region or other universe seen through it because the negative surface energy density and negative surface tensions of the exotic matter threading its throat is zero as seen and felt by light and matter passing through it.[489]

Davis provided an illustration in his paper of a ring-shaped stargate with a flat-faced entrance depicted in New York City's Times Square, with the destination site on another planet seen through the hypersurface. In addition to the stargate being a transportation device, it could also act as a "chronovisor", or "looking glass", where one could see into the past, future, or alternate realities.[490] This would make stargate technology and its derivatives a powerful intelligence-gathering tool as claimed by different whistleblowers and witnesses.[491]

The conclusion to be drawn from Davis' leaked report is that traversable wormholes, or stargates, are a form of space-time travel technology that is scientifically feasible, without requiring either singularity points or blackholes. Ostensibly, the advanced technologies required for generating exotic matter are not yet known or made available to the general public, yet whistleblower testimonies have revealed that exotic matter technology has indeed been developed in classified programs! The scientific validity of wormholes has been evaluated by multiple mainstream scientists who have proposed various designs for building stargates. The similarity between Davis' ring-shaped stargate design incorporating a flat hypersurface and the one depicted in the *Stargate SG-1* series is noteworthy. When combined with the USAF's sponsorship and active participation in the series, along with several insider testimonies, the conclusion that emerges is that *Stargate SG-1* was a form of soft-disclosure of one of the transportation modalities used by the USAF secret space program.

Figure 52. Artistic illustration of a Stargate in NYC displaying a destination world

# CHAPTER 17

## The Shock of the Navy's Parallel Secret Space Program

*They told us we were the tip of the spear, but we found out we're just the f---ing Coast Guard!*

—Senior Air Force SSP officer "Sigmund"

A s NASA's Apollo missions successfully placed men on the Moon, followed by Skylab and other civilian space missions, the US Air Force was secretly developing a parallel military space program. By the mid-1980's, the USAF had assembled several manned space stations using Manned Orbiting Laboratory (MOL) modules covertly sent into space with Titan III rockets, which were serviced by squadrons of antigravity craft delivering astronauts and supplies. All of this was accomplished while most of NASA's scientists, members of Congress and the general public remained completely unaware of what the USAF was secretly achieving in space. In turn, USAF leaders running their growing arsenal of secret space assets were confident that they had the most advanced space program on the planet – aside from the German program established in Antarctica. USAF leaders were convinced that the technological gap with the Antarctic Germans would eventually be closed due to the vastly superior industrial infrastructure possessed by the United States, and with the help they were receiving from different extraterrestrial groups.

That confidence was severely shaken in late 2015 when USAF leaders began to seriously consider whistleblower claims that the US Navy had also secretly developed a parallel space program, reportedly with massive kilometers-long "space carriers" accompanied by naval battle groups capable of deep space operations. Such claims of a highly advanced Navy program had been dismissed as disinformation, or misidentification of Antarctic German or extraterrestrial spacecraft. Witnesses such as Michael Relfe (2001) and Randy Cramer (2014), who claimed they had been involved in a Navy-run space program were simply ignored.[492] That dramatically changed after both Corey Goode and William Tompkins' emerged publicly in 2015.

Goode's first audio interviews appeared in late 2014 on the *Project Avalon* website. I conducted a series of email interviews with him next which were published in April 2015.[493] However, it was Goode's appearance on Gaia TV's new series, *Cosmic Disclosure,* which began on July 21, 2015, that attracted large audiences who were electrified by his startling new information about multiple secret space programs and various extraterrestrial alliances. Tompkins public emergence began with the publication of his book, *Selected by Extraterrestrials*, in December 2015.[494] Tompkins presented many documents substantiating his remarkable claims of a Navy secret space program (SSP), which corroborated much of Goode's earlier information. When Tompkins read volume one of my own "Secret Space Program series" which outlined Goode's claims, he was shocked since he thought he was the first to come forward with information on what the Navy had secretly built.[495] In turn, when Goode learned about Tompkins' testimony, he realized that much of the information he had read on SSP smart glass pads from 1987 to 2007 was based on Tompkins' briefing packets created between 1942 to 1946.

As Goode and Tompkins' remarkable claims steadily reached larger audiences, their testimonies came to the attention of leaders within the highly classified Air Force SSP, which closely cooperated with other US military entities such as the National Reconnaissance Office (NRO), the National Security Agency (NSA), and the Defense Intelligence Agency (DIA). Goode calls this interagency collaboration the "Military-Industrial Complex Secret Space Program" (MIC SSP), but I will continue to refer to it simply as the USAF SSP since much of its military infrastructure and leadership originate from the Air Force.

Goode's information, in particular, drew close scrutiny when he began talking about USAF space assets in sufficient detail to warrant an investigation to find out who was responsible for this potential security breach. Goode says that he first encountered a senior official from the USAF SSP, whom he dubbed "Sigmund", during a series of military abductions. These began in January 2016 when a triangular-shaped antigravity spacecraft landed near his Texas home, and two Air Force personnel from the vehicle forcibly took him into the craft.[496] Goode was medically interrogated during his abduction, a procedure in which chemicals were injected into him to force him to reveal the source(s) of his information. Eventually, "Sigmund" revealed himself as the officer in charge of the multiple interrogations Goode was being subjected to.

During the initial 2016 abductions, Goode described how Sigmund and his subordinates chemically forced him to expose several operatives working for the "SSP Alliance", a group comprising elements from the Navy's Solar Warden program and defectors from rival space programs:

> A tablet with a camera was held in front of my face and academy type military photos were shown to me. The camera monitored my eyes and marked a photograph when it was detected that I recognized the person. This incident caused the outing of 3 high ranking SSP Alliance individuals and caused a further rift between myself, Gonzales and the SSP Alliance. Because of the chemical interrogation and the attempted blank slating of my memories of the

incidents, I didn't remember the full details until I was informed later of the security breach.[497]

Figure 53. Courtesy: Cosmic Disclosure/Gaia TV

The nature of Sigmund's interrogations of Goode began to significantly change when Sigmund determined that Goode's information was accurate, at least in part, and that he was not a part of a disinformation program belonging to some rogue element secretively working within the USAF SSP. In a December 11, 2016 update on his website, Goode described how laboratory tests conducted by USAF scientists had repeatedly confirmed some of his claims of being taken off-planet by the SSP Alliance and the highly advanced extraterrestrial group Goode was working with, known as the "Sphere Being Alliance":

> The lab results once again confirmed that I had been in the approximate off-planet locations I had claimed. The MIC [USAF] space program people do not have any intelligence suggesting I had ever actually visited these locations. I was never a part of their program.

> This led to Sigmund concluding that he was being deliberately misled by his superiors about the

existence of more advanced programs to his own. My tests proved, beyond any shadow of doubt, that there was much more to the secret space program than he knew. As a high-ranking superior officer, this naturally came as quite a shock to him. He was led to believe that he had access to all relevant compartments of the UFO cover-up.[498]

Subsequently, Sigmund began an "information exchange" with Goode. "Meetings" began to occur during the latter part of 2016, and Sigmund learned about the 'alleged' Navy SSP and corroborated rumors of it cooperating with Nordic and other human-looking extraterrestrial groups.[499] In turn, Goode was given information about Antarctica and the intense power struggle within the US national security system involving the 2016 Presidential election. Goode wrote a report about one of these information exchanges that was published on December 11, 2016.[500]

On March 16, 2017, Goode gave me a detailed briefing about his ongoing meetings with Sigmund. These meetings sometimes included two of Sigmund's subordinates from the USAF SSP and were occurring regularly at the time.[501] The scope of the information revealed by Goode was breathtaking in its national security implications for the U.S. and the rest of the planet. The overall context driving these "information exchanges", according to Goode, was Sigmund's anger over the fact that his USAF SSP had been kept out of the loop concerning activities in deep space involving extraterrestrial life and technologies. The possible existence of a more powerful Navy program operating in deep space, with technologies far more advanced than anything possessed by the USAF SSP, was of particular concern to Sigmund and his colleagues.

Sigmund's USAF SSP was largely confined to near-Earth orbital operations – utilizing at least two operational space stations capable of hosting several hundred personnel, and several squadrons of TR-3Bs and reverse engineered flying saucers (aka Alien Reproduction Vehicles "ARVs") that serviced the stations and were capable of deep space operations. Goode also claims that the USAF SSP maintained small bases on the Moon and Mars. However, these bases were secondary in relation to the USAF SSP's primary mission of near-Earth surveillance and operations. Goode further describes how USAF space operations involving its TR-3Bs and ARVs had strictly been limited to our solar system:

> They have black triangular craft that can travel around our solar system, as well as other stealth-looking varieties. This technology makes them feel that if anything else was going on in our solar system, they would be able to see it.
>
> When they see the crafts from other programs, or any of a number of ET races, they are simply told that these are ours. They are told that they are not on a need-to-know basis about that particular program and not to speak about their sighting with anyone....
>
> The MIC [USAF] SSP are told that we cannot travel outside our solar system... This is due to gravitational and energetic conditions at the boundary of our heliosphere that [makes] any escape impossible with their current technology.[502]

Goode's testimony has revealed how USAF officials were duped about the Antarctic German, Navy, and corporate-run secret space programs, even when these other programs' assets were encountered or monitored on USAF surveillance devices. Having been told that such craft were part of a highly classified program requiring need-to-know access proved a very effective means for silencing USAF witnesses and leaders. The assumption was that once the different corporations completed such classified projects, the USAF would then receive these aerospace technologies for eventual deployment. According to Goode, this was not going to happen for decades, not until the technologies had become redundant as in the case of the TR-3B. Sigmund and his peers were infuriated when they learned what was happening.

Goode says that Sigmund and USAF SSP personnel were all told that they were "the tip of the spear" when it came to the deployment of advanced military technologies in space, but now they had learned that they were just the "f---ing Coast Guard!" Sigmund's fury over the lies told to him and other USAF SSP personnel led to angry confrontations with his superiors. This resulted in Sigmund initiating a serious investigation into Goode's claims, along with those provided by Tompkins through publicly released interviews and his book *Selected by Extraterrestrials*, which specifically detailed the US Navy's SSP and its cooperation with Nordic extraterrestrials. [503]

According to Goode, Sigmund began conducting a similar type of investigation to the one I had already completed for my March 2017 book, *The US Navy's Secret Space Program and Nordic Extraterrestrial Alliance*. Sigmund's investigation involved historical document searches, along with the interviewing of Navy personnel that had potential knowledge of a Navy space program that had been covertly constructed simultaneously with the creation of the USAF SSP. After eliminating a number of possible sources for Goode's impressive body of information, Sigmund concluded that there were two remaining explanations for Goode's revelations. The first was that a visiting group of "Nordic extraterrestrials", known to the USAF SSP, were manipulating Goode through mind control and feeding him accurate information for an unknown agenda. The second explanation meant that Goode was genuinely associated with a Navy-run secret space program, which in turn was part of a "Secret Space Program Alliance"; and Goode worked with a highly advanced visiting group of extraterrestrials called the "Sphere Being Alliance". [504]

Sigmund began his information exchanges with Goode, in part, to further investigate Goode's claims, thus helping him to determine which of the two explanations was accurate. Sigmund was also concerned that USAF plans for a "limited disclosure" of its SSP would only reveal part of the truth about deep space activities involving lesser classified compartments within the US military. He felt that the truth needed to be determined and exposed in order to maintain morale in each military branches' SSP. Was the USAF SSP the "tip of the spear" as they had been told, or "the f---ing Coast Guard" as Sigmund now feared?

## Support for Tom DeLonge's "Sekret Machines Initiative" Pulled

The highly-charged discovery made by Sigmund and his peers that they might well have been misled about the existence of a more advanced Navy SSP, along with German and corporate-run SSPs, would later play a role in the USAF's decision to continue or withdraw its covert support of Tom DeLonge's multimedia "*Sekret Machines* disclosure initiative". This initiative began in 2015, around the same time that Goode appeared in the public arena. DeLonge, a former lead vocalist for the rock band 'Blink 182', had enlisted top writers such as A.J. Hartley and Peter Levenda to co-author up to six books for his project. In interviews, DeLonge described how ten senior military and corporate officials backed his initiative.[505] By early 2016, as Goode's abductions by Sigmund and his team began, the first book in DeLonge's "Sekret Machines initiative" was published detailing the help his advisory team had given him.

> I've had meetings in mysterious rooms far out in the desert. I've had meetings at the highest levels of NASA. I have had conversations at research centers, think tanks, and even on the phone connected to

secret facilities. I've been introduced to a man whom I call "the Scientist," and another whom I call "the General." And there are many more of whom I cannot say much about, but some have become true friends, and all have become close counselors. Each of these men has all held, or currently holds, the highest offices of the military and scientific elite. The point is, I have done it. I have assembled a team of men and women "in the know." And they all believe I am doing something of value, something worth their time and yours.[506]

On the back cover of *Sekret Machines*, an endorsement appeared, presumably from one of the ten officials, Maj. General Michael Carey (USAF ret.), whose last military assignment was special assistant to the commander of Air Force Space Command:

Sekret Machines scratches at the surface of "who do" we trust with our classified technology – certainly our adversaries are aware of our undertakings, as they are doing the same, but what of our citizens, our politicians, even our own military. Tom DeLonge and A.J. Hartley create a convincing narrative describing the "cat and mouse" game that is timeless between strategic adversaries. It has existed under the sea, on the surface of the earth and in its skies, why wouldn't we believe it occurs in space? Our military leaders have been saying space is a contested environment for years now, perhaps we should believe them! – Maj. Gen. Michael J. Carey.[507]

General Carey's support for DeLonge's initiative is particularly significant because his work as special assistant to Air Force Space Command would have made him very knowledgeable of their secret space program. Carey was not the only USAF general supporting DeLonge's "Sekret Machine's initiative".

In late 2016, Wikileaks released emails that confirmed DeLonge's claims and outed two of the ten officials involved in his initiative.[508] One of the hacked emails was addressed to John Podesta, chairman of the Hillary Clinton Presidential Campaign, and dated January 25, 2016, with the subject header "General McCasland". In it, DeLonge stated:

> He mentioned he's a "skeptic", he's not. I've been working with him for four months. I just got done giving him a four hour presentation on the entire project a few weeks ago. Trust me, the advice is already been happening on how to do all this. He just has to say that out loud, but he is very, very aware – as he was in charge of all of the stuff. When Roswell crashed, they shipped it to the laboratory at Wright Patterson Air Force Base. General McCasland was in charge of that exact laboratory up to a couple years ago. He not only knows what I'm trying to achieve, he helped assemble my advisory team. He's a very important man.
> Best, Tom DeLonge[509]

Here is a brief biography of McCasland prior to his retirement:

> Maj. Gen. William N. McCasland is the Commander, Air Force Research Laboratory, Wright-Patterson Air Force Base, Ohio. He is responsible for managing the Air Force's $2.2 billion science and technology program as well as additional customer funded research and development of $2.2 billion. He is also responsible for a global workforce of approximately 10,800 people in the laboratory's component technology directorates, 711th Human Performance

Wing and the Air Force Office of Scientific Research.[510]

McCasland's assignment at Air Force Research Laboratory at Wright-Patterson AFB holds particular significance given what was presented in earlier chapters about the historical role of this major USAF research facility in studying captured German and extraterrestrial flying saucers. McCasland's assignment clearly establishes that he had the necessary scientific and technical background to be very familiar with the topic of 'advanced aerospace technologies' related to secret space programs and the UFO phenomenon. Because of the leaked emails, it was now confirmed that two senior USAF officers were supporting DeLonge's initiative.

DeLonge and his advisory group believed that their disclosure of an Air Force-run space program comprised of antigravity TR-3B vehicles and reverse engineered flying saucers would usher in a new age of space exploration. TR-3B squadrons operated out of Area 51, classified spaceports on the island of Diego Garcia, and a secret location in Antarctica which according to Sigmund was also a major R&D facility. Initially, there was widespread support among USAF leaders for DeLonge's effort. However, that was no longer the case by early 2017, according to Goode. After Sigmund and his team had thoroughly investigated Goode's claims, he and other USAF officials had re-assessed and now believed that DeLonge's efforts would indefinitely perpetuate an information gap between the Air Force SSP and the Navy's Solar Warden program, and in turn, the SSP Alliance.

Notably, DeLonge's initiative has changed quite rapidly from its initial focus upon disclosing the truth about what the Air Force and corporations had secretly designed, built and deployed – to concentrating on designing its own proto-type spacecraft through DeLonge's start-up organization, "To The Stars Academy of Arts & Science".[511] Today, there is controversy and a growing perception that DeLonge's initiative has been co-opted by the Deep State (aka "the Cabal") and is planning to disclose very little about what has secretly been developed in corporate laboratories. Instead, "To The Stars Academy" is pushing to merely reinvent the wheel with civilian funded projects to design and build a new generation of antigravity craft, which predictably would be far less advanced than anything built in the past by Lockheed, Boeing, and other aerospace corporations. Goode claims that Sigmund told him at a meeting in January 2018 that DeLonge's initiative had been "compromised beyond any type of repair".[512] Consequently, the once strong Air Force support for DeLonge's disclosure initiative had evaporated, with DeLonge and other figures having to increasingly deal with accusations of them promoting disinformation and a "Deep State" agenda. [513] Importantly, Air Force leaders wanted to get to the bottom of claims of Navy, German, and corporate-run SSPs before moving forward with officially disclosing their own SSP as *the only* secret space program known to exist.

## USAF SSP Takes Steps Toward Full Disclosure

Goode has claimed that Sigmund organized many meetings for him to provide briefings for select VIPs from the aerospace community in early 2017. Sigmund's staff arranged Goode's travel when necessary and provided the security for these meetings. Leading engineers, scientists, industrialists, and others allegedly listened to Goode's information about the Navy's Solar Warden program to see if any of it seemed possible to them and if they could confirm key elements. Goode said that in the end, he gave a series of briefings to a total of 28 VIPs at similar confidential meetings consisting of one to eight participants.

One specific meeting involved three notable people who were attending the "Human Spaceflight and Exploration Forum" organized by a small aerospace company, Special Aerospace Services.[514] This forum took place at a hotel where Goode was staying in late January 2017. Goode said that during his briefing presentation to the three VIPs, one of them got up and left in disbelief over the idea of a secret space program such as Solar Warden existing. The remaining two VIPs stayed until the end of the briefing but maintained bemused expressions which Goode interpreted to mean they didn't believe a word of what he was saying. Goode reported that in all the briefings he gave, one or more of the participants similarly walked out in disbelief, while the remaining VIPs largely maintained quizzical expressions. Consequently, Sigmund ended the VIP briefing sessions since they had not produced the desired effect of informing the VIPs about a possible Navy SSP, nor had they uncovered any further information that could help Sigmund in his investigation.

During a conversation I had with Goode on March 16, 2017, I pointed out to him the irony of his situation. Something very similar had happened to William Tompkins during World War II when he was tasked to deliver highly sensitive material in the form of briefing packets which contained information about a Nazi flying saucer program that used antigravity technologies and that exposed how the Nazi program was aided by a Reptilian extraterrestrial race.[515] Tompkins said that on many occasions when he delivered those briefing packets from Naval Air Station San Diego to the different aircraft companies' experimental test facilities, think tanks, and university departments, the lead scientists and engineers would dismiss the information as impossible. According to Tompkins, he had to constantly battle the closed-minded skepticism of those who had been trained to accept conventional scientific laws as immutable.

Soon after the VIP briefing sessions came to an end, Sigmund disappeared according to Goode. Sigmund had to go on the run from the Cabal/Deep State because he had finally succeeded in confirming Goode's claims of a far more technologically advanced Navy-run SSP. Eventually, Sigmund began working with the SSP Alliance after members of it rescued him from pursuing agents of the Cabal/Deep State. The events surrounding Sigmund's alleged investigation and disappearance illustrate how the Cabal/Deep State had expertly manipulated senior Air Force officials to keep them in the dark over the extent to which extraterrestrial technologies had been reverse engineered.

Put simply, the USAF had been tricked into believing its SSP was using the most advanced aerospace technologies available, which was believed to be a direct result of its decades-long cooperation with the Antarctic Germans, Grays, and Reptilian extraterrestrials. When Sigmund and other USAF officials learned the truth about the far superior Navy SSP, they realized that a major strategic realignment was necessary to rectify and redeem their state of affairs. In a momentous turnabout, the USAF began cooperating with Nordic extraterrestrials who had given such impressive technological assistance to the Navy's SSP. It is not coincidental that soon after Air Force officials confirmed the existence of the Navy's SSP, a Defense Intelligence Agency (DIA) document was leaked to the public. The release of this particular document provided startling evidence that the USAF and its NRO, DIA, and NSA partners had made a major strategic realignment, and were now depicting Nordic extraterrestrials as reliable allies and future friends of humanity.

# CHAPTER 18

# The History of Human-Looking Extraterrestrials on Earth

> Do not neglect to show hospitality to strangers, for by this some have entertained angels without knowing it.
>
> *– Hebrews 13:2*

In the December 2018 edition of the *MUFON Journal*, the first detailed analysis of a leaked 47-page Defense Intelligence Agency (DIA) document was published. For everyone debating over the validity of the document, the analysis presented a startling conclusion – the DIA document was *authentic*.[516] Released on June 14, 2017, this document provided a broad overview of the official history of extraterrestrial contact with Earth and credited the beginning of the modern UFO era with the Colorado Springs radio transmissions made by Nikola Tesla, which led to extraterrestrials sending an interstellar craft to investigate its origin.[517] This culminated in extraterrestrial vehicle crash incidents such as Roswell (1947) and Aztec (1948), and the establishment of formal diplomatic relations with a human-looking group of extraterrestrials during the Eisenhower administration. The US Air Force, in particular, benefited from this situation since extraterrestrials "living among us" offered an opportunity for the

USAF to gain help in understanding the principles and technologies necessary for future space travel.

The author of the *MUFON Journal* article, Dr. Robert Wood, is the world's foremost authenticator when it comes to the controversial Majestic Documents.[518] These are leaked documents concerning the activities of Operation Majestic 12 which was formally set up on September 24, 1947, by an Executive Order issued by President Harry Truman to manage the UFO/extraterrestrial issue. [519]  Dr. Wood, a former aerospace engineer with McDonnell Douglas whose career spanned 43-years, began his investigation and authentication of the Majestic Documents in 1995 with his son, Ryan, a fellow authenticator and respected author of *Majic Eyes Only*. Before discussing Dr. Wood and Ryan Wood's authentication of the DIA document, it is worth briefly reviewing its contents.

## Majestic Document Reveals US Diplomatic Relations with Human-Looking Extraterrestrials

According to its cover page, the leaked document is a "preliminary briefing" created by the Defense Intelligence Agency's Office of Counterintelligence on January 8, 1989. The full title of the briefing package is "Assessment of the Situation/Statement of Position on Unidentified Flying Objects", and it is addressed to the Office of the President. The controlling office in charge of preparing the briefing is designated as "Operation Majestic/MJ-1" which is a reference to the principal officer in charge of Operation Majestic 12.

Another page of the leaked DIA document identifies MJ-1 as the officer personally in charge of the preliminary briefing. Based on historical precedent, MJ-1 denotes the sitting CIA Director.[520] At the time, this was William Hedgcock Webster who is the only person to have held the positions of both Director of the FBI (1978 – 1987) and Director of the CIA (1987-1991). He currently is the

Chair of the Homeland Security Advisory Council whose primary function is "to provide the Secretary (of Homeland Security) real-time, real-world, sensing and independent advice to support decision-making across the spectrum of homeland security operations."[521]

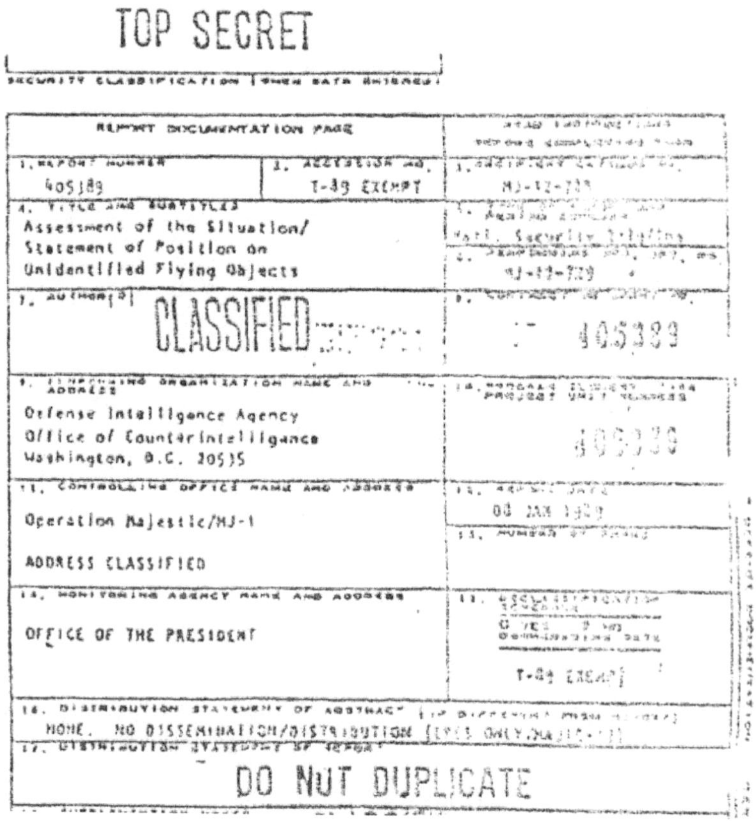

Figure 54. Cover page of leaked "Assessment of the Situation/Statement of Position on Unidentified Flying Objects" document

The briefing document refers to four groups of extraterrestrial visitors which are listed in order of importance to our planet, and cites whether they are friendly or not:

There are four basic types of EBEs so-far confirmed. And they are listed here in descending order of their influences on our planet.

A.    Earth-like humanoids. There are several variations more-or-less like ourselves. The majority of these are friendly and are the bulk of our EBE contacts. Most have a high degree of psychic ability and all use science and engineering of an advanced nature.

B.    Small humanoids or "Grays". The Grays, so–called for the hue of their skin possessed by most of this type, are a sort of drone. They are not unlike the worker ants or bees.... They are mostly under the psychic control of the Earth-like humanoids who raise them like pets (or a kind of slave). Assuming the Greys are under benign control, they are harmless.

C.    Non-humanoid EBEs. These are in several classes and come from worlds where dominant morphology took a different evolutionary course. Many of these are dangerous not for organized hostile intentions, but because such creatures do not hold human life as sacred.... Thus far, contact has been minimal with only a handful of unfortunate encounters.

D.    Transmorphic Entities. Of all the forms of EBE studied so far by Operation Majestic, these are the most difficult to understand or even to give a description of. Essentially, such entities are not "beings" or "creatures" ... exist in some either dimension or plane which is to say not in our space or time. They do not use devices or travel in space.... In essence these entities are composed of pure mind energies. ... They are said (by other EBEs) to be capable of taking on any physical form that they "channel" their energy ... as matter. [522]

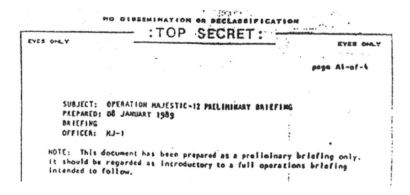

NO DISSEMINATION OR DECLASSIFICATION
: TOP SECRET :

EYES ONLY                                                    EYES ONLY

page A1-of-4

SUBJECT: OPERATION MAJESTIC-12 PRELIMINARY BRIEFING
PREPARED: 08 JANUARY 1989
BRIEFING
OFFICER: MJ-1

NOTE: This document has been prepared as a preliminary briefing only. It should be regarded as introductory to a full operations briefing intended to follow.

Figure 55. Page from DIA document identifies MJ-1 as officer in charge

The assessment depicts most extraterrestrial interactions as friendly, particularly those involving the human-looking entities. This group has engaged in a number of non-hostile interactions with humanity that date back to Nikola Tesla's pioneering radio and energy broadcasts in the late 1800's.

The briefing document continues:

In 1899, the Yugoslavic electrical scientist, Nikola Tesla, most noted for his introduction of alternating current to electrical power transmission and a laboratory device named after him (the Tesla coil) embarked on a number of researches that have made this the saucer century. Tesla had long proposed that it was possible to directly broadcast pure electrical energies at a distance without loss of power and without wires. By 1899, and with ... government and private scientific backing, Tesla had chosen a site near Colorado Springs, Colorado to conduct a massive and never repeated experiment.... Tesla's purpose to gather the Earth's own magnetic field and to use the Earth as a huge transmitter to send signals to outer space in an attempt to contact whoever might be living there. Tesla had no idea that the specific type of power he had generated was coursing through space and caused great havoc many light years away.[523]

Without realizing it, Tesla's use of the Earth's magnetic field to direct radio transmissions into space had created a disruptive weapon – essentially the world's first "Directed Energy Weapon", but at a planetary scale! The effect of doing this was analogous to the development of warp drive in the fictional *Star Trek* series, and humanity was now on the Galactic radar for the first time.

The briefing document describes what happened as a result of Tesla's experiments:

The extraterrestrial intelligence (EI's) attempted to respond to his transmissions in a form of binary code that they routinely use for long range communications (evidently these energies act instantly at a distance and are not limited to the speed of light) and asked that he cease sending. Of course, Tesla had no way of understanding the message he received back from space. Fortunately, the anger of local residents at the side-effects of his research forced him to shut down the Colorado Springs experiments in the same year he began them. [524]

Significantly, the briefing document is acknowledging that the communications were instantaneous and that the speed of light is not an absolute limit, at least for the kind of communications used by the aliens.

Critically, it was Tesla's experiments, according to the briefing document, which ushered in the "flying saucer age" due to the powerful effect of his energy transmissions which had used the Earth's magnetic field as a booster:

> This, then, was the actual start of the so-called "flying saucer age" in our times. As it became clear that our people were on the verge of an explosion of technical progress, the EI's decided on a long term program of carefully calculated and seemingly random contact with the eventual goal of raising our awareness of our place in the galactic community. With the advent of the Atomic Age, this program was escalated to include eventual diplomatic contact with many of Earth's governments. The same approach of staging apparent "accidental" contact was chosen for its low psychological impact on the human race. This was the situation in the case of the United States of America (re: Roswell and Aztec, New Mexico files, this briefing).[525]

The briefing document refers to the famed 1947 Roswell UFO crash and the national security mechanism that had been developed to investigate it, which subsequently cultivated a secrecy system. While the Roswell crash is well known, the next case described in the document is not.

There was a "controlled landing" of a flying saucer on March 25, 1948, in Aztec, New Mexico, which the briefing document describes:

> The controlled landing occurred in a small desert
> canyon on the private grazing land of a local farmer
> and rancher... about 12.2 miles northeast of Aztec
> New Mexico. The ship was determined to be 99.983
> feet in diameter. Inside the upper cabin... the team
> found the bodies of two (2) small humanoids about
> four feet in height strapped into seats like those in a
> jet cockpit... The extraterrestrials were dead. [526]

The briefing document reports that the extraterrestrials had died because of an atmospheric leak caused by impact with a nearby cliff face. The report then describes the discovery of cryogenic tubes with human-looking bodies in the lower level, four of which were successfully resuscitated:

> A closer look at the sealed tubes which looked like
> the doors of clothes dryers at a laundromat revealed
> that they were a complex form of refrigeration
> system. Two were empty and twelve contained the
> bodies of what looked like human adults and small
> children as well as infants, all frozen as if preserved
> for specimens.... Eventually the medical team was
> able to resuscitate one adult Earth-like humanoid
> male and three (3) Earth-like humanoid infants, all
> about six (6) months of age: two male and one
> female. The rest of the infants and one more short,
> gray-skinned, large-headed EBE, perished. [527]

The adult survivor, according to the briefing document, looked human but was clearly an extraterrestrial who, despite appearances, had physiological differences which included having two livers. The document describes how diplomatic negotiations began with the surviving adult extraterrestrial:

After a hasty tele-conference with President Truman, it was explained to the visitor that if his intentions proved to be non-hostile and he cooperated in an information exchange, he would be granted diplomatic status and soon be repatriated to his own kind when the arrangement could be made. To this he readily agreed, provided he was not asked to give away any scientific secrets that could alter the course of our natural cultural development. [528]

What comes next in the brief details the year-long diplomatic exchanges between the extraterrestrial visitor and the Truman administration, before he was returned to his people:

> Altogether, the Aztec EBE lived under our protective custody on the Los Alamos complex for nearly a full year, from late April 1948 – until March of 1949. After that, he was sequestered at a private safehouse set-up by Army Intelligence ... during which time he met with the President and other top government and military administrators, prior to his being returned to his people in August of 1949. He gave the scientists and military debriefers a great deal of mostly non-technical information about his civilization and its motives for being on our Earth; a total of six hundred and eighty-three (683) pages of transcripts were made of recorded conversations.[529]

The reference to Army Intelligence almost certainly would have involved the Interplanetary Phenomenon Unit (IPU) that was set up after the 1942 Los Angeles Air Raid Incident, and was eventually absorbed by the USAF as confirmed in the replies from Freedom of Information Act requests. [530]

The briefing document goes on to explain how technology given to US military authorities would also be given to the Soviet Union by the extraterrestrials:

> The EBE saw little harm in allowing us to keep the remains of his spacecraft for study, since he felt our understanding of it would only gradually develop. He did suggest that his people would "probably have to drop one in similar condition in the laps of the Soviet Union – just for the balance of things to be sustained," and "You are welcome to take this up with any higher authority you can find willing to listen to you, if you do not approve of this ...."[531]

In 1948, the United States and the Soviet Union were the world's only known nuclear powers and leaders of the two most powerful political blocs, so it makes sense why they were specifically helped in this way. This means that it was not only the U.S. which had extraterrestrial craft in its possession, which it was desperately attempting to reverse engineer to deal with the Antarctic Germans and extraterrestrial visitors, but also the Soviet Union had them as well.

More startling is what the extraterrestrials said about "human-like" infants that were to be dropped off and raised on our planet:

> The team of scientists were told that the human-like infants were destined for our world anyway, and we were welcome to keep them. On 21 August 1949, the Aztec EBE was returned to his own kind at a meeting site southwest of Kirtland Air Force Base, ... and arrangements were made for a future meeting at the same location, to open diplomatic relations.[532]

In gifting three human-looking extraterrestrial infants to US authorities, the apparent intent was for these offspring to be raised in human society, and in the future, they could help establish friendly relations between our two worlds. There was also the possibility that the infants innately possessed higher understanding and knowledge (much like Nicola Telsa exhibited) which could later unlock scientific advances. Therefore, the extraterrestrials would be raised in or brought regularly to classified USAF facilities to study and work with scientists on understanding how alien technologies worked, and the scientific principals behind them. Presumably, the same thing happened in the USSR. By 1989, when the briefing document was prepared, the extraterrestrial children were 40+ year old adults.

The briefing document refers to President Eisenhower traveling to Kirtland Air Force Base In 1954 where he established full diplomatic relations with the human-looking extraterrestrials:

> The preceding diplomatic treaty was drafted by the director of the Majestic-12 operation and a joint committee of extra-terrestrial visitors and representatives of the U.S. Diplomatic Corps, as a statement of intent. It was ratified and signed at Kirtland Air Force Base … on July the eighteenth, 1954 by President Dwight D. Eisenhower and an individual on the behalf of the EBEs.[533]

If the content of this DIA document is correct, then Eisenhower held additional meetings with human-looking "Nordic" extraterrestrial visitors besides the one that had already occurred at Edwards AFB on February 20, 1954, wherein a group of Nordics attempted to persuade President Eisenhower not to move forward with the US thermonuclear weapons program.[534] Their attempt was not successful, and the Castle Bravo nuclear test occurred just over a week later on March 1 with a 15-megaton hydrogen bomb.

In chapter ten, it was explained that the meeting between Eisenhower and the occupants of a flying saucer that landed at Holloman Air Force Base on February 9, 1955, involved the Antarctic Germans who were allied with Reptilian extraterrestrials. So the diplomatic meeting at Kirtland AFB very likely involved a different group of human-looking extraterrestrials to those who had landed at Edwards AFB, and with whom a diplomatic agreement failed to materialize due to differences over the US thermonuclear weapons program.[535] Apparently, the Nordic group that landed at Kirtland AFB was not deterred from cooperating with the U.S. in various technological areas despite the continuation of its thermonuclear weapons program. They readily left behind their damaged spacecraft for the U.S. to study, but they did not give technical guidance for understanding its workings at the time. This means that, in addition to the cooperation with the Grays, Reptilians, and the Antarctic Germans, the USAF also began cooperating with a human-looking extraterrestrial group in order to advance their study and reverse engineering efforts on captured extraterrestrial craft and related technologies.

This is not all that surprising given what William Tompkins revealed about a Nordic group of extraterrestrials who had infiltrated the Douglas Aircraft Company and other leading aerospace companies to help the US Navy design and build future space carrier battle groups. Tompkins described at length how these human-looking Nordics would assist him with design problems, but without ever revealing their true identity to him or the other scientists and engineers working in Douglas' "Advanced Design" think tank.[536] Tompkins is not the only insider/whistleblower who says he encountered Nordic extraterrestrials in classified facilities helping to bring military plans for future space programs to fruition.

According to Emery Smith, a former USAF Surgical Assistant who was stationed at Kirtland AFB from 1990 to 1995, he met different types of extraterrestrials working within highly classified facilities on the base.[537] Smith claims that among the different extraterrestrials he encountered were a human-looking group that was living amongst the Earth's population. This is a powerful corroboration that Kirtland AFB was at the epicenter of meetings and cooperation between the USAF and human-looking extraterrestrials, as clearly stated in the DIA briefing document.

Smith asserts that in addition to Kirtland AFB, he worked at other highly classified facilities which included Sandia National Laboratory and Los Alamos National Laboratory, where he met and worked with human-looking extraterrestrials. In a series of interviews completed on the popular online streaming show "Cosmic Disclosure," Smith says that the human-looking extraterrestrials were indistinguishable from modern humans. In the August 7, 2018 episode, he responded as follows to a question about whether extraterrestrials really live amongst us:

> Well, extraterrestrials also come to watch over us and look at things on the Earth. And they're here to gather information to make sure we don't blow up the Earth or kill each other. So they're here just to get information and have it and to also experience the life of a human, which is quite preferably like a vacation for them.
>
> So, it's a very interesting scientific job for an extraterrestrial to have. It's kind of an honor to come here, live amongst us, and do things, just living like a normal human being, actually, and interacting with human beings... It's like a universal Peace Corps. [538]

Smith says that in addition to encountering such extraterrestrials in classified projects run by corporations within USAF facilities, he had also recently been briefed that up to 100,000 extraterrestrials are presently living all over the planet.[539] Smith's testimony is highly significant given the contents of the leaked 1989 DIA document which identifies the start of human-looking extraterrestrials living among us in 1954.

The contents and implications of the DIA document, and its release during the first year of the Trump administration is astounding. This is especially so given the role of William Webster, Chairman of Homeland Security Advisory Council in the Trump administration, and the possibility that Webster was involved in leaking the DIA document. The question that now urgently needs to be addressed is whether the document is authentic or not.

## Authentication of the DIA Document

Dr. Robert Wood found many reasons that point to the leaked 1989 DIA document's authenticity and why it is an error to simply dismiss it as many UFO researchers have done to date.[540] His detailed analysis of the typing, spelling mistakes, signatures, patents referenced, individuals mentioned, etc., in the document, led to his conclusion that it was a briefing dictated by a member of the Majestic 12 Group (most likely William Webster) to two typists. They typed up the 47 pages, a copy of which was preserved on microfilm by the DIA as Dr. Wood explained:

> If one contemplates why there were so many errors made in a document officially recorded on a microfilm, and you pronounce the word or phrase on the left and look at the correct one on the right, usually they sound essentially the same. This would be consistent with the document having been created as a result of taking dictation and having two different typists implement the words on paper...[541]

Essentially, MJ-1 dictated a one-time introductory level summary for the recipient (who was new to his post) since one apparently was not available from prior written records at the time.

Since the DIA document's cover page refers to itself as a "preliminary briefing" and it was created on January 8, 1989, the natural assumption to make is that the briefing was intended for President-Elect George Bush, who at the time was the Vice President and winner of the 1988 Presidential election. Presumably, he was to be briefed by the then sitting CIA Director, William H. Webster. However, by analyzing the only signature appearing on the DIA document, Dr. Wood concluded that the briefing was actually intended for a distinguished astrophysicist at MIT, Dr. Philip Morrison:

> This is the only signature on the document, and the first question might be whether it was the entry level person being briefed or the briefer. It seems much more reasonable that it was the person being briefed. [542]

Dr. Morrison began his career as a nuclear physicist working on the Manhattan Project, and then moved his area of expertise to astrophysics to express his disapproval of the nuclear arms race. He became famous for his popular books and documentaries dealing with astrophysics and would continue to be a professor at MIT. In 1987, Dr. Morrison was the host of a six-part miniseries for PBS called "The Ring of Truth", which examined a number of astrophysics topics.[543]

There is important circumstantial evidence indicating that the DIA document was a briefing dictated by the head of Majestic 12, presumably the sitting CIA Director William Webster, for Dr. Morrison. In his MUFON article, Dr. Wood mentions that Dr. Morrison was on friendly terms with Carl Sagan, whom some believe replaced Dr. Robert Menzel as a member of the Majestic 12 committee upon the latter's retirement and death shortly after that, in December 1976:

> Philip Morrison was a very distinguished professor, who was a protégé of Oppenheimer and very likely was "in the know" on security issues, although I have no evidence of this. There is evidence, however, that he was very collegial with Carl Sagan during his career, helping to arrange a [UFO] symposium at the conclusion of the Colorado study in Boston. [544]

Notably, Dr. Menzel was named as MJ-10 in the Eisenhower Briefing Document.[545] In his book, *Top Secret/MAJIC,* veteran UFO researcher Stanton Friedman provided detailed evidence that Dr. Menzel was indeed a member of the Majestic 12 committee, even though Menzel wrote several books debunking the UFO phenomenon.[546] Dr. Menzel is best known for his popular astronomy books such as *Field Guide to the Stars and Planets.*[547] Either before or shortly after Dr. Menzel's death, the Majestic 12 committee was looking for an astronomer/astrophysicist who could replace him. The replacement would have needed a firm scientific grounding in astronomy/astrophysics, broad public outreach, and Dr. Menzel's recommendation of approval prior to his death.

Dr. Carl Edward Sagan fit the bill with his scientific pedigree. While a Fellow at Berkeley University, his work on NASA missions to Venus and Mars received widespread scientific recognition. Most importantly, his scientific work came to the attention of Dr. Menzel who arranged for Sagan to be given an Assistant Professor position at Harvard University (1963-1968). Sagan moved to Cornell University after he was denied tenure at Harvard, ironically due to the growing popularity of his generalist approach to science. Nevertheless, Menzel remained a firm supporter of Sagan.

Sagan achieved celebrity status through his popular books and documentaries, including his award winning *Cosmos* that aired on PBS in 1980 which was seen by 500 million people across 60 countries.[548] If a position was reserved on the Majestic 12 committee for a leading astronomer/astrophysicist with outreach to a global audience through popular books and television documentaries, and required the approval of the previous position holder, then Sagan was the natural replacement for Menzel.

Similarly, by January 1989 when the Majestic 12 Group was looking for a replacement or-an alternate for Dr. Sagan due to his impending retirement (he had been in the position since at least 1976) or some other reason (Sagan died on December 20, 1996), Dr. Morrison was a solid choice given his scientific pedigree, broad public outreach, and prior friendly relationship with Sagan. This is important circumstantial evidence in support of the authenticity of the DIA briefing document.

Also, the fact that the 1989 document was merely a preliminary briefing dictated by the head of the Majestic 12 Group for a new member taking up Dr. Sagan's position meant that little in the way of resources would have gone into preparing the briefing document. The ad hoc nature in which the document was assembled helps to explain the notable security-marking inconsistencies, spelling errors, cutting-and-pasting of different documents together, etc., as many critics have pointed out.[549]

In his MUFON article, Dr. Wood favored the authenticity of the DIA document with several compelling reasons, despite its multiple errors not uncommon in documents of this type. By identifying Dr. Morrison as the recipient of the briefing, rather than Vice President (President-Elect) Bush, Dr. Wood has provided the means of corroborating the document's authenticity. Was Dr. Morrison being briefed as Dr. Sagan's replacement on the MJ-12 committee due to the latter's retirement after 13 years or more of service as MJ-10?

If indeed genuine, this document's content offers a wealth of information on instrumental topics ranging from the 1948 Aztec

UFO crash; diplomatic relations between the Eisenhower administration and extraterrestrial life; the role of Nikola Tesla in starting the modern UFO era; the working relationship with human-like extraterrestrials at classified military facilities such as Kirtland AFB; and perhaps most important in shaping future public perceptions, the fact that human-looking extraterrestrials are friendly and have been secretly living amongst humanity for decades.

## Behind the Leak

What conclusions can be drawn from the leaking of the 1989 top secret DIA briefing document? First, the DIA document provides a startling version of suppressed history: friendly human-looking extraterrestrials began interacting with humanity in the early 1900's and observing our technological progress without any outright interference in human affairs. One important objective in the leaking of the DIA document was that it would help greatly in realigning public perception over "human-like" extraterrestrials, so that any future official disclosure of such visitors would not come as a great shock to the American public and the rest of the planet.

Second, the DIA document shows how a group of "Nordics" decided to assist human evolution by sending one of their spacecraft bearing extraterrestrial babies as "gifts" to the U.S. in order to gradually stimulate technological and social development in certain positive directions. The USAF was the military institution identified in the DIA document as the main beneficiary of this arrangement, which was later corroborated by whistleblower Emery Smith, who had many first-hand experiences while working at Kirtland AFB from 1990 to 1995 and other USAF facilities where classified projects involved human-looking extraterrestrials assisting.

Third, it is more than coincidental that the most likely briefer in the 1989 document was CIA Director William Webster who at the time of its leaking in June 2017 chaired the Homeland Security Advisory Council, which provides detailed advice on homeland security issues to the Council's Secretary. It's very plausible that issues discussed in the DIA document were deemed important for public dissemination for homeland security reasons, and approval was given for it being leaked by the Secretary of Homeland Security. At the time, this was none other than retired four-star US Marine Corps general, John F. Kelly, who became President Donald Trump's Chief of Staff on July 31, 2017, just over a month after the DIA document's release. This raises the intriguing possibility that Trump may have been briefed on what had happened and what the plan was for future revelations.

Finally, it is more than coincidental, given the USAF's reassessment of their decades-long cooperation with the Reptilians, Grays, and Antarctic Germans beginning with the 1955 Holloman AFB meeting, that the 1989 DIA briefing was leaked. The leak happened in June 2017, only a few months after Sigmund began having "information exchanges" with SSP whistleblower Corey Goode in late 2016. The USAF had learned for the first time about the Navy's Solar Warden program and other secret space programs, and the high-level diplomatic meetings occurring between these programs and visiting extraterrestrial groups. Due to what Sigmund and other USAF SSP leaders had learned from Goode, William Tompkins and other sources, the USAF had decided it was going to align itself to a much greater degree with human-looking extraterrestrial visitors who had previously been suppressed.

While the Aztec flying saucer crash and the 1954 meeting at Kirtland AFB had led to an agreement with one particular group of Nordic visitors, other human-looking extraterrestrial visitors were severely limited or hampered in performing any activities on Earth. Indeed, the degree to which most human-looking extraterrestrial visitors were suppressed only became clear in an interview that Goode and Smith gave on *Cosmic Disclosure* on August 14, 2018. The program host, David Wilcock, began the episode by asking

Goode to explain what he knew about the secret "Intruder Intercept and Interrogate Program", which Goode had briefly participated in during one of his SSP assignments:

> It's a program that is similar in part to the Men in Black. What this Intercept and Interrogate Program does is that if an intruder flies into our [solar] system, doesn't give a friend or foe signal and is intercepted, they are taken from their ship. Or, what happens in most cases is that ETs have made it to the Earth and are here secretly blending in as one of us.
>
> This group will go in retrieve the people that are here, we call the ETs people, the people that are here without permission, bring them up to a certain station, and their interrogation would begin to find out what was going on.
>
> This group also acted as a police force for non-terrestrial groups that are here with permission. They helped keep them apart, and when they (ETs) had issues they acted just like police.[550]

After describing the extraterrestrials, who were seemingly identical to humans and could easily blend into our society, Smith described encountering some of them during autopsies he would conduct for classified programs at Kirtland Air Force Base and other facilities around the country.[551] He emphasized that they showed obvious signs of torture:

> Emery: ... Some of the extraterrestrials that we were working on in the lab, doing dissections and what not, some of these full bodied human extraterrestrials would come in and they would be

very badly beaten, where subdural hematomas are being beaten in the back of the head. It looks like they were tortured – sometimes many broken bones.

I didn't find out until later on that these beings were coming from a prison or a withholding cell of some sort and were being interrogated.... After hearing your [Corey's] testimony, this is what we were receiving. Because usually they are in pretty good shape unless they were shot down or something. These had obviously been beaten to death.

Corey: ... A lot of the times the individuals that were being interrogated just died of stress and trauma ... from the process. It was very disturbing. They were tagged and sent for study afterwards.

Emery: We had a few that they actually starved to death. I'm not sure what happened....

Corey: There are prisons that they have for them. Some of them, where they bring them like a conventional prison. That's where they need to further interrogate them for a while to get more information, or have them to trade off, or to tip us off. [552]

The USAF decision to break from this earlier drastic policy of intercepting, interrogating and even torturing many human-like extraterrestrial visitors was momentous. It's important to keep in mind that USAF leaders who had participated under the long-standing harsh policy had done so based on the limited information provided to them by their allies and superiors within the Deep State and the Antarctic German-Reptilian alliance. Sigmund's discovery of the Deep State's deception over the technological advances that were kept from the USAF SSP led to the profound policy shift. The June 2017 leaking of the 1989 DIA document was a clear sign that the USAF SSP and its DIA, NSA, and NRO partners were intent on officially disclosing the existence of human-like extraterrestrials.

The policy shift had immediate results. The USAF SSP began to allow photographs to be taken of their antigravity craft coming out of sensitive facilities such as MacDill Air Force Base, home of Special Operations Command. This offered another signal that the USAF SSP was planning to disclose itself soon to the American public. In addition, the USAF SSP began to cooperate with the Nordic visitors, and at the same time turned their space assets against the Deep State and their now defunct German/Reptilian/Gray allies. The most dramatic result came when the USAF SSP partnered with the Nordics in intercepting and destroying a ballistic missile attack on Hawaii launched by the Deep State that was intended to start World War III. In the next two chapters, I will discuss each of these two major developments involving the USAF SSP, and what it portends for the future.

# CHAPTER 19

## USAF SSP Photo Disclosure of MacDill's Antigravity Vehicles

If you reveal your secrets to the wind, you should not blame the wind for revealing them to the trees.

—Khalil Gibran

On March 3, 2017, I received the first in what would become a series of photographs taken over seven months of differently shaped UFO's seen in the vicinity of MacDill Air Force Base in Tampa, Florida.[553] Making circumstances even more significant, the photographer, publicly known only as "JP" (a pseudonym), was prompted on several occasions by different strangers in passing cars, vans, or via phone calls to look up into the sky at exactly the right time to see the UFO's while also often being encouraged to take the photos. An organization had singled out JP to take these shots which would reveal some of the different antigravity craft operating in the vicinity of, or out of, MacDill AFB. All of this happened around the same time that Corey Goode's former interrogator-turned-source, "Sigmund", was conducting information exchanges with Goode (from late 2016 to mid-2017) after learning about the Deep State's deception of the USAF; as well as the leak of the DIA briefing document that had taken place in June 2017, both suggesting that the USAF SSP wanted to prepare the public for future announcements concerning human-looking extraterrestrial visitors.

Later in early 2018, JP moved to Orlando, Florida, where he continued to have sightings and take more photos, including those of a spacecraft allegedly belonging to a "Nordic" extraterrestrial who was working with the USAF. According to JP, the Nordic sought him out and relayed the message that his people were helping the Air Force's secret space program, which demonstrates how the USAF was now implementing a major policy shift to work with this group of extraterrestrials instead of their former Deep State allies. These events suggest that the photos were part of a plan by the Air Force to prepare the general public for official disclosure of the USAF SSP and the reality of human-like Nordic extraterrestrial visitors.

In early September 2017, as JP continued to supply me with photos of the different antigravity craft flying near MacDill AFB, Hurricane Irma approached Florida and headed directly towards Tampa and MacDill. The base had to be shut down and personnel evacuated as the hurricane made landfall at several Florida locations, including Tampa. As I presented the photographic and testimonial evidence for the events around MacDill in articles during the latter part of 2017, it was hard to avoid making the striking conclusion that, because the USAF SSP was deliberately allowing their craft to be photographed by a member of the general public, MacDill was being targeted by the Deep State in retaliation against the USAF's new policies and disclosure initiatives.

While JP's claims are sensational, it's important to emphasize that they are backed up by his excellent photographs taken with the tacit approval of the USAF for public release. These photos have provided hard evidence that a USAF SSP exists and has begun to actively work with Nordic extraterrestrials, while also indicating that Air Force leaders want this information released to the public. What follows is the chronology of the photos supplied to me, my analysis of them, and their policy implications.

## JP's Photos of Craft Near MacDill Air Force Base

In 2008, I first began communicating with the photographer "JP" after he phoned me claiming he had recently experienced contact with a group of Nordic extraterrestrials while visiting family in Brazil. JP said that he had been given many ideas for advanced technology through downloads placed into his mind as a result of his first contact experiences. He then began designing and building some of them upon his return home to Tampa, Florida, where he resided at the time. Before his Brazil experiences, he was an aspiring musician with no previous technical background. Married with two children, his wife is a popular Brazilian singer with a flourishing career which includes singing at many Christian devotional events. She has had great difficulty in accepting JP's experiences as either genuine or positive given her strong Christian faith, but she acknowledges their significance for her husband. I have spoken with her about JP's experiences and explained how these are not uncommon, certainly not a case of demonic possession, and that JP has likely been chosen by a group of extraterrestrial visitors to play an important role in future events. Given his wife's successful musical career and her sensitivity to JP's extraterrestrial contacts, he has chosen to remain anonymous and to use the "JP" pseudonym.

After contacting two university professors about some advanced technology inventions he was building, JP claims operatives subsequently approached him from one or more military agencies who wanted to work with him in building the devices. I gave JP advice on dealing with his sensitive situation – given that his extraterrestrial contacts wanted him to build free energy and other technologies helpful to humanity – because it was unlikely that these covert military operatives would allow this technology to be shared under current government secrecy policy. JP says that after a period of working with the covert operatives in building several of the devices, he decided to stop. His extraterrestrial contact experiences nevertheless continued, and he next began undergoing military abductions. This "re-abduction" experience is very common with individuals having extraterrestrial contact, and extensive research has been conducted on the phenomenon.[554] An example that has already been discussed is Corey Goode who was subjected to multiple military abductions after going public with his experiences concerning different extraterrestrial groups and secret space programs.

MacDill AFB is the home of US Central Command and US Special Operations Command, which are two of the ten combatant commands under the Department of Defense (DoD) that are responsible for coordinating and deploying of US military forces drawn from all four military services: Navy, Army, Marines, and Air Force. One of Special Operations Command's constituent units is Air Force Special Operations (Special Ops), which officially trains Air Force personnel for covert air operations but also secretly trains qualified personnel for classified operations in space. As will be shown, JP was being encouraged to take the photos of the flying craft by personnel connected to Air Force Special Ops and to have these disseminated to the general public.

After receiving JP's photos on March 3, 2017, of triangular-shaped craft near MacDill AFB on March 3, 2017, I took no action and archived these for further study. Then on August 31, 2017, I received a new set of photos of a triangular-shaped vehicle taken near MacDill. The photos were captured at approximately 8 am that day by JP and show a UFO that is similar in shape to the TR-3B, but JP estimated it to be much smaller – about 60 feet (20 meters) across. As stated in a previous chapter, the TR-3B was built in two sizes: a prototype of 250 feet (83 meters) and a working model of 600 feet (200 meters). Given that the TR-3Bs were built in the 1980's, it would not be surprising that by 2017 there would be operational models of different sizes, as now indicated by JP's photos. Due to the clarity of the photos and the significance of where they were taken, I decided to publish them and publicly announce JP's existence (with his consent) for the first time.[555] The *Tampa Bay Times* subsequently published a story discussing the

August 31 photos, along with later photos taken by JP.[556]

The six photos from August 31 show a triangular-shaped craft that "powers up" with a circular energy pulse. The powering up is consistent with what we know from whistleblowers such as Edgar Fouche, who have described how the TR-3B operates using highly pressurized mercury-based plasma at high temperatures rotating around a circular ring at 60 thousand rotations per minute (60k RPM) or more. Significantly, the six successive photos by JP show a triangular craft "powering up" as if on cue when he started taking the photos. How the craft moved as the photos were taken and the energy pulse it created indicate it was not an airplane, but an antigravity craft similar to a TR-3B described by Fouche. This is what JP wrote to me in his Skype communication on August 31, 2017:

> An amazing experience today[.] I felt the urge of looking up and snapping a couple pictures of clouds, beautiful evidence, spectacular. I saw a Triangle UFO ship flying left to right, I guess it noticed that I saw it, it stopped, it generated some sort of bright energy ring then it shot up to space into the blue Sky, nice technology ... if I did get picked up, I don't remember...

I've created two composites of the six photos he sent. The first composite shows the first three photos of a triangular-shaped craft that could simply be an exotic plane of some kind. The third in the sequence shows the beginning of the "power up" described by JP. However, it is the composite of the last three photos taken which show a spectacular circular ring of energy appearing around the craft that remarkably demonstrates it is an antigravity vehicle of some kind, powered by a glowing plasma ring similar to the one a TR-3B generates for propulsion.

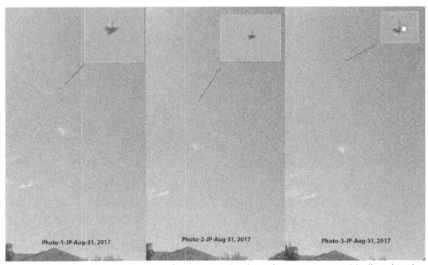

Figure 56. First 3 photos in series show start of pulsing (zoom inserts added). Taken by JP on Aug 31, 2017

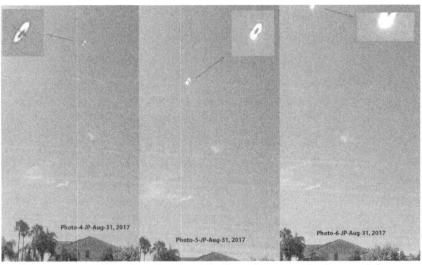

Figure 57. Last 3 photos in series show circular pulse emanating from craft (zoom inserts added). Taken by JP on Aug 31, 2017

JP described the craft's movements as darting around, then becoming stationary, and finally shooting up. These are distinct characteristics of the maneuverability of antigravity craft, not conventional airplanes or military fighter jets.

On September 4, 2017, JP sent me three more impressive and clear photos of a triangular UFO taken near MacDill AFB. The photos were taken around 9:50 am, about 10 miles from the Air Force base. His estimate of the vehicle's size was between 50-70 feet (17-23 meters). JP says that he quickly snapped the photos over a three second period and observed the craft for about 9 seconds in total. During this time period, the craft hovered and slowly moved towards his right. JP says that the craft then disappeared in an odd way, like an invisibility field sweeping over it from right to left over it. This suggests that the triangular craft either used "cloaking technology" that would quickly move from one end of the vehicle to the other, or JP saw the craft enter into the cloaked field of a larger vehicle.

**Figure 58. Photos of flying triangle (zoom inserts added) taken by JP on Sept 4, 2017**

The three September 4 photos show a different type of triangular craft to those photographed by JP on August 31. The August 31 craft appeared to have a large exaggerated tail fin of some kind that was very visible before it powered up with a luminous circular field generated by its propulsion system, and shot away into the upper atmosphere. Conversely, the September 4 craft disappeared using some kind of cloaking technology, and JP claims it looked similar to ones that had come to abduct him in the past.

## Flying Triangle Sightings Continue as Hurricane Irma Approaches

On the morning of September 5, 2017, I was contacted by JP about the unusual circumstances that led him to take more photos, again of yet another type of triangular-shaped craft in the vicinity of MacDill AFB. JP said he was prompted to look in the direction of the UFO by a stranger who drove up to him in an unmarked black car and instructed for him to see what was happening in the sky, and to take photos. Here is what JP said in his Skype communication to me about 30 minutes after the incident which had occurred at 3:45 pm EDT:

> 3:45 today a guy in a black car dress[ed] in white with a black hat, sunglasses. Put his tinted windows down, drove towards me and told me to look up. I saw a jet pass by, the guy in the black car was already … [at] the stop sign making the right to leave, after I saw the jet there was a ship, no noise, interesting pictures. I even think it has a little bit more details.

Figure 4 is the clearest of the sequence of five photos JP took after the incident with the mysterious man in white driving the black car. The September 5 incident suggests that a covert operative out of MacDill was sent to get JP to photograph the flying triangle very likely associated with Special Operations Command.

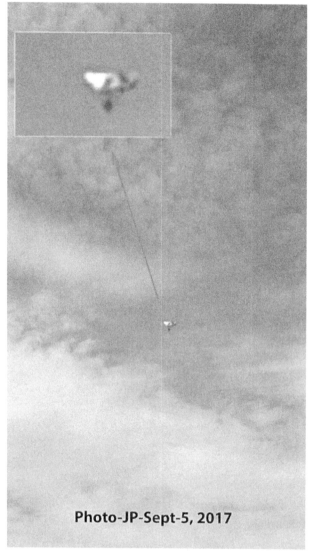

**Photo-JP-Sept-5, 2017**

Figure 59. Photo of triangular craft (zoom insert added) taken by JP on Sept 5, 2017

The next day, on September 6, 2017, between 7:00 to 7:15 am (EDT), JP took a series of nine photos which very clearly depict a triangular-shaped craft that again appears to be a smaller version of a TR-3B antigravity craft. Three photos from the series of nine are shown that most clearly present the triangular craft, which JP reports was strangely weaving in and out of the clouds in different directions. Significantly, these flying triangle antigravity sightings over successive days were happening at the same time as Hurricane Irma was heading directly for Florida.

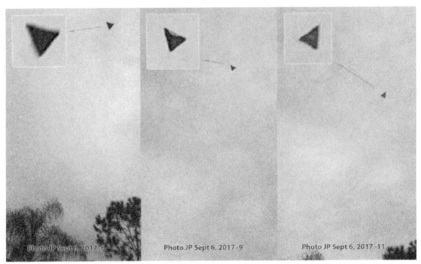

Photo JP Sept 6, 2017 -6    Photo JP Sept 6, 2017 -9    Photo JP Sept 6, 2017 -11

**Figure 60. Three photos of triangular craft (zoom inserts added) taken by JP on Sept 6, 2017**

On September 7, 2017, JP sent me six more exceptional photos in a series showing a flying triangle sighted again near MacDill, which was being shadowed by a military helicopter in what appeared to be a sanctioned military exercise or escort. Notably, JP described that the triangular-shaped craft was at times stationary. Photo analysis has shown that the triangular craft was not a propeller driven drone or kite of any kind.

Here is what JP said in his Skype message that was sent the next day at 9:23 am EST on September 8:

> WOW. Almost a whole week of sightings, I know a lot more people are seeing these, this sighting is amazing … At 8:00 am I saw a black helicopter circling the neighborhood where I was at. Then around 8:25 am, 4 miles away from MacDill Air Force Base I saw the TR-3B interacting with this

helicopter, some sort of exercize. You can clearly see the size comparisons confirming the size that many whistleblowers have talked about. The ship was on its side, top, side, facing the helicopter, like communicating with it. You could tell the helicopter [was] stationary. Then it zips to the top on its regular position that flew over head ... The helicopter went towards MacDill Air Force Base ... the last photo is spectacular just like the last one seeing the details ... not sure if they equipped it already with weapons.

JP estimated that the triangle craft was 100 yards from the helicopter, 500 yards away from him, and 300 yards in elevation. He again estimated the triangle craft's size as around 50-70 feet (17-23 meters). All this occurred about 4 miles (6.5 km) away from MacDill.

Figure 61. Three of the photos by JP of flying triangle and helicopter (zoom inserts added). Taken on Sept 8, 2017

In my discussion with JP, I pointed out that the TR-3B was approximately 600 feet (200 meters) wide according to reports by

Edgar Fouche. JP responded that the craft he saw was much smaller but very similar in size and shape to the ones that had picked him up in military abductions related to his extraterrestrial contact experiences. JP explained that the interior of the craft seated three people in capsules that were designed to stay upright no matter what position the ship was in, like a gyroscope. This would enable the pilots to see forward in terms of where the craft was traveling or move to whatever direction they wanted to see, no matter the ship's orientation to external objects. When carrying a passenger, one of the capsules would stay in a horizontal position. In addition, JP described the capsules as having "a mind of their own" because they would automatically adjust to a person's body shape. The capsules had a control panel that one or more pilots could use to navigate the craft. The craft did not have windows but instead used a technology which allowed the pilots to see through the metal alloy.

The day after JP took the flying triangle with military helicopter photo series, a mandatory evacuation order was issued for MacDill AFB, in response to the imminent impact of Hurricane Irma which was projected to make landfall in southern Florida on Sunday morning, September 10.[557] The evacuation order went into effect at 12 pm EST on September 8. Significantly, the evacuation order was mandatory for all personnel except for those involved in "mission essential" activities:

> Personnel who are not designated as members of a ride-out team but who are deemed mission essential or required to remain past 1200L 8 Sep 2017 by their commanders to perform mission related duties will be excused from this evacuation order until they are cleared for release by their commander.[558]

DEPARTMENT OF THE AIR FORCE
6TH AIR MOBILITY WING (AMC)
MACDILL AIR FORCE BASE, FLORIDA

8 Sep 2017

MEMORANDUM FOR MACDILL AFB PERSONNEL

FROM: 6 AMW/CC

SUBJECT: CONUS Limited Evacuation Order (LEO) of MacDill AFB, FL

1. This is a Limited Evacuation Order (LEO) for MacDill Air Force Base. Evacuees are authorized/ordered to move from a CONUS residence to the nearest available accommodations (which may be Government quarters) outside of evacuated areas.

2. My authority to order the LEO is Joint Travel Regulations (JTR), par. 6080-B (uniformed services eligible dependents) and par. 6580 (civilian employees/eligible dependents). This order applies to all individuals assigned to, residing on, or present on MacDill AFB.

**Figure 62. Evacuation Order for MacDill AFB issued September 8, 2017**

MacDill's US Special Operations Command possesses significant underground facilities which allow covert operations to continue regardless of a hurricane's approach. Given the approach of Hurricane Irma, it appears that Special Operations Command was conducting surveillance and possible "mitigation operations". It's highly likely that Special Operations Command was conducting covert operations in relation to Irma, but for some unknown reason they chose to alert JP to photograph their craft in operation. Are they sending a public message, and if so, what did Special Operations Command want people to know at the time of the impending chaos to be caused by the hurricane?

Landfall for Florida happened early on September 10, 2017. Here is what ABC News had to say about the hurricane situation just after midnight on September 11 at 12:21 am (EST) in a story titled: "Irma approaching Tampa Bay area with hurricane-force winds":

> Hurricane Irma is barreling up the Sunshine State, bringing heavy rainfall and powerful winds. The deadly hurricane, which is now a Category 2 storm with sustained winds of 100 miles per hour, was

moving east of the Tampa metro area.... Tampa, which is now experiencing wind gusts of over 50 mph, is in the predicted path of the storm, and the city's mayor Bob Buckhorn did not mince words when he warned residents on Twitter earlier today.

"We are about to get punched in the face by this storm. We need to be prepared," he wrote. [559]

Note the date of Hurricane Irma hitting Tampa – September 11, 2017 (the 16th anniversary of the 9/11 terrorist attack). Tampa's proverbial "punch in the face" as stated by the Mayor raises the intriguing possibility that Hurricane Irma was linked in some way with the flying triangle UFO sightings, which appear to be sanctioned by US Special Operations Command as a means of disclosing the existence of a USAF secret space program. As a result, the question becomes whether MacDill AFB and Tampa had been deliberately targeted using weather modification technology.

## Did Maser Satellites Steer Hurricane Irma in Weather War Against U.S.?

Weather modification technologies date back to the 1940's, when it was discovered that chemicals could be seeded into the atmosphere to influence weather. A 1978 US Senate committee report details how the US Congress has been involved in reviewing weather modification technologies as far back as 1947; and since 1953, six acts of Congress have been passed dealing with the subject matter.[560] In the early 1970's, it was discovered that high frequency radio waves would provide the next major technological evolution in weather modification, according to Nick Begich and Jeanne Manning in their 1995 book, *Angels Don't Play This Haarp*.[561] They traced the technological developments that led to the High Frequency Active Auroral Research Program (HAARP) and its multifaceted uses in weather manipulation and mind control.

While ostensibly designed as a communications system, powerful radar arrays like Alaska's "HAARP" Research Station near Gakona could be used to bombard the Earth's ionosphere with extremely low frequency (ELF) radio waves that would cause the Earth's surface to heat up tremendously. HAARP produces weather modification by artificially inducing high pressure systems which can be created to steer moisture laden air from one area to another. Dane Wigginton from GeoEngineering Watch described how this has occurred to create drought in the Western United States:

> When the ionosphere is bombarded with the massively powerful RF / microwave transmissions from installations like HAARP, high pressure heat domes can be created and maintained. In the case of the current record shattering high pressure heat dome over the US West, the clockwise spin of upper level air currents around the high pressure zone allow the weather makers to steer cooler air from the Pacific into the center of the country while completely bypassing the West.[562]

HAARP's ability to create high pressure systems can be used to steer hurricanes in whatever direction the controllers of HAARP desire. Wigington claims multiple HAARP facilities have been used in the creation and directional path taken by hurricanes such as Harvey and Irma.[563]

In the 1970's, another step in the evolution of weather modification technologies occurred with the Air Force/NRO deploying military satellites with dual purpose surveillance and electronic weapons functions. These were later incorporated into Ronald Reagan's Strategic Defense Initiative (Star Wars) as a futuristic space-based defense system against intercontinental ballistic missile attacks using electronic weapons such as "masers"

(Microwave Amplification by Stimulated Emission of Radiation). It was found that masers loaded into satellites could then generate powerful microwave radiation with great accuracy. Maser satellites were allegedly tested in the creation of many crop circles that first began appearing in the English countryside in the late 1970's, with large numbers appearing by the late 1980's.

According to many scientists who have investigated the crop circle phenomenon, there is extensive evidence showing that microwaves were used to generate some of the elaborate patterns. For example, Professor Richard Taylor, head of the Materials Science Institute from the University of Oregon identified satellite-based microwave technologies as the most likely explanation for crop circles.[564] Professor Taylor and others conclude that many crop circles have been created as a means of testing and refining space-based maser satellites, which can direct microwaves anywhere on Earth with pinpoint accuracy. Here is how another crop circle researcher who mentions patents on maser technologies summarized his analysis of crop circle evidence:

> At many crop circle locations, especially the more elaborate ones, it appears that microwaves have been used to create the designs. The evidence of stalks burnt from the inside out exists. Small animals have been discovered at these places that look exactly how an animal would look like if you shoved one of them in a microwave oven and blasted them until dead. Burned from the inside out. This fact to me, along with my knowledge of what is currently on the patent books, seems to indicate the use of military microwave weapons being directed from orbiting satellites. The evil outgrowth of Ronald Reagan's Star Wars Program. Masers! ... Crop circles are, in my educated opinion, test firings of military weapons.[565]

This is where crop circle researchers make the critical link to a key ingredient for weather modification:

> Perhaps they are testing newly launched satellites to see if they work properly. Perhaps they are aligning the beam or beams. One thing is for certain, microwaves produce some mean heat! Stick a bowl of water in a microwave oven. Hit the start button. Three minutes later, you've transformed a cool bowl of water into a hot bowl of bubbling, boiling excitement.[566]

The work of crop circle researchers in identifying maser technologies as responsible for creating many crop circles through their ability to heat water molecules rapidly provides the critical element for understanding how hurricanes can be generated or steered anywhere around the world. Maser satellites can be used with pinpoint accuracy to heat up bodies of water anywhere in the world to augment a developing tropical storm into a hurricane, and then steer it in any direction.

Maser producing satellites are classified military technology. However, it is likely that the Raytheon Corporation is involved in their use through its Joint Polar Satellite System and Common Ground System which uses a network of satellites to survey weather systems around the world. Raytheon's website states:

> The Joint Polar Satellite System (JPSS) Common Ground System supports the latest generation of U.S., European and Japanese polar-orbiting satellites designed to monitor global environmental conditions and collect and disseminate data related to weather, atmosphere and oceans.[567]

Raytheon provides the data needed for forecasting models used by the National Oceanic and Atmospheric Administration (NOAA).

Raytheon has long been suspected as a major player in weather modification due to its ownership of key HAARP patents.[568]

Identifying the different kinds of weather modification technologies in terms of their historic development and use helps clarify some key questions associated with Hurricane Irma. First, why would the Strategic Defense Initiative military controllers in charge of maser satellites direct a major hurricane like Irma directly into a specific area of Florida to directly impact a critical military installation like MacDill Air Force Base, which is home to two major US commands: Central Command and Special Operations Command?

Second, why would the managers of a covert aerospace program at MacDill, using advanced antigravity craft, allow these to be on full display in broad daylight for the general public and to be photographed by a witness who was even prompted to take the photos? For answers, we can turn to Corey Goode who received briefings about the satellite weapons used in steering Hurricane Irma.

## Q & A Session with Corey Goode on Hurricane Irma, Maser Weapons & Flying Triangles

Over September 9 and 10, 2017, I asked Corey Goode a series of questions about Hurricane Irma and the "flying triangle" sightings near MacDill AFB that were photographed by JP.[569] The following text is a portion of that exchange:

MS – Michael Salla / CG – Corey Goode

MS: I know Raytheon and Lockheed are part of the weather forecasting used by NOAA. I assume they are involved in or getting tipped off by the Cabal about the strength and direction of these hurricanes. Are they involved in the Maser satellites being used, or is this more the ICC [Interplanetary Corporate Conglomerate]?

CG: It's closer to MIC [Military-Industrial Complex] level with heavy corporate ties of course, built by both Raytheon and Lockheed and a few others. Placed in orbit in the 90's by Shuttle and booster rocket tech. There are lots of classified satellites up there that aren't even on the books with fake names. There are a lot of weapons platforms orbiting the planet right now that are a part of the MIC program.

Some of the satellites that use microwave lasers were put up under the strategic defense initiative mostly put in orbit in the 90's. Classified documents showed that the satellites were for taking out missiles on Earth as well as defending earth from [the] unknown.

MS: Makes sense they would have classified satellites doing that. So, do you think the Raytheon Joint Polar Satellite System might be used in some way with this?

CG: Yes, it's all about "Environment" aka weather, in my opinion, based on briefings and other available exotic weapons platforms in orbit that I was briefed on years ago. Also, some remote imaging satellites that penetrate hundreds of miles into the Earth.

MS: No wonder the Inner Earth groups were concerned about us surface dwellers.

CG: Exactly, we have satellites that can not only

cause EQ's [earthquakes] and volcano eruption but also cause deep, deep earth penetrating waves that cause the implosion or explosion of deep facilities. Most likely 'Rods from God' will be used in conjunction with the Triangle Craft, in a first strike on N. Korea, thus disclosing part of their program to the public.

These Air Force base sightings are exactly what I told people about around 6 months ago when I said people would start seeing triangle craft in the skies more and more leading up to the imminent revealing of a TR3B-like craft. Most likely, these exotic weapons platforms will debut during a first strike on N Korea. Then they will disclose the MIC SSP in stages based off of that showing of tech. Similar to how the Stealth planes were introduced.... They said the MIC SSP would start accidentally showing their triangle craft and that we would see white orbs that are the corona of these craft (at night and day)... Glad I'm on record reporting that.

MS: Might there be a connection between the HAARP facilities and the MASER satellite systems working in tandem with Hurricane generation and steering, or is HAARP the older more antiquated weather modification system?

CG: I'm told they don't need the HAARP facilities... They are used for several purposes (HAARP arrays) they can do all weather modification from orbit now.

MS: So, the increased microwave activity in Antarctica may not be related to Irma at all in terms of a HAARP type radar system? Basically, does Antarctica just provide the control and command system of the ICC/Cabal that is directing these MASER satellite systems to manipulate the weather?

CG: The Command and Control can be from anywhere. Just a simple tasking of a satellite. This is mostly on the MIC (AF/DIA) SSP end.

MS: So, basically, one faction of the MIC (AF/DIA), which is still Cabal aligned, is directing IRMA to hit Florida, thereby taking out another group, Special Operations Command, MacDill, that has become a threat? MacDill has now been evacuated.

CG: I believe so. Major distraction for the Military and maybe a bit of punishment for states that supported Trump.

MS: Makes sense. The flying triangle sightings are therefore likely to increase as the Special Operations people hit back using information warfare tools against the Cabal.

CG: Some of these Air Force assets (MOST) are Cabal controlled.

MS: The most recent hurricane forecast shows how the eye of Irma passes directly over MacDill Air Force base, Tampa, in the early morning of September 11. Is this coincidence or a veiled declaration of war from those steering Hurricane Irma, which is derived from the name of the Old Germanic War Goddess, Irmin?

CG: Hard to tell with this group. They could be directing the storm and showing the craft is a part of disclosure of MIC SSP. I am hearing that the Cabal controlled MIC SSP is directing this for several reasons.

MS: Two main reasons you've mentioned so far is: 1. The hurricane is steered to preoccupy the US military to delay/forestall a military coup to take down the cabal. 2. A way of revealing the USAF/DIA SSP through the flying triangles that have been photographed. What other reasons might there be for the MIC SSP doing this?

CG: 3. Punish the alliance for its victory and those who supported it in electing Trump. For each one of those there are at least a few dozen "Sub Agendas".
570

Goode's responses make clear the significant role space-based masers have played in augmenting and steering Hurricane Irma and others after meteorological processes naturally form them. His statement that HAARP technologies are an outdated technology for weather modification raises two key interconnected questions. First, do those controlling the maser satellites have conflicting agendas with those running HAARP facilities around the planet? And second, was Irma a false flag weather attack on the U.S.?

Concerning the first question, if Hurricane Irma was an example of a weather war attack against the U.S. using space-based masers, then one or more factions within the US military may have used some of the HAARP facilities to mitigate this attack. Certainly, US Special Forces Command at MacDill would have done all in its power to diminish Hurricane Irma, which was predicted to make landfall near its headquarters. Indeed, after forcing the mandatory evacuation of MacDill on September 8, there was much expectation of a devastating hit to Tampa Bay on September 10/11, 2017. [571] However, Irma was downgraded to a Category 1 when it finally impacted Tampa at the start of September 11, and MacDill suffered only minor damage and was quickly reopened later that day.[572]

The antigravity flying triangles photographed by JP near MacDill from August 31 to September 7, 2017, might have been part of the mitigation effort directed at Irma. Additionally, any HAARP facilities cooperating with Special Operations Command could have been used in the effort. This raises the issue of Hurricane Irma being a false flag attack. Goode pointed out that the maser satellites are under the control of what he calls the "Military-Industrial Complex Secret Space Program" (MIC SSP) involving USAF Space Command, the Defense Intelligence Agency, and the National Reconnaissance Office.

In addition, there may also be secret HAARP facilities in Antarctica that are being used in a weather war against the United States. In an article on September 11, 2017, former *Forbes* magazine writer Benjamin Fulford cited unnamed Navy sources saying that electromagnetic waves were detected coming from Antarctica.[573] This coincided with mainstream news reports about large subterranean caves being discovered in Antarctica that were thermally heated to 77°F (25°C) and could support abundant life.[574] Both William Tompkins and Corey Goode have said there are secret R&D facilities scattered under the kilometers deep Antarctic ice sheets that are controlled by various US corporations, which have historically cooperated with the German Antarctic-based secret space program that escaped World War II largely unscathed.[575]

Corporations such as Raytheon and Lockheed have each, in turn, led the logistical operations of the US Antarctic Program, currently provide the weather modeling for NOAA, and have allegedly worked closely with the German Antarctic-based SSP.[576] It's also important to note that the MIC SSP, which controls US military satellites including those with maser weapons forming a space defense grid, also have historically cooperated with the

Antarctic-based Germans. This points to programs in Antarctica playing a significant role in weather war attacks against the U.S., partnered in collusion with some factions of the US Military-Industrial Complex to manipulate hurricanes such as Irma.


## Cylindrical UFO near MacDill AFB Affirms USAF-Nordic Alliance

Only three days after MacDill AFB reopened from a mandatory evacuation due to Hurricane Irma, UFO's were once again photographed near the base by JP. This time the vehicles were "cigar-shaped" and at least 325 feet (100 meters) in size according to JP's estimates. On September 14, 2017, at around 3 pm EDT, JP took a series of 11 photos and also shot video footage of the cigar-shaped UFO which appeared approximately two miles from MacDill. Several orb-like objects were also moving around the cigar-shaped ship, which first appeared coming out of the clouds according to JP's eyewitness account. Further, he claims that during the six minutes of filming and photographing the incident, all city noises ceased around him and it became very quiet and peaceful. The din of the city quickly returned after the cigar-shaped UFO disappeared back into the clouds.

A composite of three images which most clearly show the cigar-shaped UFO is shown with added close-ups (see Figure 63). The first image displays approximately five small orb-like objects which are likely scoutcraft that can enter and exit the cigar-shaped craft. JP emphasized that he felt "very calm and peaceful" during the sighting as everything around him turned quiet. This made him suspect that the craft was extraterrestrial in origin since the quieting of the environment was very unusual and he had not encountered it while photographing the flying triangles between August 31 to September 7, 2017.

**Figure 63. Three photos of cylinder-shaped craft (zoom inserts added) taken by JP on Sept 14, 2017**

At the start of this chapter, I mention that JP first got in touch with me in 2008 to talk about the extraterrestrial contact experiences he had while in Brazil. Later, it became clear that a human-looking Nordic group of extraterrestrials was in communication with him. During his earlier contact experiences, JP told me that he had unusual feelings of "peace and calm", and now his encounter with the cigar-shaped craft had generated a similar emotional impact on him. This is why he believes this particular craft is extraterrestrial in origin rather than part of the USAF SSP.

However, something threatening occurred after JP took the eleven photos and video. Once the cigar UFO departed, he says the ambient noise returned and he received a phone call. The person on the line was very abrupt and menacing. The male voice told JP, "don't put [out] no videos, delete [them]", then he hung up. Immediately after receiving the call, JP said it began to rain and two lightning bolts struck the ground in succession about 20 feet from him. He felt very intimidated by the events and decided to delete the video, but to keep the eleven photos which he quickly sent to me.

So, who called JP? A clue comes from another incident JP related to me which occurred the day before on September 13, 2017. JP was scheduled to visit a private residence for his job, and at this location, he and the customer began talking. During this conversation, JP received a veiled warning as he explains:

> I was at a customer's house and we were talking. We were talking about weather manipulation. I never actually talked to this customer before. He told me something really, really interesting. The house that is in front of his is the house of the commander of the CIA at MacDill Air Force Base... I got cold feet, cold hands, the customer said, "we're watching you" ... He went inside.

What this exchange suggests is that CIA operatives are closely monitoring JP, likely because he has been chosen to photograph the antigravity craft from MacDill overseen by Special Operations Command.

Importantly, JP claims that he has also been abducted and brought aboard the "flying triangles" on various occasions, and then taken to unknown locations where he encountered human-looking extraterrestrials. Since a group of extraterrestrials is also cooperating in allowing JP to photograph their vehicles near MacDill, this indicates that Nordic extraterrestrials are actively working with Special Operations Command to disclose the existence of antigravity vehicles and the presence of human-looking extraterrestrial allies.

Such a scenario is bolstered by the leaked 1989 Defense Intelligence Agency document discussed in the previous chapter, which reveals how human-like Nordic extraterrestrials are friendly and can be relied upon, as opposed to some of the other calculating and deceptive extraterrestrial groups encountered by the military.[577] The leaked DIA document marks the first time that any Majestic document has ever referred to a "human-looking" extraterrestrial group. The DIA document and JP's experiences add further weight to the conclusion that the USAF is now actively working with a Nordic group.

This Nordic/USAF/DIA alliance is staunchly opposed by other factions within the US Military-Industrial Complex, which helps to explain why Hurricane Irma was steered towards Tampa and directly aimed at MacDill AFB, home of Special Operations Command. Goode has confirmed that such a factional war exists within the Military-Industrial Complex, which provides the infrastructure and corporate resources for the USAF SSP. This Nordic/USAF/DIA alliance would also explain why JP, who was chosen to take and pass along the photos of various craft in the vicinity of MacDill for public release, has also been experiencing intimidation by another CIA-linked group that wants to slow or prevent such disclosures.

## Covert Disclosure of Antigravity Craft near MacDill AFB

On October 19, 2017, I was sent another series of photos by JP. This time nine shots were showing two UFOs in the vicinity of MacDill. One craft looked like a rectangular platform while the other exhibited the now familiar TR-3B triangular-shaped design. Both appeared to use antigravity technology and were interacting with one another. Before the October 19 sighting, JP says he received a phone call instructing him exactly when and where the UFO's would appear. This is what he wrote in a Skype message:

> 11 a.m. I got a call to look at the skies... like around 11:02 I see this rectangle platform black ship... it came out from the clouds 6 seconds after seeing the

ship... TR-3B shows up, you can see the characteristics of it, the circle in the middle still glowing a little bit, in a blink of an eye it was there interacting with the black triangle platform ship similar to when it was interacting with the helicopter... this was quite fast, 2 minutes after the call lasting a minute and 20 sec.[578]

He added in a later Skype message:

The call was really fast like 10 seconds, the guy said (I'm calling to inform you to look at the skies clock view (10:30) in between East and North have a nice day. Click...

Next, JP writes that he has never encountered a rectangular-shaped UFO before, which flew in from the direction of the Gulf of Mexico:

I never seen those platform looking UFOs
Rectangle
It was quite interesting
I was hearing something weird this time
like a high pitch sound my ears were kind of like
ringing after the event..
8 miles from MacDill this time
MacDill was in the Direction behind me
The platform ship looks like it came from the Gulf
of Mexico
And I'm thinking the tr3b came from MacDill
They met up.[579]

Here is how JP describes the sequence of photos and the interaction between the two vehicles in another message:

I saw the platform ship and out of nowhere the TR-3b showed up in a blink of an eye... then started interacting with the platform ship
I don't know the exact size but all I can tell you, it was the same height where the clouds were
Me knowing that there are three levels of clouds, I'm thinking this is the lower cloud... 6500 feet
Those are regular cumulus clouds
They're at least 6,500 ft from sea level.

A close-up of the first photo in the sequence shows the rectangular-shaped platform UFO taken by JP directly after he received the phone call telling him to look up into the sky. (see Figure 64)

Figure 64. Photo of rectangular-shaped craft (zoom insert added) taken by JP on Oct 19, 2017

A close-up of the third photo in the sequence displays the rectangular-shaped platform and the TR-3B-like craft, both appearing to interact with one another while JP watches and hears high pitched sounds. (see Figure 65)

Figure 65. Photo of flying rectangle and flying triangle (zoom insert added) taken by JP on Oct 19, 2017

Based on JP's account of how the photos were taken and their estimated elevation, size, sounds, and maneuvers, they, like his previous photos, do not appear to be merely balloons, drone aircraft or aircraft using conventional propulsion systems as skeptics may automatically assume. The fact that JP received a phone call to alert him is very significant. It clearly suggests that the UFO's presence was deliberately brought to his attention to display classified technologies for photographing. The phone call he received on October 19, 2017, and the photos themselves are further evidence that JP is being used to facilitate a covert disclosure of antigravity craft operating in the vicinity of MacDill AFB. Four days later, JP was to witness the flying rectangle platform again, and this time he claims the encounter became much more personal.

On the morning of October 23, 2017, JP claims he was once again instructed to look up into the sky, this time by two men in a nearby van who pulled up next to him. Immediately, he saw a rectangular-shaped platform UFO and used his camera phone to take five photos. He then noticed that he had missing time and when he tried to recall what had happened, he had memory flashes of being inside the flying rectangle he had just photographed. Here is the Skype message I received from JP at 10:49 am EDT:

> 9:30 am 30 minutes lost time... it was 10:08 when I get back in the same position where I was doing the pictures today. a white mini work van with two guys black baseball hats Black sunglasses... He pulled over in front of me and told me to look up and they left really fast like you can really hear the tires... a platform ship really near, it swooped up fast when it knew it got my attention...

> I was getting flashes on and off of me being inside of this ship, and looking down on the area where I was... in these flashes of remember[ing] being cold really cold inside... the guys told me to look up, I saw the ship really close, then I saw the guys leaving really fast and I saw the ship swoop up... when they told me to look up it was like 9:30, I'm guessing when I was taking the pictures of the ship it was already 10:05 10:08 am

> I could still smell inside the ship... it smelled similar when you have a lot of coins in your hands the copper smell... copper, copper, copper. Beautiful ship amazing maneuvering. Cradling, it went back into the clouds...[580]

JP estimated that the flying rectangle was about four school buses in length (approx. 40-50 yards), and one and a half school buses in width (approx. 15-20 yards). He said the distinct copper smell was likely related to its antigravity drive or some other aspect of its propulsion system. JP specifically recalls being inside of the flying rectangle and walking down a corridor with windows. Outside the windows it was completely black, and he got the impression that the rectangle was flying in space. He has provided a rough sketch of the corridor of the craft with a bank of windows on one side.

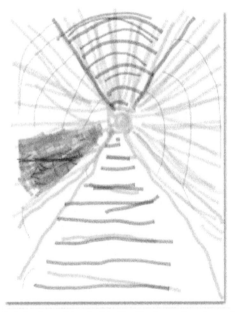

Figure 66. JP's sketch of corridor inside the Rectangular-shaped Platform UFO

JP remembers being injected with something, presumably a mind control substance so he would cooperate. He shared a photo with me of an inflamed reddish area on his upper forearm where he says he was injected. He also reported feeling some dizziness afterwards. However, JP reported this effect quickly wore off as did the reddish discoloration on his arm.

According to JP, the interior of the flying rectangle craft looked and smelled like a military vehicle. Most importantly, he recalled a military patch worn by his abductors which he described as having a red dot and two wings. He first sketched the patch he witnessed and passed on a copy to me. We discussed this further and hours later he found the specific patch online. The patch he identified belongs to Air Force Special Ops. The Air Force patch suggests that the occupants of the craft were part of a squadron of vehicles operating out of MacDill, the headquarters of Special Operations Command which provides the command structure for Air Force Special Ops.

Figure 67. Patch used by Air Force Special Operations

It is rare for an experiencer or abductee to take photos of the craft which has abducted him/her. Often people are incapacitated due to the effects of sedatives or mind control used on them. In the case of JP, however, he was instructed to look in the sky by two operatives while still under the influence of the substance injected into him. What this indicates is that the operatives were part of an officially sanctioned covert program run by USAF Special Ops which was helping JP to photograph the rectangular-shaped craft, and further, to recall his abduction experience. JP strongly believes that the operatives came from MacDill AFB, which is very close to where he was living at the time he took the photos.

According to SSP insider Corey Goode, the rectangular-platform craft are weapons platforms whose genesis he explained in a personal correspondence:

> I think similar platforms or UFOs have been reported in the 80s and 90s as they were first building out of these platforms that have modular weapons and personnel delivery systems or configurations. They were built up on that superstructure type frame a long time ago and they [have] been improved upon since but these are pretty old technologies both the TR 3B and the square platform... It's crazy all of this is happening not six months after I announced on *Cosmic Disclosure*.

Based on the circumstances surrounding JP's abduction and him being prompted by covert operatives to look to the sky at the right time, it can be concluded that the rectangular-shaped platforms photographed by JP are part of a USAF-run secret space program, which is deployed out of MacDill. JP's military abduction experience aboard the craft he photographed clearly suggests that these "flying rectangles" have been covertly deployed in squadrons which are fully operational and deployed by Air Force Special Ops in space. The disclosure of these fully operational space-based weapons platforms appears to have been sanctioned by senior officials from Air Force Special Ops through Special Operations Command.

## Contact with a Nordic Extraterrestrial Working with USAF SSP

In late 2017, JP and his family moved to Orlando, Florida, where his sightings and extraterrestrial contact experiences eventually resumed. After a hiatus of several months, he again took photos of "flying triangles". Incidents occurred on March 16 and 3March 23, 2018, which once again clearly displayed TR-3B-type craft.[581] However, it is what occurred later on May 24, 2018, that holds special significance for understanding the extent to which the USAF SSP had begun working with Nordic extraterrestrial visitors.

On May 24, JP claims he encountered a human-looking "Nordic" extraterrestrial wearing a USAF uniform who invited him to go for a ride in a landed saucer-shaped craft in a secluded wooded area of Orlando, Florida. JP declined the offer but took photos with his cell phone of the spacecraft as it was departing (see Figure 68). His conversation with the Nordic revealed that this group was directly cooperating with the USAF and abiding by the terms of an agreement concerning interactions with the general public.

What follows is the Skype conversation I had with JP which began at 10:32 EDT (I have removed minor grammatical errors and added missing words in square brackets):

> JP: 9am. Nice contact today.
>
> MS: What happened?
>
> JP: I went Into the Woods. I was hearing a ringing in my ear. I'm thinking it was like a calling. Like I got from the last Contact. I met up with this man. Wearing a United States Air Force suit. Dark blue. His face was different. Possibly a Nordic.
>
> Light blue eyes, blond hair, wearing a United States Air Force Jumpsuit. He was really kind. He said a lot of these craft are going to be seen all over. He invited me to go into the craft. I said no. Yes [he asked] me again if I was sure....
>
> I told him that we're gonna start needing video evidence. He told me [that] because [of] a deal they

have with the United States and other countries, they can't.

This is consistent with the information contained within the leaked 1989 DIA document, as well as the testimonies of William Tompkins, Corey Goode, and Emery Smith presented in earlier chapters concerning human-looking extraterrestrials blending into human society and working in classified USAF and corporate facilities. Additionally, early contactees such as George Adamski, Howard Menger, and others have said that they learned about, and even helped, human-looking extraterrestrials blend into local communities to evade recognition by the general public, and when necessary, national authorities.[582]

It's worth noting that the current Russian Prime Minister, Dimitry Medvedev, revealed in December 2012 that when he became President, he was given a highly classified file about how Russian and global authorities register and track extraterrestrials secretly living among the human population.[583] This all suggests that secret agreements have indeed been reached between public authorities and Nordic extraterrestrials, where any video recording of their existence is not allowed by those they contact, just as JP claims he was told.

I return now to my Skype conversation with JP where I next queried him on his decision not to accept the Nordic's offer to go into his craft:

> MS: Surprised you said no. Did you get a bad feeling at any point?
>
> JP: No, I actually felt happiness, but I was kind of mad because of not getting video.
>
> MS: Hmmm, would you have been allowed to take photos?
>
> JP: He told me: This kind of ship gonna be seen all over. After he talked to me. He touched me. I had a

Flash that I was inside the ship. Floating around. It was a split second. Then he told me, you're back.

Here, JP is illustrating that the "Nordic" had arranged for him to view the interior of the craft without physically going inside. Perhaps astral projection or some other kind of non-physical encounter had been made possible for JP through the Nordic's more highly developed psychic abilities or through some altered state of advanced consciousness. As our Skype conversation continued, JP went on to write about the Nordic's cooperation with the Air Force:

> JP: I ask him what's going on with you guys and the United States Air Force. He told me that they're working on a big project together. A couple of projects in the Gulf of Mexico Involving a couple of MacDill Air Force personnel.
>
> MS: Why was he wearing a US Air Force uniform?
>
> JP: All these sonic booms are part of the project that they're doing.
>
> MS: Did it have the USAF Special Operations patch?
>
> JP: Another type of space insignia.

In the earlier October 23, 2017, military abduction incident, JP described the occupants on board a rectangular-shaped spacecraft wearing USAF uniforms bearing the patch of Air Force Special Operations. However, the patch worn by the Nordic in the May 24, 2018, encounter according to JP was different and contained the symbols of a star and crescent. Later, JP explained to me that the uniform itself featured the acronym "USAF" with these letters also spelled out in a foreign language he did not recognize. Only after some internet research did JP discover that the letters looked very similar to ancient Sumerian. This patch enabled the Nordic to walk undisturbed inside USAF bases:

> MS: What made you think it was a USAF uniform and not a Nordic uniform?
>
> JP: To camouflage [himself] in the [USAF] base when they're walking. It [the patch] had United States Air Force.
>
> MS: Did he tell you he walks around on USAF bases, MacDill?
>
> JP: ... Yes, in other bases around the world
>
> MS: That makes sense. Did he say how long his group has been working with the USAF in joint projects? When did it start?
>
> JP: We never got far in that conversation. I'm thinking a long time Michael. When he left... I ran to my truck.
>
> MS: Was the conversation done in English or telepathically?
>
> JP: English. European type. Like people talk in England. The perfect English. Not American.
>
> MS: Were you able to take any photos?
>
> JP: Yes.

JP was only able to take photos of the craft once it took off, but not while it was on the ground because his cell phone did not work at the 25 meter distance he stood from the craft.[584] He estimated the size of the vehicle to be about 20 meters in diameter and it had three windows which he could see through into the craft. I next asked JP questions about the Nordic's cooperation with the USAF:

MS: What kind of projects were the Nordics working on with the USAF?

JP: Training how to maneuver the ships. Invasive [evasive] maneuvers all around United States, in Europe, in Middle East, Russia, Asia.

MS: The Nordics were teaching the USAF invasive [evasive] maneuvers? Can you explain more?

JP How another craft chasing another craft. Is not the same as a jet chasing a jet.

MS: Were the Nordics sharing technology with the USAF or just know how?

JP: The United States Air Force has the technology but does not know how to correctly use it. Is [like] the comparison of dropping a smartphone to an Indian in the Amazon. The Indian would not know how to charge [the] phone, will not know how to search through ... So basically, they're teaching the technology, the physics. When to use the weapons because the weapons sometimes affect the ability of the craft when it's flying. The weapon that the craft has is a pulsing weapon energy superheated [or superheated energy weapon]. Beams of light, Radioactive, how to interact with thunderstorms to conserve and gain energy.

It's worth pointing out here that based on the testimonies of Tompkins, Goode and other whistleblowers, the US Military-Industrial Complex needed to work with German scientists after World War II to learn about the technology of crashed extraterrestrial spacecraft that had been recovered from Roswell and elsewhere.[585]

## Conclusion

Based on JP's testimony, the USAF still needs assistance to understand Nordic technologies which are reportedly more advanced than those found aboard the recovered spacecraft belonging to the small Gray extraterrestrials who have worked alongside USAF engineers and scientists at classified facilities for decades, as presented in chapter 11. It's worth emphasizing that in many of the incidents involving JP taking photographs of USAF aerospace vehicles, he was prompted by covert operatives to look in the direction of the craft. This indicates that JP has been selected for some reason to be part of a sanctioned disclosure initiative where he is being helped to reveal the existence of a USAF SSP and their growing alliance with Nordic extraterrestrials.

JP's testimony, backed by multiple photographs, is concrete evidence that the USAF has made a dramatic decision to work with a previously shunned group of Nordic extraterrestrials, and to break away from its earlier alliance with the Deep State/Antarctic-based German/Reptilian faction. These former allies promised much in the way of advanced technologies but only delivered these technologies decades after they had become redundant. Once leaders and personnel within the USAF SSP discovered that an advanced Navy SSP existed, which was built with Nordic assistance, it became a determining factor in this dramatic turnaround. The Deep State retaliated against the USAF SSP by using its maser satellite system to direct a hurricane against MacDill AFB, home of a critical element of the USAF SSP – Air Force Special Operations. Thankfully, the Deep State did not succeed, and the hurricane threat was significantly mitigated by the time it hit Tampa, Florida. Even more significantly, the USAF SSP has been able to turn the tables on the Deep State by using its space-based weapons systems, including some of the same type of maser satellite systems used by the Cabal to neutralize a false flag operation involving a ballistic missile attack over Hawaii intended to start a Third World War.

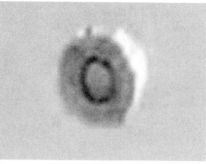

This is the fifth in a sequence of six photos taken by JP on May 24, 2018. The image on the left is the original and the upper right portion shows a close up of the flying saucer craft as it was departing, after having earlier been on the ground where JP met its Nordic extraterrestrial occupant.

Figure 68. Photo courtesy of "JP" (zoom insert added)

# CHAPTER 20

## USAF SSP Shoots Down Deep State "False Flag" Missile Attack on Hawaii

> When the will defies fear, when duty throws the gauntlet down to fate, when honor scorns to compromise with death – that is heroism.
>
> – Robert Green Ingersoll

### State of Hawaii Issues Ballistic Missile Alert on January 13, 2018

On a typically warm and sunny morning in the state of Hawaii, as island residents and a multitude of vacationing tourists on holiday began their Saturday activities, cell phones across Big Island to Kauai suddenly blared an emergency alert message. It was January 13, 2018, at 8:07 am when the unthinkable happened: an alert that warned of an incoming nuclear missile heading straight for America's Pacific jewel. Stunned people, young and old, trembled as they read the terrible message that said in all caps: "BALLISTIC MISSILE THREAT INBOUND TO HAWAII. SEEK IMMEDIATE SHELTER. THIS IS NOT A DRILL". In addition, television broadcasts by affiliates of CBS (KGMB) and NBC (KHNL) in Hawaii were interrupted by an emergency message stating that the missile alert emanated from the US Pacific Command. A KGMB television video clip was uploaded to Twitter, and the red ticker-tape alert at the top read as follows:

> A civil authority has issued A CIVIL DANGER WARNING for the following counties or areas: Hawaii: at 8:07 AM on JAN 13, 2018. Effective until 6:07 PM. Message from IPAWSCAP. The U.S. Pacific Command has detected a missile threat to Hawaii. A missile may impact on land or sea within minutes. THIS IS NOT A DRILL. If you are indoors, stay indoors. If you are outdoors, seek immediate shelter in a building. Remain indoors well away from windows. If you are driving pull safely to the side of the road and seek shelter in a building or lay on the floor. We will announce when the threat has ended. Take immediate action measures. THIS IS NOT A DRILL. Take immediate action measures ...[586]

Residents near Honolulu, the expected ground zero of any nuclear attack, fled for their lives to more remote areas of Oahu expecting the worst. Then after 38 minutes of sheer terror for many all over the Hawaiian Islands, residents were notified by the emergency alert system that it was all a false alarm.

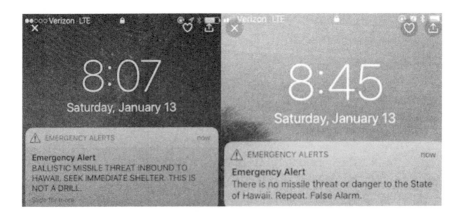

Figure 69. Emergency Alert messages issued by the civil authority to Hawaii residents

Someone within the Hawaii Emergency Management Agency (HI-EMA) had pressed the wrong button according to local State authorities.[587] On January 30, the Governor of Hawaii David Ige gave a press conference about the results of an official investigation into the January 13 ballistic missile alert that was allegedly *mistakenly issued*.[588] Brigadier General Bruce E. Olivera (ret.) wrote the report for the investigation which was issued to Major General Arthur J. Logan, Director of HI-EMA. The synopsis said:

> At approximately 8:06 am, the SWP (HI-EMA State Warning Point) mistakenly issued a BMA [Ballistic Missile Alert]. As the investigating officer and carefully considering the facts, I find a preponderance of evidence exists that insufficient management controls, poor computer software design, and human factors contributed to the real-world BMA and the delayed false BMA correction message that was issued over the WEA/EAS [Wireless Emergency Alert/Emergency Alert System] on January 13, 2018.[589]

According to the investigation, a drill was spontaneously scheduled by overnight SWP shift supervisors for the beginning of the morning shift (shortly after 8:00 am) and proposed the following timeline for what happened:

> 0800 – In preparation for the BMA [Ballistic Missile Alert] response drill, Employee 4 and Employee 2 discussed their plan to conduct a BMA at the end of their shift. Employee 4 prepared a recording of PACOM BMA notification script from the BMA Checklist manual. Employee 4 discussed leaving the SWP area while Employee 2 told the dayshift about the previous shift change drill.

It's important to emphasize here that the investigation directly implied that no actual warning came from Pacific Command (PACOM). Instead, the PACOM notification was said to be merely a "notification script" written in-house by Employee 4. This means that as far as PACOM was concerned, there was no direct connection between the HI-EMA drill and the security measures implemented at Pearl Harbor and other PACOM facilities. The timeline continues:

> 0803 – The incoming day shift entered the SWP. Employee 4 met with Employee 5 and discussed the BMA drill.
>
> 0806 – Employee 4 initiated the BMA drill using a phone outside the SWP area calling into the SWP STE [State Warning Point Secure Telephone Equipment]. Employee 2 activated the STE speaker. At this time, it was announced loud and clear, "EXERCISE, EXERCISE, EXERCISE," and then concluded with, "EXERCISE, EXERCISE, EXERCISE," which is a normal procedure for all drills including this BMA drill....
>
> 0806 – Employee 1 logged into the AlertSense system and waited for Employee 3's announcement of the simulated siren warning activation. Following the simulated siren announcement, Employee 1 erroneously activated the real-world Alert Code.
>
> - There is a drop-down menu that includes:
> - Test missile alert (this sends message internally to agency)
> - Missile Alert (this sends message to public)
> - Missile alert was selected.
> - The computer asks to confirm choice.

  o Employee 1 clicked yes.

 0807 – SWP began receiving WEA/EAS message on their personal smartphones

The official timeline of events went on to state that US Pacific Command notified Hawaii State authorities that there was no missile launch at 8:10 am, 3 minutes after the warning went out. Local authorities attempted to inform the public via a variety of means that the ballistic missile alert was a false alarm. But it was only at 8:45 am, 38 minutes after the initial alert, that a second emergency alert was sent over the Public Alert and Warning System announcing the mistake. The official investigation concluded that the emergency alert was a false alarm due to human error caused by a HI-EMA employee who had confused a hastily organized drill with a real-life ballistic missile attack. Indeed, the report emphasized that the responsible employee "has confused real life events and drills on at least two separate occasions".[590] Additionally, the report attributed the 38 minute delay between the alert being issued and the cancellation notice going out due to further human error:

> HI-EMA mistakenly believed that it had to consult with FEMA [Federal Emergency Management System] to issue an event code which contributed to the 38 minutes to issue an official notification that the threat was false. HI-EMA has corrected this concern and understands that it is the HI-EMA's authority to issue an official notification.[591]

There are a number of problems with the official Hawaii State narrative which tells us it was merely a HI-EMA in-house drill without any link to an actual notification received from PACOM about an inbound ballistic missile. First, there are eyewitness reports of Naval Station Pearl Harbor activating its emergency sirens and evacuating personnel. For example, one eyewitness staying next to the Pearl Harbor naval base described in a video how the base's siren system went off during the alert, and he watched personnel being evacuated.[592] This was later confirmed with the release of an email from Admiral Harry Harris to a senior Air Force officer about the missile alert in which he described the siren (aka Big Voice) going off:

> "Just for my education and edification, when the Big Voice went off at Pearl Harbor-Hickam this morning, there was no indication that this was a drill; in fact just the opposite... So, what happens on the flight line, and what message, if any was passed to aircraft in the air?"[593]

The siren going off at Pearl Harbor contradicts the Hawaii investigation's report that PACOM did not issue the alert and it came only from local Hawaiian authorities. The conclusion emerging here is that PACOM had issued a missile alert to its facilities, the message was then immediately relayed to HI-EMA which put out its public alert message at 8:07 am. Afterward, PACOM issued the 8:10 am cancellation that once again went to HI-EMA, which then began declaring a false alarm.

Second, the information provided during the interrupted television broadcasts by CBS and NBC affiliates contained several more details about the threat received from PACOM, which the stations used in their public alert messages. Details such as "a missile may impact on land or sea within minutes" was missing in the official Hawaii State alert message that only stated: "BALLISTIC MISSILE THREAT INBOUND TO HAWAII", and had no information about it making its impact "within minutes", as announced on the television broadcasts. This means that the television networks acquired their information from sources other than HI-EMA. It also points to PACOM as the original source for putting out the alert information that was picked up by television and radio sources. The additional information relayed in the television broadcasts suggests that PACOM was tracking an inbound missile and was informing military personnel and members of the media that impact was only minutes away. This incongruity between the additional information transmitted by the television stations and the role played by PACOM has been completely ignored in the Hawaii State report.

A third problem with the Hawaii Report is that an anonymous Airman working as a Fusion Analyst at Joint Base Pearl Harbor-Hickham, who was on duty at the time of the alert, posted his/her experiences on the *4chan forum*. S/he pointed out the great amount of confusion over the alert which was at first considered by US Pacific Command personnel to be a drill, and then soon after, a genuine attack. This would explain why the television broadcasts had additional information: PACOM personnel had shared it immediately after they were themselves informed of an "official attack". Significantly, s/he points out that it was only after the White House replied "Negative" to a request for a military response to the presumed attacker, North Korea, that local authorities in Hawaii were instructed by the Governor to say it was all a false alarm. This is what the Airman posted at 7:15 pm on the day of the alert:

> Hickam AFB Fusion Analyst here. The false alert was not a mistake. It was ordered. We were informed it was to be a drill, but then all information was put out that the threat was real. Immediate contact was to the White House and requests were made for retaliation. I think there was a push to have the White House approve an attack on foreign vessels off the coast. When the White House replied "Negative" and demanded further information, my superiors called Gov. David Ige where he ordered us to contact media outlets of there being a false alarm. One of the Gov. Aids was present and

speaking with who looked to be a federal investigator. I overheard them state this "Demonstrated weakness in the Trump admin and a refusal to protect the people."[594]

The Hickam airman's information reveals that the alert was not merely an in-house drill hastily concocted by shift coordinators at the HI-EMA facility, as indicated in the official report. The drill was coordinated with PACOM all along, and PACOM personnel were notified at some point that the drill had metamorphized into an actual ballistic missile attack that was only minutes away. This would explain the television messages announcing a missile impact in only a few minutes, whereas the HI-EMA alert at 8:07 am had only the original drill message warning of an inbound ballistic missile. Put simply, the television broadcasts were reporting on further information provided by PACOM after the HI-EMA alert had gone out.

Furthermore, the confusion over whether the missile alert was only a drill or a genuine strike has all the tell-tale characteristics of a "false flag" attack. These are known to be implemented under exactly these same circumstances. Similar confusion arose amongst security personnel in past false flag events taking place on September 11, 2001, (New York) and July 7, 2005, (London), etc. In these alleged terrorist attacks, military and government authorities were first told a drill was taking place where security forces were instructed to stand down, and then everyone was informed that a real attack was occurring by which time it was too late to prevent the attack. Analysis of this pattern observed in past alleged false flag attacks has led to the conclusion that: "The exercise or drill – at the same time, at the same place – has become the *sine qua non* or indispensable element of the recent false flag operation."[595]

The final problem with the official Hawaii report is that eyewitnesses claim they saw a meteorite, or something like it, exploding high in the sky over the Hawaiian Islands around the time of the alert. Several witnesses also reported watching a television story about an exploding meteorite or object before the news story was pulled. Such reports were ignored by the mainstream media which exclusively focused on the Hawaii State authorities' explanation for the problem occurring due to one person pressing the wrong button. This is why alternative media reports of an intercepted nuclear missile attack need to be considered. Importantly, they raise another scenario that better fits the true sequence of events which took place on that Saturday morning: the USAF SSP had intervened to neutralize a false flag attack orchestrated by the Deep State, and this classified information was covered up.

## Alternative Media Reports & Witnesses Point to Genuine Missile Attack

A number of alternative media outlets citing anonymous sources reported that there was a real missile launched against Hawaii. Among the first was a news report appearing on the *Operation Disclosure* blogsite on January 13, 2018. It was titled *RV/Intelligence Report* and said:

> Missile launches were detected in the Pacific Ocean off the coast of Hawaii. The launches originated from the same anomaly detected yesterday, Jan. 12. The missiles were immediately intercepted and destroyed. The anomaly was revealed to be a nuclear stealth submarine. The nuclear stealth submarine was located and destroyed shortly after the attempted attack.[596]

On January 16, a still active US military officer known to former USMC and CIA intelligence analyst, Robert David Steele, came out under the pseudonym "DefDog" to state publicly that the Hawaii event was an intercepted nuclear missile attack.[597] After explaining how the emergency alert system is designed to prevent the kind of human error claimed by state authorities, DefDog wrote:

> There was a missile. Probably fired from a submarine under the control of individuals loyal to the Deep State ... The missile was not fired from North Korea or by North Korean forces. The missile was intercepted. Then the coverup began.[598]

An additional alternative news media source corroborating key aspects of the thwarted missile attack against Hawaii is a veteran journalist and orthopedic surgeon, Dr. Dave Janda.[599] He revealed in a radio interview what a "Deep Level source" known to him for years had told him about the Hawaii missile incident: "What this source told me was that, and again I've never found this source to be wrong, was that there was actually a missile fired."[600]

There are also witness reports of something being destroyed high over the Hawaiian Islands during the morning of the alert. In a video published on January 14, 2018, Marfoogle Watutu (a pseudonym) says he was told by his sister, a Maui resident, about a tourist boat which carried multiple eyewitnesses to the exploding meteorite:

> My sister has lived in Maui for eight years and what she is finding out is that a group of boaters that actually run a tour company say now that **they saw something get blown out of the sky**.
>
> So, they said it looked like a meteor and then all of a sudden there was a big boom and **it lit up the entire sky**. It was 8 o'clock in the morning...
>
> **Maybe this was not a drill after all**. Maybe our Boys in Blue and Boys in Green shot it down. And I would think that the U.S. would not want to cause more

panic if they did try to launch something at Hawaii. Then, of course, they would say it was a mistake. They took responsibility very quickly that it was a mistake... This could be bullshit but I just had to relay that...[601]

Some of the tourists were interviewed by a local news channel that purportedly aired the story once before it was pulled.

In the 500+ comments to Marfoogle's video, various people claimed they were also an eyewitness to the exploding object in the sky or saw the television news story. What follows are three independent comments that cast light on what really happened during the missile alert.

> I was awakened by a Thundering sound, thought it was 7:00AM but it was 8:00AM. My clock is shaded from where I sleep. I live on the N.Shore of Kauai. We did not receive a cell phone call, my Son called to inform us, he lives in town. We should have had a Civil Defense Siren sound but there was None! Definitely something was Not normal in the Civil Defense Warning System. Many People do not have cell phones in rural area's. These siren's are tested every 1st business day of Every Month, for as long as I have lived here in Hawaii, 32 year's. (Janie Langley)

> So I live on Island and our family has a lighthouse, we all saw a bright flash that morning as well. There is much buzz about this ... Thank you for actually putting what islanders are all talking about. (Ashly P)

> My cousin lived there, she came back to the states because of this, and she told me the same thing ... that something was shot out of the sky that morning. (Jaycen Leapline)[602]

The next four comments focus on an alleged KHON2 news story about the incident, the first of which refers to a tourist boat and an exploding object that was covered in the story.

> You are right. I did see a video of the tourists on [the] boat and there was something blown out of sky. But it was deleted right away. I stay [on] Molokai, Hawaii. Also there is more than one button to push. Seen that video also and now it is gone... [Yoshi Copeland]

> I'm on Oahu, there was a news story and then it cut straight to commercial, I am a sceptic usually, but I did see this my self. [Themexican21491]

> I don't know why people are doubting this, it WAS on the local news and then the feed was cut ... at least 10 fellow islanders have commented here agreeing with me... [Alex Woo]

> I saw something about it on the news, but it has since been taken down ... no surprise there [Kevin Brown] KHON2 had a snippet of this but its no longer available? A lot of people on Island are saying they saw lightning in the morning of the false alarm, very weird... the sky was blue as ever. [Darren Cultis] [603]

Since I have resided in Hawaii since 2004, I investigated Marfoogle's claims using my own local network. Local residents told me that witnesses were being visited and intimidated by police and other authorities who told them to remain quiet. In addition, I spoke to a Maui boat captain who believed that something had been shot down because the week following the alert, she had never seen so many US Coast Guard vessels scouring the area. Such reports help to explain why the only sources to emerge so far have been anonymous.

Let's come back to the anonymous Fusion Analyst airman at Hickham AFB who worked in a sensitive military intelligence position. The Airman described how the Trump White House requested further information rather than choosing to retaliate against the alleged source of the ballistic missile attack. It appears that cooler heads at the White House did not order military retaliation against North Korea after the missile was shot down. President Trump and his national security advisors likely suspected that a false flag attack had been attempted. Indeed, this is exactly what another anonymous military intelligence source was telling the public via online postings, operating under the pseudonym "QAnon".

## "Q" Exposes False Flag Attack Meant to Start WWIII

"QAnon" (aka "Q") is widely regarded to be a group of military intelligence officers working within the Trump White House who have been given the authorization to leak classified information to expose the Deep State and those seeking to undermine the Trump administration.[604] QAnon's posts first appeared on the *4chan forum* on October 28, 2017, and then moved to the *8chan forum* just over a month later when the original site was compromised.[605] Q appears to have predicted an upcoming false flag missile attack in six posts on January 7, 2018, six days before the Hawaii missile alert. One post referred to DEFCON 1 and five others to DEFCON [1], which is a defense readiness condition used by the US Department of Defense.[606] DEFCON 1 signifies nuclear war is imminent and is the most serious state of defense readiness. In post 500, Q writes:

> DEFCON 1 ...
> [non-nuclear]
> [1] OWL [1]
> Q[607]

Q is suggesting that a staged or false flag nuclear attack against the U.S. will be thwarted because of the intelligence gained through the US military intelligence community's eyes (the "OWL" reference likely refers to the NSA, NRO, etc.) about such an upcoming event.

On the evening of January 13, after the false Hawaii missile alert, Q linked the alert to the previous DEFCON 1 or [1] posts that had been made on January 7. Here is the first part of Q's post #520:

> BIG news week?
> Future proves past.
> What news was unlocked?
> Do you believe in coincidences?
> >U1
> What public disclosure occurred re: U1?
> >DEFCON 1 **[non-nuke FALSE]**
> COMMAND?
> WHY?
> NECESSARY?
> NO SUCH AGENCY.
> Where did POTUS stop **[post]** ASIA?
> IT WAS NECESSARY.
> FOR GOD & COUNTRY.
> IT WAS NECESSARY.
> NO OTHER VEHICLE TO REGAIN ENTRY.
> :**[AGAIN]** direct pre-knowledge.
> :**[AGAIN]** warning ALERT.[608]

Q's reference to a "Big News Week" and "Future proves past" is informing us that issues raised in previous posts had just been proven true. One of these involved the "DEFCON 1 [non-nuke FALSE]" issue. Q was linking directly to the earlier January 7 posts and saying that the event that occurred in Hawaii on January 13 was related to the predicted "DEFCON 1 [non-nuke FALSE]" event. One way of interpreting Q's usage of a double negative "non-nuke

FALSE" is that the "false missile alert" was, in fact, a real nuclear missile attack on Hawaii. This is consistent with the whistleblower and witness reports discussed earlier and is also substantiated by later Q posts, which will soon be discussed. First, however, I need to analyze an alternative interpretation of Q's posts on the Hawaii missile alert to get a more comprehensive picture of what really happened.

An alternative explanation focuses on Q's reference in post #520 to the NSA ("No Such Agency") and Trump's inability to get access to classified files being held at the NSA's Honolulu facility, which allegedly was controlled by the Deep State. According to several Q posts, Edward Snowden is really a CIA asset whose mission has been to penetrate the NSA, and to eventually leak files from its Honolulu facility to the global media to undermine the NSA's surveillance of the Deep State/CIA.[609] Apparently, Trump's short visit to Honolulu, which occurred while *en route* to Asia for his November 2017 tour, was to specifically go to this secure NSA facility and use his Commander in Chief authority to get access to all the files exposing Deep State malfeasance.[610] Q's references to "Command?" and "BDT" (Bulk Data Transfer) in a January 7 post suggests that Trump had been denied access and a workaround was put into place involving a security protocol involving "BDT". [611]

Consequently, one interpretation of the January 7 post is that the Hawaii missile alert was an attempt by the Trump administration to implement the BDT security protocol so that all the NSA files held at the Honolulu facility could be transferred back to NSA headquarters at Fort Meade, Maryland. This interpretation was first raised by Jordan Sather, who runs the popular "Destroying the Illusion" YouTube channel, where he was among the first to analyze Q's posts. Sather's interpretation has been summarized as follows:

> Documents necessary for freeing America from the
> iron grip of the Deep State were being held hostage
> in Hawaii. It is there that the NSA maintains one of

their most important facilities. Because of their hostility and the MAGNITUDE of their corruption, the documents requested were not being surrendered to the Trump Administration, and were most likely in danger of being destroyed.

The only way to obtain these documents was to trigger a bulk data transfer (BDT). This could happen if the Emergency Missile System were activated. In this scenario, a fail-safe measure comes into play: When the alert is activated, all of the data is transferred to the Pentagon, in the form of a BDT.

Is it possible that the fake missile alert (that occurred in Hawaii on January 13th), was a necessary event, enacted to trigger the bulk data transfer from the NSA in Hawaii, to the Pentagon...into the hands of the White Hats?[612]

The big problem with this interpretation is that, if correct, the Deep State would have been more than happy to leak this information concerning what the Trump administration had done to major media. The Deep State's control over the mass media would have ensured round the clock coverage of the scandal, which would have surely led to Trump's impeachment.

While insufficient on its own, this interpretation raised by Sather helps to understand some important background to the Hawaii missile alert. When combined with all the evidence offered so far about a real ballistic missile being sent and shot down while *en route* to Hawaii, another motivation crystalizes to further expose the most likely actor behind the attack – the Deep State/CIA.

The Deep State was well aware that Trump had failed in his first attempt to get access to the NSA facility in Honolulu and feared that he would succeed on his next try. Sather pointed out that the files had all the incriminating evidence of what the Deep State/CIA had been doing through its multiple assets, including Snowden. The Deep State's solution was to destroy the evidence. A nuclear attack on Honolulu would not only destroy all the accumulated NSA evidence that the Deep State needed to keep out of the hands of Trump's White House, but it would also serve as a pretext to instigate a Third World War.

## "Q" Pulls the Curtain on the Puppet Masters Controlling the False Flag Attacks

The day after the Hawaii missile alert, Q's post #538 included a warning about an upcoming week of false flag attacks by the Deep State: "BEWARE of MAJOR FALSE FLAG attempts this week". [613] This implied that the Hawaii incident was part of a series of false flag attacks and not a mere false alarm as concluded by the official Hawaii State investigation. This post is particularly significant given that it appeared only two days before what in all likelihood was an attempted false flag attack against Japan.

On Tuesday, January 16, 2018, a *New York Times* report referred to another nerve jarring false alarm taking place, this time in Japan where residents were warned by a major TV broadcaster to take shelter from an incoming ballistic missile:

> Japan's public broadcaster on Tuesday accidentally sent news alerts that North Korea had launched a missile and that citizens should take shelter — just days after the government of Hawaii had sent a similar warning to its citizens. [614]

The false missile alert incident in Japan has many parallels to the one three days earlier in Hawaii which was thwarted. Q's January 14 post #538 established a link between the Japan and Hawaii missile alerts as false flag events attempted by the Deep State.

Further support for this linkage comes from Pentagon sources as reported on by former *Forbes* magazine bureau head for Japan, Benjamin Fulford, who described what the Deep State/Cabal was trying to do with their false flag attacks:

> Our here-now Pentagon sources admit that "the Secret Space Program may have been used to down nuke missiles aimed at Hawaii and Japan and to stop the cabal from leaving earth."[615]

QAnon next alluded to the Hawaii and Japan ballistic missile alerts on February 11, 2018, in a long post (#725) referring to a recent false flag missile attack intended to start a war with North Korea. A section of this post appears below:

> Ask yourself, do we want a WAR?
> Ask yourself, who is trying to start a WAR?
> Ask yourself, if a missile was launched by rogue actors, what would be the purpose?
> Ask yourself, what would/should immediately start a WAR?
> Ask yourself, would the PUBLIC understand the following statement: "Rogue actors (Clowns/US former heads of State) initiated a missile launch in order to 'force' the US into a WAR/conflict against X?"
> Be the autists we know you are....
> Q[616]

It is worth examining QAnon's questions in light of what has been presented about the Hawaii and Japan missile alerts, and an Executive Order issued by President Trump on December 21, 2017. QAnon wrote: "Ask yourself, who is trying to start a war?" The answer is individuals and groups tied to the Deep State who were directly targeted by the extraordinary Executive Order issued by

Trump on December 21 declaring a national emergency and a freeze on the financial assets of anyone involved in human rights abuses and corruption anywhere around the world.[617] Individuals and groups responsible for human rights abuses and corruption in the U.S. and globally had the most to lose if their financial assets were frozen. A nuclear war escalating into a world war would have neutralized Trump's Executive Order.

QAnon then wrote: "Ask yourself, if a missile was launched by rogue actors, what would be the purpose?" The purpose was to create a false flag attack where the blame would be pinned on a credible state capable of launching a ballistic missile that could hit Hawaii and Japan: North Korea. The "rogue actors" really responsible for the attack had created a scenario where the US military would be forced to respond, thereby opening the door to rapid military escalation as other major nations intervened.

QAnon next pointedly quizzes: "Ask yourself, what would/should immediately start a WAR?" We know from World War II history that the December 7, 1941, surprise attack on Pearl Harbor, Hawaii, home of the US Pacific Fleet was the catalyst for America entering the war. The next day, the US Congress overwhelmingly voted to declare war on the Empire of Japan. QAnon is telling us here that a successful nuclear missile attack on Pearl Harbor would have triggered a similar war declaration by Congress against the alleged perpetrator – North Korea.

Were the "rogue actors" behind the attacks also hoping that even an unsuccessful nuclear missile attack in the event the missile was shot down would trigger war? This possibility arises from what the anonymous Fusion Analyst serviceman at Hickham Air Force Base said happened after the missile was shot down. Most significantly, the analyst claims s/he overheard a Federal Investigator and a Hawaii state official say Trump's refusal to order a retaliation "Demonstrated weakness in the Trump administration and a refusal to protect the people". [618] Clearly, several actors involved in the Hawaii ballistic missile alert were trying to manipulate the Trump administration and US Pacific Command.

QAnon's next question is very important for identifying who was behind the missile attacks against Hawaii and Japan:

> Ask yourself "would the PUBLIC understand the following statement: Rogue actors (Clowns/US former heads of State) initiated a missile launch in order to 'force' the US into a WAR/conflict against X?"

In QAnon posts, the CIA is referred to as "Clowns in Action", and former Presidents Bush (Snr), Clinton and Obama have been categorized as Deep State/CIA assets. Thus, in post #725, "Clowns" is QAnon's codeword for the CIA. In a January 20, 2018, article published online at exopolitics.org, I provided documentary evidence supporting insider claims that the CIA's clandestine services division had created a shadowy naval fleet with its own ballistic missile carrying submarines, aircraft, and even an aircraft carrier battle group, which altogether comprise a secret "Dark Fleet".[619] Therefore, QAnon was suggesting that a CIA-controlled submarine was directly involved in staging a false flag attack against Hawaii and Japan using nuclear missiles. This finally brings me to the role played by the USAF SSP in shooting down the Deep State/CIA ballistic missile.

## USAF Space-Based Weapons Platform

Initial speculation concerning who was responsible for intercepting the ballistic missile targeting Hawaii on January 13 focused on an anti-ballistic missile defense system run by the US Navy. Only five months earlier, the Navy had successfully intercepted a medium-range ballistic missile target in a test off Hawaii that was launched from the Pacific Missile Range Facility on the island of Kauai.[620] This test followed a similar failed test that had been launched from the same facility in June 2017. Then, on

January 30, 2018, the Navy announced that it had conducted another exercise in Hawaii to intercept a ballistic missile using its Aegis class anti-ballistic missile defense technology, which once again failed.[621] Since the exercise marked the second failure by the Navy to intercept a ballistic missile attack in the past year, the Navy's antiballistic missile defense was shown to be unreliable. Given that the exercise was conducted less than three weeks after the January 13 incident, it is clear that the Navy did not currently have the capacity for reliably intercepting ballistic missiles. The announcement of the failed exercise so soon after the Hawaii attack raises an intriguing question. Were Naval leaders signaling they were not responsible for shooting down the January 13 missile? If so, then who did shoot down the rogue missile?

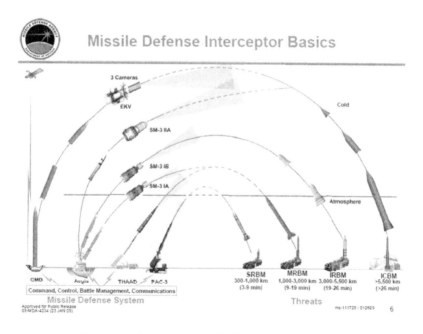

Figure 70. Comparison of Aegis and other defense systems. Source: Missile Defense Agency

On January 17, 2018, I published an article on the Hawaii missile attack focused on the likelihood that an SSP run by the US

Air Force had used space-based weapons technologies to shoot down the ballistic missile.[622] Photos taken on October 23, 2017, showed a rectangular-shaped UFO in the vicinity of MacDill AFB in Tampa, Florida. The photographer JP claims he had been taken inside the rectangular-shaped UFO and informed that it was a weapons platform.[623] Further, JP states that he witnessed uniformed military personnel wearing the insignia of Air Force Special Ops who were aboard to operate the weapons platform.

It is highly probable that the kind of weapons in use on the platform were directly linked to the technology of the MASER satellites used by the Deep State in steering Hurricane Irma. Importantly, these microwave technologies are capable of being used for both offensive and defensive military purposes. The existence of the large rectangular USAF SSP space-based advanced weapons platforms, likely including MASER technologies, has incredible implications. A "Directed Energy Weapon" system placed on a space-based weapons platform would have the capacity to reliably shoot down any ballistic missile, and thwart Deep State efforts such as those intended to simulate an attack from North Korea. Unlike the "anti-missile" missiles used in the Navy's Aegis defense system which have failed twice in tests against simulated ballistic missile attacks, a Directed Energy Weapon could send a pulse of laser energy at the speed of light. Therefore, a Directed Energy Weapon could easily intercept a ballistic missile in the terminal or re-entry stage of its flight trajectory when it travels at hypersonic speeds (5x speed of sound).

What gives further credence to the possibility that Maui residents and tourists witnessed a missile being shot down by a space-based Directed Energy Weapon is the USAF surveillance installation on the summit of Haleakala, Maui. The capabilities of the Maui Space Surveillance Site (MSSS) is described by GlobalSecurity.org as follows:

> The Maui Space Surveillance Site (MSSS) includes the Air Force Maui Optical Station (AMOS) is an

asset of the US Air Force Materiel Command's Phillips Laboratory, the Maui Optical Tracking and Identification Facility (MOTIF), and a Ground-based Electro-Optical Deep Space Surveillance (GEODSS) site operated by US Air Force Space Command....

The site is the only one of its kind in the world, combining operational satellite tracking facilities (MOTIF and GEODSS) with a research and development facility (AMOS). The MSSS operates primarily at night, but performs many of its SOI missions 24 hours a day.[624]

It is very likely that the Maui Space Surveillance Site was used in the tracking and shooting down of the ballistic missile launched against Hawaii. This would help explain why the television broadcasts included the information: "A missile may impact on land or sea within minutes". The incoming missile was being tracked at the time by the MSSS. If indeed so, this would further confirm why many Maui residents, in particular, saw the explosion. It was only during the final stage in the trajectory of the attacking missile that a Directed Energy Weapon could have been used, and the Maui facility likely helped to triangulate the missile's position accurately.

## Did the USAF Receive Help from Nordic Extraterrestrials in Shooting Down the Missile?

There are alternative scenarios to be considered because of what appeared onscreen during a live CNN broadcast covering the false alarm in Hawaii. Viewers took photos of their screens during the coverage which displayed several UFOs over Hawaii thirty-one minutes after the second emergency alert had been issued. One photo and a close-up presented here shows five UFOs with at least one exposing an advanced technology in the form of a circular pulse, or plasma ring, similar to the one made by a TR-3B antigravity craft (see Figure 71). This makes it possible that these craft belonged to the USAF SSP that had broken away from Deep State control. Another explanation is the craft are extraterrestrial in origin and are following a decades-long policy held by the Nordics – secretly intervening in human affairs to forestall nuclear weapons incidents. This is what the person who provided the photos and their genesis, Solaris Modalis, contends.[625]

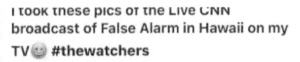

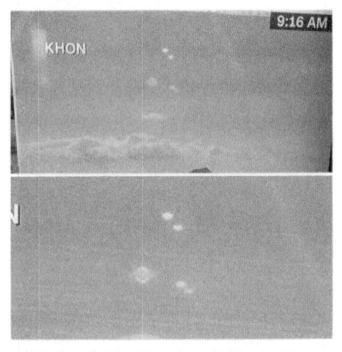

Figure 71. Screenshot of UFO's seen during live CNN broadcast
from Hawaii. Courtesy of Solaris Modalis

The evidence of UFOs, whether they are craft belonging to a USAF SSP or extraterrestrial in origin, is very significant given the newly formed alliance between the Nordics and the USAF SSP. It is very likely that the Nordics coordinated with the USAF SSP in thwarting the intended false flag attacks against Hawaii, and soon after, Japan. Indeed, this UFO incident may offer the key to how the incoming ballistic missile was shot down.

It's highly possible that the Deep State responded to the defection of the USAF by launching a nuclear weapons attack against Hawaii, which hosts the largest concentration of US military assets anywhere in the world. After the failed attempt against Hawaii, another attempt may have occurred with Japan if Q and other sources are to be believed. In the end, the evidence presented suggests a high likelihood that the attacks were thwarted, and the culprits taken out by advanced weapons and craft belonging to or affiliated with the USAF SSP, which ironically, the Deep State played a hand in creating.

## Conclusion

In contrast to the official State of Hawaii report on the January 13, 2018 incident, many other alternative conclusions can be drawn.

- First, a ballistic missile likely carrying a nuclear warhead was destroyed during the final stage of its trajectory by the USAF SSP using a space-based "Directed Energy Weapon" deployed from a rectangular-shaped weapons platform, and/or satellite.

- Second, the Maui Space Surveillance Site helped in the tracking and destruction of the ballistic missile as it neared the Hawaiian Islands to ensure it would not reach its most likely target – Pearl Harbor, Honolulu.

- Third, the drill scheduled for the morning of January 13 was intended to create confusion so that the real ballistic missile "false flag" attack could succeed, and the missile's trajectory would point to North Korea as the culprit.

- Fourth, the party responsible for the false flag attack is a rogue Deep State/CIA group with its own ballistic submarines, which are part of a secret naval force called the "Dark Fleet".

- Finally, the US Navy's January 31 anti-ballistic missile test was intended to let the general public know that its Pacific Missile Range Facility on Kauai did not yet have the capability to intercept ballistic missiles reliably, thus steering attention away to look for the real heroes in tracking and intercepting the nuclear attack on Hawaii — USAF Special Operations.

Further information is needed to determine the accuracy of these conclusions. However, one thing is pretty clear at this point given the number of insider sources and witnesses that have come forward so far. The official State of Hawaii investigation is a cover-up for a ballistic missile attack, which evidence suggests was a false flag attack using one or more nuclear weapons by a mysterious naval force desiring to start a regional nuclear war. Thankfully, this did not occur. A classified space-based weapons system was most likely employed by USAF Special Operations, thereby giving US national security planners time to develop an appropriate response to those behind the false flag attack on Hawaii, and the one that followed only three days later in what may have been a similar attack on Japan. The possible involvement of a USAF SSP, acting alone or with extraterrestrial assistance in thwarting false flag attacks against Hawaii and Japan, provides confidence that future false flag attacks using conventional military technologies are unlikely to succeed.

# CHAPTER 21

# Space Force Opens the Door for USAF SSP Disclosure

> Good courage in a bad affair is half of the evil overcome.
>
> —Titus Maccius Plautus

## President Trump Proposes Space Force as New Branch of US Military

On June 18, 2018, President Donald Trump gave a speech in which he recognized space as a "war-fighting domain" and called for the development of a United States Space Force (USSF) which will take over the current space functions of the US Air Force and other military services. The proposed Space Force is to become the sixth branch of the US military and will be placed under the administrative control of the Department of the Air Force, which is said to possess approximately 80% of all the US military space assets.[626] At the National Space Council meeting held at the White House, Trump said:

> We must have American dominance in space... I'm hereby directing the Department of Defense and Pentagon to immediately begin the process necessary to establish a Space Force as the sixth

branch of the armed forces…. We are going to have the Air Force, and we are going to have the Space Force. Separate, but equal. It is going to be something so important.[627]

Trump then commanded General Joseph Dunford, Chairman of the Joint Chiefs of Staff, to "carry that assignment out."[628]

Trump's stance appeared to be at odds with previous statements made by the US Air Force Secretary, Air Force Chief of Staff, and Secretary of Defense James Mattis, who all came out in opposition to a similar proposal by Republican members of Congress in early 2017 to establish a Space Corps.[629] On October 17, 2017, Mattis wrote in a letter to the Senate's Defense Committee explaining his opposition:

> I oppose the creation of a new military service and additional organizational layers at a time when we are focused on reducing overhead and integrating joint warfighting efforts.[630]

Even Trump's White House was opposed to the creation of a "Space Corps", saying back in July 2017 that it was "premature at this time."[631] Yet by March 13, 2018, Trump had reversed himself and said in a speech at the USMC Air Station Miramar, San Diego:

> My new national strategy for space recognizes that space is a war-fighting domain, just like the land, air, and sea. We may even have a Space Force. We have the Air Force. We'll have the Space Force. We have the Army, the Navy.[632]

Something had apparently changed in the time between the emergence of the Congressional initiative in early 2017 and President Trump's surprising speech on March 13, 2018, during which he proclaimed with rallying confidence that the creation of a

"Space Force" was a good idea. In fact, in his speech, Trump referred to off-handedly suggesting the "Space Force idea" himself in a discussion that took place at the White House, which then led to its serious consideration. This means the people in his circle of advisors concluded that Space Force *would* serve Trump's administration and USAF interests.

Clearly there is more to this story, and the dramatic change came about more likely due to USAF SSP leaders realizing over this period the full extent of the deception conducted by the Deep State. Not only did the Deep State hide the truth of its own and other space programs, but it also concealed the role major aerospace corporations had played in making all this possible through an almost impenetrable compartmentalization process implemented for "Deep Black" aerospace projects. Also, Trump and USAF SSP leaders were well aware of the Deep State's use of Maser satellites and other weather modification technologies used in steering Hurricane Irma towards MacDill AFB in September 2017; and the Deep State's nefarious role in the Hawaii ballistic missile attack on January 13, 2018.

Critically, however, the full extent of the space assets belonging to the Deep State and its Interplanetary Corporate Conglomerate (ICC) partners was not known. Something had to be done to assert US military authority over these rogue elements to eliminate future threats by the Deep State and its Antarctic German/Draconian/Corporate allies. Consequently, the advantages of creating a "Space Force" escalated. Now seen with new eyes, it represented a means of allowing the USAF to assert authority over all US military-related activities in space by coordinating all of these under one authority. This alone could lead to the elimination of the decentralized compartmentalized system that had evolved over decades in the Military-Industrial Complex, which had greatly benefited the Antarctic Germans' "Dark Fleet" and led to the creation of the ICC that remained closely tied to the Deep State.

USAF officials and Secretary of Defense James Mattis quickly re-aligned themselves with Trump's June 18, 2018, order to actualize his Space Force vision.[633] While Trump did not have the executive authority to create a new military service, since only the US Congress possesses the power to do so, he did have the authority to order the Pentagon to develop a comprehensive plan for the creation of such a service.[634] Thus, Mattis and the Pentagon launched into putting together an all-encompassing plan for the creation of a Space Force, which now importantly carried the backing of the President and military leaders for its upcoming submission to Congress.

## Department of Defense Report for the Creation of a Space Force

On August 9, 2018, the Department of Defense (DoD) released a final report to the Congressional Defense Committees outlining the steps to be taken in implementing Trump's proposal. The "Final Report on Organizational and Management Structure for the National Security Space Components of the Department of Defense" identified two main phases for developing the Space Force through its four constituent components:

> Establishing the Space Force will be multi-dimensional and phased. In this first phase, using existing authorities, the Department of Defense will establish several of the component parts of the Space Force. The second phase requires Congress to combine these components into the sixth branch of the Armed Forces. The Department of Defense is immediately pursuing four components:
>
> - Space Development Agency – capabilities development and fielding,
> - Space Operations Force – developing space leaders and joint space warfighters,

- Services and Support – leadership and support structures, and
- Space Command – developing Space Force warfighting operations to protect U.S. national interests.[635]

The first component, a "Space Development Agency", looms as a dramatic break from past practices in which the Pentagon largely relied upon corporations for building their new generations of aerospace vehicles and weapons. The Final Report states that the proposed Agency will:

- Identify opportunities to move from dependence on a few independent assets to a proliferated architecture enabled by lower-cost commercial space technology and access,
- Shift from an acquisition organization and mindset to a development organization focused on experimentation, prototyping, and accelerated fielding, and
- Change from a matrixed and overlapping structure to a concentrated and decoupled structure to generate speed. [636]

In moving from "an acquisition organization and mindset" to a "development organization", Space Force will essentially assert its authority over corporate contractors building advanced spacecraft components. The ground shaking result will be that corporations can no longer get away with secretly building components for other space programs without informing the Air Force. The Space Development Agency will enable the Air Force to assert its primacy through Space Force over the development of future space technologies with military applications.

According to the Final Report, the second component of Space

Force, the "Space Operations Force", will:

- Include the space personnel from all Military Services, including Guard, Reserve, and civilians,
- Create a clear career track for the space community containing all relevant space specialties including operations, intelligence, engineering, science, acquisition, cyber, etc., and
- Be overseen by U.S. Space Command with civilian oversight from the Office of the Secretary of Defense, while personnel remain in the Services, until the establishment of the Space Force.[637]

The establishment of a Space Operations Force will make it very difficult, if not impossible, for rival space programs led by the Antarctic Germans, Deep State and corporations to secretly recruit US military personnel and civilians. It will establish "space specialties" that allow for career advancement, thereby eliminating the need for practices such as the "20 and back" programs described by whistleblowers such as Corey Goode, Michael Relfe, and Michael Gerloff.[638] These and other insiders have described how mind control and advanced technologies were used to wipe their minds of all memories of their covert space service. They explained how they were age-regressed and returned back in time to when they began their service, without the skills and benefits they had acquired, which subsequently led to them struggling with post-traumatic stress due to the suppressed memories.

The fourth component of Space Force involves the establishment of a "Space Command", which the Final Report also describes in detail:

To further accelerate warfighting capability, the

Department recommends creating a new U.S. Space Command to become a unified combatant command. U.S. Space Command will be responsible for preparing for and deterring conflict in space and leading U.S. forces in that fight if it should happen.

U.S. Space Command will:
- Lead U.S. warfighting activities in space and establish unity of command for operational space forces,
- Integrate space planning and operations across military campaigns and contingency plans,
- Simplify the command structure by aligning operational forces to the commander responsible for joint space warfighting,
- Develop space doctrine, concepts of operation and space tactics, techniques and procedures (TTPs),
- Establish enterprise space standards to be adopted by the Military Services, ensuring interoperability of the joint force, and
- Utilize commercial practices and digitization to streamline the footprint and automate labor-intensive operations.[639]

The key benefit in creating a Space Command is that this one institution will have operational oversight for all US military space activities, including control over the corporations that have developed advanced space technologies with military applications. This institution's powers will make it very difficult, if not impossible, for space asset technologies created within and for the different military services to be secretly siphoned off into black projects run by the Deep State, corporations, and the Antarctic Germans, without the knowledge of the USAF SSP.

On December 18, 2018, this vital component of Trump's future Space Force moved forward with a Presidential Memorandum authorizing the creation of a United States Space Command:

> Pursuant to my authority as the Commander in Chief and under section 161 of title 10, United States Code, and in consultation with the Secretary of Defense and the Chairman of the Joint Chiefs of Staff, I direct the establishment, consistent with United States law, of United States Space Command as a functional Unified Combatant Command. I also direct the Secretary of Defense to recommend officers for my nomination and Senate confirmation as Commander and Deputy Commander of the new United States Space Command.
>
> I assign to United States Space Command: (1) all the general responsibilities of a Unified Combatant Command; (2) the space-related responsibilities previously assigned to the Commander, United States Strategic Command; and (3) the responsibilities of Joint Force Provider and Joint Force Trainer for Space Operations Forces. The comprehensive list of authorities and responsibilities for United States Space Command will be included in the next update to the Unified Command Plan.[640]

As the forerunner to the creation of a Space Force, Space Command will allow the necessary integration of, and command over, space assets from all five military services to take place after its Congressional approval. Most importantly, Space Command will incorporate the USAF Space Command, which until now has been subordinate to the Commander of US Strategic Command. It will also include USAF Special Operations, which provides personnel for the squadrons of antigravity spacecraft deployed in Earth-orbit and deep space, as shown in chapter 19.

Eventually, after undergoing all the stages of Congressional approval, Space Force will rise up as a new military branch placed under the Department of the Air Force. This relationship will replicate the one held by the Marine Corps, which is an independent military service under the Department of the Navy. Nostalgically, it echoes the historical evolution of the Air Force itself which first gained autonomy from the Army Air Force under General Hap Arnold during World War II; then became a *de facto* separate military service under the administrative control of the Army immediately after World War II; and soon after culminated in the creation of the Department of the Air Force after passage of the September 1947 National Security Act.

On February 19, 2019, President Trump released Space Policy Directive-4 affirming that the creation of Space Force will be a two-step process. It will be "initially established within the Department of the Air Force", which is vital to gaining Congressional and Air Force support for its legislative passage.[641] His directive only implies the next step will involve the creation of a wholly independent Department of the Space Force in order to bypass any initial resistance to the second phase. What has been left unsaid in Trump's Directive is that the second step is critical because it is the only way in which the Navy will agree to hand over authority for its deep space battle groups, which have been operating since the early 1980's. Basically, Space Force first being under the Department of the Air Force will aggregate all military space assets deployed in near-Earth orbit. After the creation of an independent Department of the Space Force, the Navy will be required to relinquish control of its deep space assets and place these under the authority of the new department.

## Space Force, Q Anon, and Secret Space Program Disclosure

A major question that arises from President Trump's Space Force initiative is whether it will lead to the disclosure of not only the USAF SSP but the existence of other space programs and extraterrestrial life in its many forms. It's important to emphasize that the idea for a Space Force emerged in White Houses discussions that took place in early March 2018, just before Trump's March 13 speech at the USMC Miramar Air Station. A clue into what Trump's military advisors have told him about disclosure comes from QAnon ('Q'), the alleged group of military intelligence officials advising Trump on the best strategy for overcoming the Deep State.

On September 19, 2018, Q dropped two bombshell posts affirming the existence of secret space programs and extraterrestrial life, thus providing insights into what the Trump administration is planning to disclose in the future. In the first post (#2222), Q responded to two questions about whether extraterrestrial life exists and whether the Roswell UFO crash really happened:

> Q !!mG7VJxZNCI ID: 98088e No.3094236
> Sep 19 2018 19:25:34 (EST)
>> Anonymous ID: 5948d8 No.3093831
>> Sep 19 2018 19:10:44 (EST)
>> Q
>> Are we alone ?
>> Roswell ?
> >>3093831
> No.
> Highest classification.
> Consider the vastness of space.
> Q[642]

Q's response makes clear that we are not alone and that the truth about the Roswell flying saucer crash has the "highest classification". This response is significant since it affirms what a senior official with the Canadian Government's Department of Communications, named Wilbert Smith, reported back in 1950 after he had inquired about the Roswell crash among senior US government scientists. Smith's official report was eventually released through the Freedom of Information Act: "The matter is the most highly classified subject in the United States government, rating higher even than the *H-bomb*."[643]

In the second bombshell post (#2225) on the same day, Q responded to questions about the Moon and secret space programs:

> Q !!mG7VJxZNCl ID: 922952 No.3095105
> Sep 19 2018 19:58:13 (EST)
>> Anonymous ID: a16b71 No.3094804
>> Sep 19 2018 19:45:34 (EST)
>> Q,
>> Did NASA fake the moon landings? Have we been to the moon since then? Are there secret space programs? Is this why the Space Force was created?
> >>3094804
> False, moon landings are real.
> Programs exist that are outside of public domain.
> Q[644]

In stating that the "moon landings are real", Q is refuting conspiracy theories that the Moon landings were hoaxed. This removes an impediment for the general public learning the truth about what has really been happening on the Moon after the Apollo Moon landings.

In the next statement, Q affirms the existence of space programs that "are outside of public domain". This means that Q wants readers to understand that the information on these "non-public domain" space programs is highly classified and restricted to those with need-to-know access. While Q did not explicitly refer to Trump's Space Force initiative, it is more than coincidental that the posts appeared only one month before the Pentagon released its "Final Report" to the Congressional Defense Committee on establishing the Space Force. [645] It can be inferred from Q's reply that information about the existence of secret space programs has been shared with Trump. Even more importantly, the fact that Q chose to answer these questions from a long list suggests that a disclosure agenda exists within the Trump White House.

## Disclosure of USAF Secret Space Program and Nordic Alliance

Ever since the creation of the US Air Force in 1947, its senior leaders have grappled with the question of how to deal with visiting extraterrestrial life and the existence of a German secret space program in Antarctica that is allied with the most aggressive of Earth's space visitors, the Draconian Reptilians. After both the US Navy and US Air Force failed in their respective efforts (from 1947 to 1952) to contain the threats posed by the Antarctic Germans through direct military confrontation, a strategy of negotiation and cooperation was taken to unlock the secrets of interstellar space travel.

Contrary to the approach taken by the US Navy, which closely cooperated with a group of human-looking Nordic extraterrestrials, the USAF worked closely with the Antarctic Germans and their Reptilian allies.[646] The USAF would also work with other extraterrestrials groups such as Grays, Tall Whites, and another human-looking group in understanding the advanced spacecraft they had in their possession. Progress was slow, but over succeeding decades, the USAF was able to assemble an impressive collection of space propulsion technologies that enabled it to travel into Earth-orbit and eventually throughout our solar system. At least three different sized saucer-shaped "Alien Reproduction Vehicles"; at least three different sized TR-3B-like "triangular-shaped craft"; "rectangular-shaped weapons platforms"; and "stealth space stations" make up the crown jewels of the USAF SSP. The USAF even possesses traversable wormhole technologies that enabled its airmen to explore different solar systems. This technology was notably revealed in a soft disclosure initiative sanctioned by two Air Force Chiefs of Staff – the *Stargate SG-1* science fiction series.

In its mission to confront any potential planetary threats, the USAF stood proudly and confidently in its belief that it was the proverbial "tip of the spear", that is until it learned differently in late 2016. It was the remarkable testimonies of Corey Goode and other insiders which led USAF SSP leaders to seriously investigate the claims of the existence of other secret space programs. Such information first became public back in 2000 with Michael Relfe's groundbreaking testimony of serving for twenty years on Mars.[647] The investigation led to the stunning discovery that the USAF SSP was very far from being the tip of the spear when it came to advanced space technologies for confronting potential planetary threats. The USAF found out it had been completely deceived by the Deep State, the Antarctic Germans and its extraterrestrial allies, who had only helped the Air Force to develop and deploy redundant space technologies that were decades behind those in use by other space programs at the time.

The USAF's institutional humiliation only increased when it was confirmed that the US Navy had developed interstellar-capable "space battle groups" that patrolled the entire solar system, and that the Navy also worked closely with a group of Nordics that the USAF leaders had been deceived into believing were a threat. Indeed, the USAF had worked with the Antarctic Germans and Deep State operatives to intercept, interrogate, and torture many human-looking extraterrestrials that were found on Earth, perhaps even some that had helped the Navy develop its more advanced space program.

After uncovering the full extent of the web of misinformation and falsehoods that had corrupted its mission and usurped its technological growth, the USAF made the momentous decision to realign its fundamental extraterrestrial alliances. Like the Navy, the USAF would now also choose to work closely with human-looking extraterrestrials willing to assist in its development of advanced space technologies that could enable it to join in the galaxy-wide confrontation already taking place with the Draconian Reptilian extraterrestrials. The Defense Intelligence Agency's "preliminary briefing" leaked by the USAF in June 2017 was a sign of this strategic change because it revealed for the first time a more accurate account of historical events than anything previously leaked by the Air Force through the Majestic Documents. Most importantly, the leaked DIA document stated that human-looking extraterrestrials were the friendliest of all of the space visitors and proved reliable partners for diplomatic agreements.

The June 2017 leaking of the DIA document was followed by USAF Special Operations covertly assisting a civilian to take photos of some of its antigravity spacecraft operating out of MacDill AFB in Florida. Chapter 19 covered in-depth JP's photos and how these have provided hard evidence for his claims regarding being taken inside differently shaped USAF antigravity spacecraft: both "flying triangles" and "flying rectangles". He even took photos of a craft belonging to one of the Nordics who said his people were helping the USAF to develop more advanced technologies and weapons for its secret space program.

This strategic realignment of extraterrestrial allies had required some important policy changes taking place within the highest levels of the USAF SSP. Among the most important was helping U.S. and global society to break free of the corrupting influence of the Deep State, and its Antarctic German/Reptilian/corporate allies. It's no accident that similar scenarios have taken place on many other worlds where the Draconians have successfully established a foothold, and where the

Nordics have stepped in to help these worlds break free of similar corrupting influences.

The dramatic change in USAF policy by September 2017 led to the Deep State's quick retaliation. Using its Maser satellite systems, the Deep State targeted MacDill AFB by steering Hurricane Irma directly towards its Tampa home. It's not coincidental that Irma made landfall near MacDill on September 11, 2017, the 16th anniversary of the 9/11 attacks, or that the hurricane was named after a Germanic goddess of war. This event was a covert declaration of war by the Deep State against the USAF SSP, and it also held a dark warning for any other facility that would be used to promote a disclosure process.

Four months later, when the Deep State attempted a false flag attack in Hawaii on January 13, 2018, in order to trigger a Third World War, the USAF and their new Nordic allies were ready. The ballistic missile launched from a CIA-controlled submarine positioned near North Korea was destroyed in mid-flight. This was followed three days later by a similar false flag event happening in Japan, suggesting that once again the USAF SSP had intervened. It would be ironic if one day it emerged that some of the same Maser satellites used to steer Hurricane Irma were used to shoot down the ballistic missiles launched against Hawaii and Japan. QAnon has raised such a possibility. Posts by Q have described the process by which the Deep State's misuse of satellite technologies have provided the justification for these space assets to be taken over by the US military.[648] In the end, the Deep State has failed miserably. USAF leaders are now working closely with the Trump administration to eliminate the power of the Deep State through the creation of Space Force.


## Official Disclosure Through Space Force

By ordering the creation of Space Force, Trump is now soundly shaking the bureaucratic and corporate trees that hide multiple space programs that have been secretly built and supplied by US corporations, both for different customers within the Military-Industrial Complex and the Antarctic Germans. Large aerospace companies such as Lockheed Martin, Northrup Grumman, Raytheon, General Dynamics, etc., have for decades supplied the technologies and components for the Air Force SSP while hiding their involvement in similarly supplying other space programs. For decades, the USAF had little to no awareness of these other SSPs due to the compartmentalization process and cunning lies told to them about spacecraft belonging to the other space programs in deep space. Consequently, the Deep State has played a major role in setting space policy due to its longstanding ability to manipulate Air Force officials through the supply and acquisition process facilitating the multiple space programs. Trump's proposed Space Force will finally enable the USAF to remedy such problems, thereby allowing the USAF to work with Navy SSP leaders in coordinating their respective space assets and move forward in a future official disclosure process.

In earlier chapters, I've described the Air Force's "limited disclosure" plan to reveal the existence of two or more orbiting space stations along with antigravity flying triangles (TR-3Bs) and other exotic spacecraft built by complicit aerospace companies. Tom DeLonge's *To the Stars Academy* started as a USAF-backed disclosure initiative, but it was abandoned after Air Force leaders learned of the Deep State's deception and recognized that the initiative had been compromised. Also, once these same leaders had learned of the Navy's more highly classified and advanced SSP by confirming Corey Goode's testimony and claims by other SSP whistleblowers, a change of strategy became necessary.

The USAF's discovery that the Navy had its own SSP with advanced space technologies which had been secretly built by various corporations, and that other space programs existed with the Antarctic Germans (Dark Fleet) and the Deep State/Corporations (the Interplanetary Corporate Conglomerate), caused Air Force leaders to appreciate the need to eliminate their dispersed development, acquisition, and recruitment systems. These administrative systems had made all of this deception possible. Therefore, modifying them became a major factor in the USAF and Pentagon officials' decision to reassess and approve the Space Force initiative after strong initial resistance, based on decades of disinformation. The degree of cooperation required by Air Force and Navy leaders to make Space Force successful will be tremendous. This fact begs the question of whether these two highly competitive military giants will publicly disclose the advanced aerospace technologies employed in their respective secret space programs, and cooperatively merge their programs under the new Space Force military service branch.

It is very likely that the advanced space technologies possessed by the USAF SSP will be the first to go public. These can be most easily disclosed since there is already public awareness of their existence due to of multiple whistleblowers, insiders, and witness reports on the existence of saucer-shaped craft (ARVs), triangular-shaped vehicles (TR-3Bs), rectangular-shaped platforms and stealth space stations. The emergence of Space Force will enable the surfacing of the myriad "deep black" technologies which have been funded by the Deep State and CIA through multiple corrupt practices. Exposure of the Deep State and CIA's complicity in international corruption and human rights abuses – used to raises billions of dollars toward a black budget estimated to exceed one trillion dollars annually – will forever end these illicit funding mechanisms.[649] Consequently, Space Force will have to be funded through more conventional sources such as Congressional appropriations. And in turn, the public will have to be persuaded to accept the wisdom of a dramatic budget increase for the Pentagon so it can match the funding levels attained by the Deep State through its illicit activities.

However, a much more dramatic disclosure event will be the revelation of the role of visiting Nordic extraterrestrials who have been helping the USAF advance its secret space program. It is very likely that the USAF SSP will reveal its relationship with the Nordics as part of its disclosure process. The first concrete sign of this took place with the June 2017 leaking of the 1989 Defense Intelligence Agency preliminary briefing document discussing human-looking extraterrestrials. By now it should be clear to readers that the DIA, NRO, and NSA are important partners in the USAF SSP. The DIA document is already publicly available and will only gain widespread attention as others authenticate key elements in it, as Dr. Robert Wood did in December 2018.[650]

The revelation of a secret space program having been developed by the USAF is big news by itself. However, adding that this was achieved with the assistance of human-looking Nordic extraterrestrials will come as a great shock to many. There are some who might question whether or not the public is ready for such dramatic announcements and if this would cause a societal collapse as warned in the 1961 Brookings Report.[651] It is highly unlikely such a collapse will occur since the public has been thoroughly acclimated to accept the existence of extraterrestrial life. Therefore, an announcement that such visitors have proved themselves friendly and are helping our world to develop advanced space technologies should be welcomed by many. Indeed, there is reason to believe that the growing QAnon movement may be destined to play a prominent role in such a disclosure process and that more Q posts can be expected to discuss secret space programs and extraterrestrial life.

When is official disclosure of a USAF SSP and the existence of Nordic extraterrestrials likely to occur? One date to keep in mind is the formal launch of the US Space Force sometime in 2020, or soon thereafter. In the interim, the newly launched US Space Command is likely to figure prominently in any official disclosure process. The disclosure of the USAF SSP, its absorption into Space Command and then Space Force, and disclosure of Nordic visitors will revolutionize our planet in many ways. This will even escalate once the Navy's SSP becomes part of the disclosure process and eventually merges its deep space assets within the proposed Space Force, which will only occur after the newly created military service moves from under the administrative control of the Department of the Air Force to become a fully independent Department of the Space Force. There will be great technological, economic, political, and cultural changes that will occur in a short period. The elimination of the Deep State and its countless corrupting practices used in secretly funding multiple secret space programs will be profound.

In the decade ahead, humanity will increasingly need to confront destructive planet-wide events that are predicted to involve an impending solar flash (aka micronova),[652] a geophysical pole shift, and potential asteroid impacts.[653] The need for disclosing the existence of advanced technologies will not only result in making space travel available for widespread commercial use; but, more importantly, it will help countless individuals and communities to avoid such impending destructive events. Similarly, the release of suppressed health technologies will help humanity to overcome the time-old enemies of disease and the aging process, and to mitigate the effects of the coming planetary challenges considerably. Finally, when these technological revelations are combined with the disclosure of human-looking extraterrestrials, to whom we are genetically related, humanity will collectively shift into a *Star Trek*-like future where space flight to distant star systems will become as commonplace as intercontinental air travel.

# ACKNOWLEDGMENTS

I am grateful to a great number of insiders, whistleblowers, researchers, and experiencers who made this book possible because of their pioneering efforts in unraveling the many layers comprising the US Air Force's historic research and development of extraterrestrial technologies. Special thanks to Dr. Robert Wood, William Tompkins, Corey Goode, Emery Smith, Alex Collier, Edgar Fouche, David Wilcock, Linda Moulton Howe, William Hamilton, Richard Hoagland, Mike Bara, and my anonymous source "JP" for their respective research, investigations, and testimonies that helped me understand the big picture behind the US Air Force Secret Space Program.

Many thanks to the offices of Senator Brian Schatz and Representative Tulsi Gabbard for their assistance in a Freedom of Information Act request to Indo Pacific Command about the January 13, 2018, Hawaii Missile Alert.

Warm thanks to Mark McCandlish for his permission to reproduce the cutaway of the "Alien Reproduction Vehicle" witnessed at Edwards Air Force Base in 1988.

Much appreciation to Corey Goode, the Sphere Being Alliance, and Gaia TV for permission to include several images depicting Corey's secret space program experiences.

Many thanks to JP for sending me dozens of photos of multiple antigravity craft he has witnessed in Florida, and permission to reprint a selection of them.

I am very grateful to Duke Brickhouse for proofreading this manuscript, and to Thomas Keller for his helpful editorial services.

My deep heartfelt thanks to Rene McCann for generously donating her time to create another wonderful book cover, and for liaising with other artists to incorporate different elements. Thanks to Rene Armata for the SSP craft, Vashta for the Draco Reptilian figure, and Matt McCann for modeling for the Nordic depicted on the book's cover.

Finally, my deepest gratitude goes to my wonderful soulmate Angelika Whitecliff who has never wavered in her support of my "exopolitics" research, has offered wise counsel using her unique intuitive gifts, and whose first-rate editing skills turned my raw manuscript into the polished product before you.

Michael E. Salla, Ph.D.
May 5, 2019

# ABOUT THE AUTHOR

Dr. Michael Salla is an internationally recognized scholar in global politics, conflict resolution and U.S. foreign policy. He has taught at universities in the U.S. and Australia, including American University in Washington, DC. Today, he is most popularly known as a pioneer in the development of the field of 'exopolitics'; the study of the main actors, institutions and political processes associated with extraterrestrial life.

He has been a guest speaker on hundreds of radio and TV shows including Coast to Coast AM, and Ancient Aliens, and featured at national and international conferences. His Amazon bestselling *Secret Space Program* book series has made him a leading voice in the Truth Movement, and over 5000 people a day visit his website for his most recent articles.

Dr. Salla's Website: www.exopolitics.org

# ENDNOTES CHAPTER 1

[1] "Wright Brothers",
http://www.daytonhistorybooks.com/the_wright_brothers_9.html (accessed 9/5/18).

[2] CPI Inflation Calendar http://www.in2013dollars.com/1909-dollars-in-2018?amount=30000 (accessed 9/1/18)

[3] Ohio History Central, "Wright Field"
http://www.ohiohistorycentral.org/w/Wright_Field?rec=1703 (accessed 9/2/18)

[4] Ohio History Central, "Wright Field"
http://www.ohiohistorycentral.org/w/Wright_Field?rec=1703 (accessed 9/2/18)

[5] , Thomas J Carey and Donald R. Schmitt, *Inside the Real Area 51* (New Page Books, 2013) Kindle Locations 299-300.

[6] See Henry Samuel, "Winston Churchill 'ordered assassination of Mussolini to protect compromising letters' https://www.telegraph.co.uk/history/world-war-two/7978285/Winston-Churchill-ordered-assassination-of-Mussolini-to-protect-compromising-letters.html (accessed 9/7/2018).

[7] Ryan Wood, *Majic Eyes: Earth's Encounters with Extraterrestrial Technology* (Wood Enterprises, 2005). 35

[8] Leonard H. Stringfield, *UFO Crash Retrievals: The Inner Sanctum, Status Report VI* (Leonard H. Stringfield, July 1991).

[9] Ryan Wood, *Majic Eyes: Earth's Encounters with Extraterrestrial Technology*, pp. 35-36.

[10] See Cantwheel Document: http://majesticdocuments.com/pdf/s-aircraft.pdf (accessed 9/5/2018). Ryan Wood discusses how the author may have innocently mistaken the southeastern location in Missouri where the 1941 crash occurred with the southwest. *Majic Eyes: Earth's Encounters with Extraterrestrial Technology*, p. 37.

[11] A short biography of Timothy Cooper is available at:
http://ufoevidence.org/researchers/detail34.htm (accessed 9/6/18)

[12] Timothy Cooper, "Analysis of the Intelligence Background of Thomas Cantwheel, Preliminary Report",
http://www.majesticdocuments.com/pdf/cantwheel_report.pdf (accessed 9/5/2018).

[13] The Woods discuss their eight factors that make up their authenticity ratings for Majestic documents in "Introduction to Authenticity Ratings",
http://www.majesticdocuments.com/documents/authenticity.php (accessed 9/5/2018).

[14] Discussion of the two page memo titled "S-Aircraft Drawing and Memo by Thomas Cantwell" is available at:
http://www.majesticdocuments.com/documents/1970-present.php (accessed 9/5/2018).

[15] See entry for White Hot Report at: http://www.majesticdocuments.com/documents/pre1948.php (accessed 9/11/2018).

[16] White Hot Report available at: http://www.majesticdocuments.com/pdf/twining_whitehotreport.pdf (accessed 9/11/2018). Printed version found in Dr. Robert M. Wood & Ryan Wood (eds), *The Majestic Documents* (Wood and Wood Enterprises, 1998) p. 77.

## ENDNOTES CHAPTER 2

[17] See Michael Salla, *The U.S. Navy's Secret Space Program & Nordic Extraterrestrial Alliance* (Exopolitics Consultants, 2017) 1-16.

[18] William Tompkins, *Selected by Extraterrestrials: My life in the top secret world of UFOs, think-tanks and Nordic secretaries* (Createspace, 2015) p. xi.

[19] "The Great Los Angeles Air Raid," http://theairraid.com/ (accessed 9/16/18).

[20] "The Great Los Angeles Air Raid," http://theairraid.com/ (accessed 9/16/18).

[21] William Tompkins covert work with the Office of Naval Intelligence will be discussed in chapter two.

[22] Phone interview with William Tompkins, September 19, 2016.

[23] Wesley Frank Craven and James Lea Cate, eds., *The Army Air Forces in World War II*, p. 284. Available online at: http://tinyurl.com/jcxxmu8 (accessed 9/16/18).

[24] According to Timothy Good, prior to the documents release, the Department of Defense denied having any further information about the Los Angeles Air Raid, see *Above Top Secret: The World Wide U.F.O. Cover Up* (Quill, 1988) p. 17.

[25] According to Timothy Good, prior to the documents release, the Department of Defense denied having any further information about the Los Angeles Air Raid, see *Above Top Secret*, p. 17.

[26] Quoted in Timothy Good, *Above Top Secret,* p. 17.

[27] "History of Helicopters", https://tinyurl.com/ycje9xoj (accessed 9/15/18)

[28] "Introduction to Authenticity Ratings", http://www.majesticdocuments.com/documents/authenticity.php (accessed 9/16/18).

[29] "Memorandum for Chief of Staff of the Army", http://www.majesticdocuments.com/pdf/fdr.pdf (accessed 9/15/18)

[30] See "Franklin Delano Roosevelt - Assistant Secretary of the Navy" https://www.nps.gov/articles/franklin-delano-roosevelt-assistant-secretary-of-the-navy.htm (accessed 9/15/15).

[31] See "The Manhattan Project" https://www.osti.gov/opennet/manhattan-project-history/Events/1942/enter_army.htm (accessed 9/15/15).

[32] Available online at: http://majesticdocuments.com/pdf/marshall-fdr-march1942.pdf (accessed 12/30/18).

[33] Available online at: http://majesticdocuments.com/pdf/marshall-fdr-march1942.pdf (accessed 12/30/18).

[34] Available online at: http://www.textfiles.com/ufo/UFOBBS/1000/1723.ufo (accessed 9/15/15).

[35] Cited by Ryan Wood, *Majic Eyes Only*, p. 4; and Timothy Good, *Above Top Secret* ((William Morrow and Company, 1988) p. 267.

[36] Available online at: http://majesticdocuments.com/pdf/marshall-fdr-march1942.pdf (accessed 12/30/18).

[37] Robert and Ryan Wood, http://majesticdocuments.com/documents/pre1948.php (accessed 12/30/18).

[38] For discussion of the report called "Estimate of the Situation", see Michael D. Swords, "Project Sign and the Estimate of the Situation," *Journal of UFO Studies* 7 (2000): pp. 27-64.

[39] Donald Keyhoe, *Aliens from Space: The Real Story of Unidentified Objects* (New American Library) p. 14.

[40] Timothy Cooper, "Analysis of the Intelligence Background of Thomas Cantwheel, Preliminary Report", http://www.majesticdocuments.com/pdf/cantwheel_report.pdf (accessed 9/5/2018).

[41] Dana T. Parker, *Building Victory: Aircraft Manufacturing in the Los Angeles Area in World War II* (Dana T. Parker, 2013) Kindle edition, location 138.

[42] See "United States aircraft production during World War II", *Wikipedia*, https://en.wikipedia.org/wiki/United_States_aircraft_production_during_World_War_II#cite_note-buildingvictory-2 (accessed 9/16/2018).

[43] Dana T. Parker, *Building Victory: Aircraft Manufacturing in the Los Angeles Area in World War II* (Dana T. Parker, 2013) introduction.

[44] Major General John W. Huston, ed. *American Airpower Comes of Age: General Henry H. "Hap" Arnold's World War II Diaries*, Volume 1. (Pickle Partners Publishing, 2014) 229.

[45] "Memorandum for Chief of Staff of the Army", http://www.majesticdocuments.com/pdf/fdr.pdf (accessed 9/16/2018).

## ENDNOTES CHAPTER 3

[46] "The Majestic Documents", http://majesticdocuments.com/documents/pre1948.php (accessed 9/19/18)

[47] Feb 22, 1944, Roosevelt Memorandum, Majestic Documents, http://majesticdocuments.com/pdf/fdr_22feb44.pdf

[48] The Navy espionage program in Nazi Occupied Europe is discussed in detail in Book 2 of the Secret Space Program series, *The U.S. Navy's Secret Space Program & Nordic Extraterrestrial Alliance* (Exopolitics Consultants 2017) pp. 17-60

[49] Major General John W. Huston, ed. *American Airpower Comes of Age: General Henry H. "Hap" Arnold's World War II Diaries*, Volume I1 (Pickle Partners Publishing, 2014) 439.

[50] Dana T. Parker, *Building Victory: Aircraft Manufacturing in the Los Angeles Area in World War II* (Dana T. Parker, 2013). Kindle Edition. Location 256. (accessed 9/17/18)

[51] Dana T. Parker, *Building Victory: Aircraft Manufacturing in the Los Angeles Area in World War II* (Dana T. Parker, 2013). Kindle Edition. Location 256. (accessed 9/17/18)

[52] For discussion of Arnolds friendship with Donald Douglas and other aircraft manufacturers, see Major General John W. Huston, ed. *American Airpower Comes of Age: General Henry H. "Hap" Arnold's World War II Diaries*, Volume I1, 144.

[53] See Michael Salla, *The U.S. Navy's Secret Space Program & Nordic Extraterrestrial Alliance*, pp. 17-60.

[54] William Tompkins, *Selected by Extraterrestrials*, p. xii.

[55] William Tompkins, *Selected by Extraterrestrials*, p. xv.

[56] Private telephone interview, September 19, 2016.

[57] Feb 22, 1944, Roosevelt Memorandum, Majestic Documents, http://majesticdocuments.com/pdf/fdr_22feb44.pdf

[58] Tompkins, *Selected by Extraterrestrials*, p.58.

[59] Phone interview with William Tompkins, September 19, 2016.

[60] "A Brief History of RAND," http://www.rand.org/about/history/a-brief-history-of-rand.html

[61] Major General John W. Huston, ed. *American Airpower Comes of Age: General Henry H. "Hap" Arnold's World War II Diaries*, Volume I1 (Pickle Partners Publishing, 2014) 439.

[62] Major General John W. Huston, ed. *American Airpower Comes of Age: General Henry H. "Hap" Arnold's World War II Diaries*, Volume I1, 439-40.

[63] Wesley Marx, "The Military's 'Think Factories'," *The Progressive*, https://www.cia.gov/library/readingroom/docs/CIA-RDP88-01315R000400280026-3.pdf See also Wikipedia, https://en.wikipedia.org/wiki/Henry_H._Arnold (accessed 10/19/16).

[64] Wesley Marx, "The Military's 'Think Factories'," *The Progressive*, https://www.cia.gov/library/readingroom/docs/CIA-RDP88-01315R000400280026-3.pdf See also Wikipedia, https://en.wikipedia.org/wiki/Henry_H._Arnold (accessed 10/19/16)

[65] According to the website, "Measuring Worth," 10 million dollars in 1945 converts to between 110 to 855 million in 2018 dollars. https://www.measuringworth.com/calculators/ppowerus/ (accessed 9/18/18)

[66] "A Brief History of RAND," http://www.rand.org/about/history/a-brief-history-of-rand.html (accessed 9/18/18)

[67] Major General John W. Huston, ed. *American Airpower Comes of Age: General Henry H. "Hap" Arnold's World War II Diaries*, Volume I1, 439-40.

[68] Tompkins, *Selected by Extraterrestrials*, p.104.

[69] William Tompkins, *Selected by Extraterrestrials*, p. xv.

[70] William Tompkins, *Selected by Extraterrestrials*, p. 192.

[71] William Tompkins, *Selected by Extraterrestrials*, p. 192.

[72] The words "from the Missouri" are blacked out in the leaked document, "Twining's 'White Hot' Report," *The Majestic Documents* (Wood and Wood Enterprises, 1998) p. 56.

[73] See Majestic Documents website, http://tinyurl.com/jt49ov3

[74] See Michael Salla, *The US Navy's Secret Space Program and Nordic Extraterrestrial Alliance.*

[75] William Tompkins, *Selected by Extraterrestrials*, p. 192.

[76] Wesley Marx, "The Military's 'Think Factories'," *The Progressive*, https://www.cia.gov/library/readingroom/docs/CIA-RDP88-01315R000400280026-3.pdf (accessed 9/19/18)

[77] Smith's memo is available online at: http://www.majesticdocuments.com/pdf/smithmemo-21nov51.pdf

[78] Major General John W. Huston, ed. *American Airpower Comes of Age: General Henry H. "Hap" Arnold's World War II Diaries*, Volume 11, 439-40.

[79] See Barrett Tillman, *LeMay* (St Martin's Press ebook) 101. Available online at: https://www.scribd.com/read/282722552/LeMay-A-Biography# (accessed 9/20/18)101.

[80] *Earthfiles*,http://www.earthfiles.com/news.php?ID=1503&category=Real+X-Files (accessed 9/24/14).

[81] *Earthfiles*, http://www.earthfiles.com/news.php?ID=1501&category=Real+X-Files (accessed on 11/14/17).

[82] *Earthfiles*, http://www.earthfiles.com/news.php?ID=1501&category=Real+X-Files (accessed on 11/14/17).

[83] See Barrett Tillman, *LeMay* 102. Available online at: https://www.scribd.com/read/282722552/LeMay-A-Biography# (accessed 9/20/18)

[84] For detailed discussion of Schumann and his role in helping develop Germany's flying saucer prototypes, see Michael Salla, *Insiders Reveal Secret Space Programs and Extraterrestrial Alliances*, pp. 60-65.

[85] For further discussion see Richard Sauder, *Hidden in Plain Sight: Beyond the X-Files* (Keyhole Publishing, 2011). See also "Vorticular Madness Of The Dark Magicians," https://truthtalk13.wordpress.com/category/operation-paperclip/ (accessed 8/11/15).

[86] Barrett Tillman, *LeMay* 103. Available online at: https://www.scribd.com/read/282722552/LeMay-A-Biography# (accessed 9/20/18)

[87] Alex Abella, *Soldiers of Reason: The RAND Corporation and the Rise of the American Empire* (Houghton Mifflin Harcourt, 2008) 24.

[88] Mark Erickson, *Into the Unknown Together: The DOD, NASA, and Early Spaceflight* (Air University Press, 2005) p. 5. Available online at: https://tinyurl.com/y76onu7q (accessed 10/31/18)

[89] "Preliminary Design of an Experimental World-Circling Spaceship," http://www.rand.org/pubs/special_memoranda/SM11827.html

[90] "Preliminary Design of an Experimental World-Circling Spaceship." Santa Monica, CA: RAND Corporation, 1946. http://www.rand.org/pubs/special_memoranda/SM11827.html. Also available in print form.

[91] Alex Abella, *Soldiers of Reason: The RAND Corporation and the Rise of the American Empire*, p. 28

[92] Alex Abella, *Soldiers of Reason: The RAND Corporation and the Rise of the American Empire*, p. 28.
[93] Mark Erickson, *Into the Unknown Together: The DOD, NASA, and Early Spaceflight*, pp. 5-6. Available online at: https://tinyurl.com/y76onu7q (accessed 10/31/18)
[94] The Navy did not completely abandon its plans for Earth orbiting satellites since in 1955 it was assigned to provide the rocket launchers for the Vanguard space satellite that was part of the International Geophysical Year, see Mark Erickson, *Into the Unknown Together: The DOD, NASA, and Early Spaceflight*, p. 15. Available online at: https://tinyurl.com/y76onu7q (accessed 10/31/18)
[95] Tompkins, *Selected by Extraterrestrials*, p.427.
[96] For detailed analysis of the conflicting Army Air Force Press announcements, see Thomas Carey and Don Schmitt, *Witness to Roswell: Unmasking the Government's Biggest Cover-up* (New Page Books, 2009)

## ENDNOTES CHAPTER 4

[97] https://apps.dtic.mil/dtic/tr/fulltext/u2/a433273.pdf
[98] Thomas Carey and Donald Schmitt, *Witness to Roswell: Unmasking the 60-Year Cover-up* (New Page Books, 2007) p. 81.
[99] Thomas Carey and Donald Schmitt, *Witness to Roswell: Unmasking the 60-Year Cover-up*, p. 81.
[100] Jesse Marcel Jnr, *The Roswell Legacy: The Untold Story of the First Military Officer at the 1947 Crash Site* (New Page Books, 2008). See also website: http://www.marceljr.com
[101] Charles Berlitz and William Moore, *The Roswell Incident* (Grosset & Dunlop, 1980).
[102] "2002 Sealed Affidavit Of Walter G. Haut", http://roswellproof.homestead.com/haut.html#anchor_8 (accessed 12/30/18).
[103] Thomas Carey and Donald Schmitt, *Witness to Roswell: Unmasking the 60-Year Cover-up*. For a comprehensive book review, see David Rudiak, "Witness to Roswell," http://ufodigest.com/news/0607/witnesstoroswell.html
[104] Haut Affidavit #8, http://roswellproof.homestead.com/haut.html#anchor_8 (accessed 12/30/18).
[105] Haut Affidavit #9., http://roswellproof.homestead.com/haut.html#anchor_8 (accessed 12/30/18).
[106] Haut Affidavit #10., http://roswellproof.homestead.com/haut.html#anchor_8 (accessed 12/30/18).
[107] Haut Affidavit #12., http://roswellproof.homestead.com/haut.html#anchor_8 (accessed 12/30/18).
[108] Haut Affidavit #13., http://roswellproof.homestead.com/haut.html#anchor_8 (accessed 12/30/18).
[109] Witness to Roswell: Unmasking the 60-Year Cover-Up (New Page Books, 3007) p. 113.
[110] Philip Corso, *The Day After Roswell* (Pocket Books, 1997) pp. 34-35.

[111] Corso's DA Form 66 shows he was stationed at Fort performing various duties from April 1947 to May 1952: http://www.cufon.org/cufon/corso_p2.htm (accessed 10/6/2018).

[112] Leonard Stringfield published his findings in Status Report II, published in January 1980. For more information see Thomas Carey and Donald Schmitt, Inside the Real Area 51: The Secret History of Wright Patterson (New Page Books, 2013) p. 39, Kindle location 430.

[113] Linda Moulton Howe, "Part 3: Army/CIA Unit Studied the Real UFO Blue Book Cases" https://www.earthfiles.com/2008/09/07/part-3-army-cia-unit-studied-the-real-ufo-blue-book-cases/ (accessed 10/15/2018).

[114] The two memos are respectively titled "Directive to General Twining by Eisenhower" and "Directive to General Twining by President Truman". They are available online at: http://www.majesticdocuments.com/documents/pre1948.php (accessed 10/7/2018).

[115] Dr Robert and Ryan Wood, eds, The Majestic Documents (Wood and Wood Enterprises, 1998) p. 20. Available online at: http://www.majesticdocuments.com/pdf/twining_truman.pdf (accessed 10/7/2018).

[116] Dr Robert and Ryan Wood, eds, The Majestic Documents, p. 25.

[117] Dr Robert and Ryan Wood, eds, The Majestic Documents, p. 25..

[118] Dr Robert and Ryan Wood, eds, The Majestic Documents, p. 26. Available online at: http://www.majesticdocuments.com/pdf/airaccidentreport.pdf (accessed 10/8/2018).

[119] See "Documents Dated Prior tot 1948", http://www.majesticdocuments.com/documents/pre1948.php (accessed 10/8/2018).

[120] "The Twining Memo", http://www.roswellfiles.com/FOIA/twining.htm (accessed 10/7/2018).

[121] Dr Robert and Ryan Wood, eds, The Majestic Documents, p. 26..

[122] While Twining's Memo denied any knowledge of crashed flying saucers, it is likely that was due to the fact that such information was classified at a higher level than Secret. For further discussion of this see, "Did UFOs prompt a secret agenda?" http://www.educatinghumanity.com/2014/11/proof-aliens-twining-memo.html (accessed 10/7/2018)

[123] See: http://www.majesticdocuments.com/documents/pre1948.php (accessed 10/6/2018).

[124] Dr Robert and Ryan Wood, eds, The Majestic Documents, p. 35. Available online at: http://www.majesticdocuments.com/pdf/ipu_report.pdf (accessed 10/6/2018).

[125] Dr Robert and Ryan Wood, eds, The Majestic Documents, p. 35..

[126] Dr Robert and Ryan Wood, eds, The Majestic Documents, p. 25..

[127] Dr Robert and Ryan Wood, eds, The Majestic Documents,, p. 37. Available online at: http://www.majesticdocuments.com/pdf/ipu_report.pdf (accessed 10/6/2018).

[128] See: http://www.majesticdocuments.com/documents/pre1948.php (accessed 10/6/2018)

[129] Dr. Robert Wood and Ryan Wood, "Documents Dated Prior to 1948", http://www.majesticdocuments.com/documents/pre1948.php (accessed 10/6/2018).

[130] Dr Robert and Ryan Wood, eds, *The Majestic Documents*, p. 77.

[131] Dr Robert and Ryan Wood, eds, *The Majestic Documents*, p. 78.

[132] Dr Robert and Ryan Wood, eds, *The Majestic Documents*, pp. 77-78.

[133] Dr Robert and Ryan Wood, eds, *The Majestic Documents*, p. 83.

[134] See entry for "Hillenkoetter Memo to Joint Intelligence Committee," on Majestic Documents Prior to 1948, http://www.majesticdocuments.com/documents/pre1948.php (accessed 10/26/2018).

[135] Friedman, *Operation Majestic-12 and the United States Government's UFO Cover-up*, 54.

[136] Available online at: http://www.majesticdocuments.com/pdf/cutler_twining.pdf (accessed 10/7/2018).

[137] Even though the document was found in a box of official records at the National Archives, it does not have the standard control number. This has led to some claiming it was planted in the National Archives and is a hoax, but that is unlikely given NARA security procedures. See: http://www.ufoforhumanrights.com/mj-12.php . For a NARA statement on the memo, go to: http://www.archives.gov/foia/ufos.html#mj12 (accessed 10/7/2018).

[138] Available online at: http://www.majesticdocuments.com/pdf/cutler_twining.pdf (accessed 10/7/2018).

## ENDNOTES CHAPTER 5

[139] "A Brief History of RAND," https://www.rand.org/about/history/a-brief-history-of-rand.html (accessed 12/30/2018).

[140] Alex Abella, *Soldiers of Reason: The RAND Corporation and the Rise of the American Empire* (Mariner Books, 2009) pp. 40-41.

[141] Private telephone interview, September 19, 2016.

[142] Tompkins, *Selected by Extraterrestrials*, p.58.

[143] Alex Abella, *Soldiers of Reason: The RAND Corporation and the Rise of the American Empire*, p. 42.

[144] Available online at: https://rense.com/general96/DouglasDocsPreface.pdf (accessed 12/30/2018).

[145] More of these Douglas Aircraft unconventional propulsion system documents available online at: https://rense.com/general96/deepspaceexmp4.html (accessed 10/10/2018)

[146] "Preliminary Design of an Experimental World-Circling Spaceship." Santa Monica, CA: RAND Corporation, 1946.

http://www.rand.org/pubs/special_memoranda/SM11827.html. Also available in print form.

[147] The following biographical information for Wolfgang Klemperer was extracted from Wikipedia, https://en.wikipedia.org/wiki/Wolfgang_Klemperer#Life_in_the_United_States (accessed 10/10/2018)

[148] According to the website, "Measuring Worth," 8 million dollars in 1924 converts to between 92 to 1,780 million in 2018 dollars. https://www.measuringworth.com/calculators/ppowerus/ (accessed 10/10/18).

[149] "USS Akron (ZRS-4), Airship 1931-1933" https://web.archive.org/web/20040218013105/https://www.history.navy.mil/photos/ac-usn22/z-types/zrs4-v.htm (accessed 10/10/18).

[150] See Michael Salla, *The US Navy's Secret Space Program and Nordic Extraterrestrial Alliance*, pp. 43-60.

[151] Dr Robert and Ryan Wood, eds, *The Majestic Documents*, p. 77

# ENDNOTES CHAPTER 6

[152] Translation supplied via Linda Moulton Howe and Duncan Roads, "Operation Highjump Photos from Anonymous Source" https://www.earthfiles.com/news.php?ID=2577&category=Science, (accessed 11/13/17).

[153] Translation supplied via Linda Moulton Howe and Duncan Roads, "Operation Highjump Photos from Anonymous Source" https://www.earthfiles.com/news.php?ID=2577&category=Science, (accessed 11/13/17).

[154] See Michael Salla, *The US Navy's Secret Space Program and Nordic Extraterrestrial Alliance* (Exopolitics Consultants, 2017) pp. 91-103.

[155] Kenneth Arnold letter to US Army Air Force, available online at: http://www.project1947.com/fig/ka.htm (accessed 10/11/18).

[156] Ben Rich in his book, *Skunk Works: A Personal Memoir of My Years at Lockheed* (Little Brown and Co., 1994).

[157] The role of the Ho 229 in the evolution of flying wing jet craft is available online at: https://www.militaryfactory.com/aircraft/detail.asp?aircraft_id=105 (accessed 10/16/18).

[158] See Michael Swords, "Project Sign and the Estimate of the Situation", http://www.nicap.org/papers/swords_Sign_EOTS.htm (accessed 10/16/18).

[159] Edward Ruppelt, *The Report of Unidentified Flying Objects* available online at: *https://tinyurl.com/yafcoxge* (accessed 10/11/18).

[160] For the two page memorandum ordering the creation of the new laboratory, see Dr. Robert Wood and Ryan Wood, "Documents Dated Prior to 1948", http://www.majesticdocuments.com/documents/pre1948.php (accessed 10/6/2018).

[161] Available online at: https://tinyurl.com/y37uh8hv (accessed 10/18/18).

[162] *The Report of Unidentified Flying Objects available online at: https://tinyurl.com/yafcoxge* (accessed 10/11/18).

[163] CIA website, https://www.cia.gov/library/readingroom/docs/CIA-RDP81R00560R000100020011-8.pdf (accessed 10/18/18).

[164] See Michael Swords, "Project Sign & Estimate of the Situation," https://www.bibliotecapleyades.net/sociopolitica/sign/sign.htm (accessed 10/15/2018).

[165] Donald Keyhoe, *Aliens from Space* (1973) p. 14.

[166] Interviewed by Linda Moulton Howe, *Earthfiles*, https://tinyurl.com/y24lbxdu (accessed 4/4/15).

[167] CIA website, https://tinyurl.com/y3z3vz8w (accessed 10/18/2018)

[168] See Edward Ruppelt, *The Report of Unidentified Flying Objects,* ch. 8, available online at: https://tinyurl.com/yafcoxge (accessed 10/11/18).

[169] Interviewed by Linda Moulton Howe, *Earthfiles*, https://www.earthfiles.com/2008/09/12/part-4-army-cia-unit-studied-the-real-ufo-blue-book-cases/ (accessed 10/15/2018).

[170] For discussion of these alleged German extraterrestrial agreements see Michael Salla, *The US Navy's Secret Space Program and Nordic Extraterrestrial Alliance*, pp. 43-60.

[171] I discuss the relationship between Gray aliens and the Draco Reptilians in Michael Salla, *Galactic Diplomacy* (Exopolitics Institute, 2013) pp. 181-84.

## ENDNOTES CHAPTER 7

[172] See Michael Salla, *Antarctica's Hidden History: Corporate Foundations of Secret Space Programs* (Exopolitics Consultants, 2018) pp. 91-103.

[173] "Las Cruces Sun-News from Las Cruces, New Mexico · Page 1" https://www.newspapers.com/newspage/35271541/ (accessed 10/17/18).

[174] Timothy Good, *Need to Know: UFOs, the Military and Intelligence* (Pegasus Books, 2007), pp. 55-56.

[175] Cited by Linda Moulton Howe in ""Peculiar Phenomena," V-2 Rockets – and UFOB Retaliation?" https://tinyurl.com/y5tj4lgj (accessed 10/17/18).

[176] Byrd's reference to hostile aircraft can be found here: "Operation Highjump Photos from Anonymous Source" https://www.earthfiles.com/news.php?ID=2577&category=Science, (accessed 11/13/17).

[177] Cited by Timothy Good in *Need to Know*, pp. 57-58.

[178] Cited by Linda Moulton Howe in "Peculiar Phenomena," V-2 Rockets – and UFOB Retaliation?" https://tinyurl.com/y5tj4lgj (accessed 10/17/18).

[179] Linda Moulton Howe in "Peculiar Phenomena," V-2 Rockets – and UFOB Retaliation?" https://tinyurl.com/y5tj4lgj (accessed 10/17/18).

[180] Cited by Linda Moulton Howe in "Peculiar Phenomena," V-2 Rockets – and UFOB Retaliation?" https://tinyurl.com/y5tj4lgj (accessed 10/17/18).

[181] Andrew Kissner cited by Tim Good, *Need to Know*, p. 101.

[182] Stein as quoted by Linda Moulton Howe"Part 2: Army/CIA Unit Studied the Real UFO Blue Book Cases", https://www.earthfiles.com/2008/09/02/part-2-army-cia-unit-studied-the-real-ufo-blue-book-cases/ (accessed 10/19/18).

[183] Stein as quoted by Linda Moulton Howe"Part 4: Army/CIA Unit Studied the Real UFO Blue Book Cases", https://www.earthfiles.com/2008/09/12/part-4-army-cia-unit-studied-the-real-ufo-blue-book-cases/ (accessed 10/19/18).

[184] Edward Ruppelt, *The Report of Unidentified Flying Objects available online at: https://tinyurl.com/yafcoxge* (accessed 10/11/18).

[185] Cited by Timothy Good, *Need to Know*, p. 105.

[186] Edward Ruppelt, *The Report of Unidentified Flying Objects available online at: https://tinyurl.com/yafcoxge* (accessed 10/11/18).

[187] Cited by Timothy Good in *Need to Know*, p. 105.

[188] Cited by Frank Fechino, "Shoot Them Down! - The Flying Saucer Air Wars Of 1952", https://rense.com/general78/shoot.htm (accessed 12/30/18).

[189] Edward Ruppelt, *The Report of Unidentified Flying Objects available online at: https://tinyurl.com/yafcoxge* (accessed 10/11/18).

[190] https://rense.com/general78/shoot.htm ((accessed 10/19/18).

## ENDONTES CHAPTER 8

[191] See "1952 Washington DC UFO Incident," http://en.wikipedia.org/wiki/1952_Washington,_D.C._UFO_incident

[192] "Conversations with Major Donald Keyhoe", https://tinyurl.com/y7ul2b3h (accessed 10/21/18).

[193] Washington Post, "50 Years ago, UFOs buzzed the nation's Capitol," reprintedWaterloo/Cedar Falls Courier:" (July 26, 2002) p A8. Available online at: https://tinyurl.com/ydxcqxnh (accessed 10/19/18).

[194] Washington Post, "50 Years ago, UFOs buzzed the nation's Capitol," reprintedWaterloo/Cedar Falls Courier:" (July 26, 2002) p A8. Available online at: https://tinyurl.com/ydxcqxnh (accessed 10/19/18).

[195] Washington Post, "50 Years ago, UFOs buzzed the nation's Capitol," reprintedWaterloo/Cedar Falls Courier:" (July 26, 2002) p A8. Available online at: https://tinyurl.com/ydxcqxnh (accessed 10/19/18).

[196] See "1952: Second American Sighting Wave" https://www.bibliotecapleyades.net/ciencia/ufo_briefingdocument/1952.htm (accessed 10/22/18).

[197] See Wikipedia, "1952 Washington, D.C. UFO incident", https://en.wikipedia.org/wiki/1952_Washington,_D.C._UFO_incident (accessed 10/22/18).

[198] Edward Ruppelt, *The Report of Unidentified Flying Objects,* ch. 12, available online at*: https://tinyurl.com/yafcoxge* (accessed 10/11/18).

[199] Edward Ruppelt, *The Report of Unidentified Flying Objects, available online at: https://tinyurl.com/yafcoxge* (accessed 10/11/18).

[200] See: Barbara LaGrange, "The Intrigue of Giant Rock" http://www.lucernevalley.net/giantrock/ (accessed 10/21/18).

[201] George Van Tassel, *I Rode on a Flying Saucer* (New Age Publishing Co. 1952) p. 30. Available online at: www.scribd.com/doc/467760/I-Rode-A-Flying-Saucer-George-Van-Tassel (accessed 10/21/18).

[202] George Van Tassel, *I Rode on a Flying Saucer,* p. 19. Available online at: www.scribd.com/doc/467760/I-Rode-A-Flying-Saucer-George-Van-Tassel (accessed 10/21/18).

[203] George Van Tassel, *I Rode on a Flying Saucer,* 20.

[204] George Van Tassel, *I Rode on a Flying Saucer,* 20.

[205] George Van Tassel, *I Rode on a Flying Saucer,* 30-32. Available online at: www.scribd.com/doc/467760/I-Rode-A-Flying-Saucer-George-Van-Tassel (accessed 10/21/18).

[206] For discussion of Oppenheimer's opposition to the Hydrogen Bomb, see http://en.wikipedia.org/wiki/J._Robert_Oppenheimer#Postwar_activities (accessed 10/21/18).

[207] See: http://en.wikipedia.org/wiki/Ivy_Mike (accessed 10/21/18).

[208] Rakesh Krishnan Simha, "Nuclear overkill: The quest for the 10 gigaton bomb", *Russia Beyond,* https://www.rbth.com/opinion/2016/01/05/nuclear-overkill-the-quest-for-the-10-gigaton-bomb_556351(accessed 10/21/18).

[209] Tom Van Flandern, "The Exploded Planet Hypothesis," https://www.bibliotecapleyades.net/sumer_anunnaki/esp_sumer_annunaki23.htm (accessed 10/21/18).

[210] See John E. Brandenburg, *Death on Mars: The Discovery of a Planetary Nuclear Massacre* (Adventures Unlimited Press, 2015).

[211] Michael Salla, *Antarctica's Hidden History: Corporate Foundations of Secret Space Programs*, pp. 295-314.

[212] "Briefing Document: Operation Majestic 12: Prepared for President-Elect Dwight D. Eisenhower", http://www.majesticdocuments.com/pdf/eisenhower_briefing.pdf (accessed 10/23/18).

[213] See Michael Salla, *Antarctica's Hidden History: Corporate Foundations of Secret Space Programs*, pp. 1-12.

[214] A video of Byrd's mysterious February 19, 1947 flight over Antarctica is available online at: https://www.youtube.com/watch?v=zjKI_LzPjeY (accessed 10/23/18).

[215] Interview with Corey Goode, May 19, 2014 "Corporate bases on Mars and Nazi infiltration of US Secret Space Program," http://exopolitics.org/corporate-bases-on-mars-and-nazi-infiltration-of-us-secret-space-program/ (accessed 6/30/15).

[216] Admiral Richard Byrd's Diary, available online at: http://www.bibliotecapleyades.net/tierra_hueca/esp_tierra_hueca_2d.htm (accessed 10/23/18).

[217] Admiral Richard Byrd's Diary, available online at: http://www.bibliotecapleyades.net/tierra_hueca/esp_tierra_hueca_2d.htm (accessed 10/23/18).

[218] Joseph Farrell, *Reich Of The Black Sun: Nazi Secret Weapons & The Cold War Allied Legend* (Adventures Unlimited Press, 2005).

[219] See Michael Salla, *Antarctica's Hidden History: Corporate Foundations of Secret Space Programs*, pp. 159-78.

[220] See Michael Salla, *Antarctica's Hidden History: Corporate Foundations of Secret Space Programs*, pp. 174-78.

[221] Justin Deschamps, "David Wilcock and Corey Goode: History of the Solar System and Secret Space Program - Notes from Consciousness Life Expo 2016 ," https://stillnessinthestorm.com/2016/02/david-wilcock-and-corey-goode-history/ (accessed 10/21/18).

[222] See Michael Salla, *The US Navy's Secret Space Program and Nordic Extraterrestrial Alliance* (Exopolitics Consultants, 2017) pp. 43-60.

[223] Private Interview with William Tompkins, April 30, 2017.

[224] Interview with Corey Goode, May 19, 2014 "Corporate bases on Mars and Nazi infiltration of US Secret Space Program," http://exopolitics.org/corporate-bases-on-mars-and-nazi-infiltration-of-us-secret-space-program/ (accessed 6/30/15).

[225] Clark McClelland, *The Stargate Chronicles*, chapter 28, http://www.stargate-chronicles.com/site/ (accessed 10/21/18).

[226] Clark McClelland, *The Stargate Chronicles*, chapter 32, http://www.stargate-chronicles.com/site/ (accessed 10/21/18).

## ENDNOTES CHAPTER 9

[227] The history of the Soviet Union's hydrogen bomb development is available online at: http://soviethistory.msu.edu/1954-2/hydrogen-bomb/ (accessed 10/24/18).

[228] "Eisenhower's "Atoms for Peace" Speech", https://www.atomicheritage.org/key-documents/eisenhowers-atoms-peace-speech (accessed 10/24/18).

[229] Bernard Brodies papers on nuclear strategy are available online at: https://www.rand.org/pubs/authors/b/brodie_bernard.html (accessed 10/24/18).

[230] Initial communist press coverage of Eisenhower's speech is available at: https://eisenhower.archives.gov/research/online_documents/atoms_for_peace/Binder15.pdf (accessed 10/24/18).

[231] "Atoms for Peace: The Mixed Legacy of Eisenhower's Nuclear Gambit", https://www.sciencehistory.org/distillations/magazine/atoms-for-peace-the-mixed-legacy-of-eisenhowers-nuclear-gambit (accessed 10/24/18).

[232] William Moore, "UFO's: Exploring the ET Phenomenon," *Gazette* (Hollywood, CA., March 29, 1989). Available online at: https://tinyurl.com/ycn9kcun (accessed 12/31/18).

[233] Bill Kirklin, "Ike and UFO's," *Exopolitics Journal*, available at http://exopoliticsjournal.com/vol-2/vol-2-1-Exp-Ike.htm (accessed 12/31/18).

[234] Private email sent to author on February 6, 2009.

[235] Transcript of Project Camelot, "Air Force One and the Alien Connection: A Video Interview with Bill Holden," June 2007, http://www.bibliotecapleyades.net/vida_alien/alien_zetareticuli03.htm (accessed 12/31/18).

236 For description of the U.S. Nuclear Testing program at the Bikini Islands, Jane Dibblin, *Day of Two Suns: Us Nuclear Testing and the Pacific Islanders* (New Amsterdam Books, 2002).

237 Available online at http://www.youtube.com/watch?v=NNV8-k5UvpY. Transcript available at McElroy's former website: http://tinyurl.com/bzlu74x (accessed 12/31/18).

238 "New gold money bill introduced in N.H. General Court," http://www.newswithviews.com/NWVexclusive/exclusive55.htm

239 See http://www.gencourt.state.nh.us/house/committees/committeedetails.aspx?id=16 (accessed 10/24/18)

240 Available online at http://www.youtube.com/watch?v=NNV8-k5UvpY. Transcript available at McElroy's former website: http://tinyurl.com/bzlu74x (accessed 12/31/18).

241 Available online at http://www.youtube.com/watch?v=NNV8-k5UvpY.

242 Available online at http://www.youtube.com/watch?v=NNV8-k5UvpY.

243 Available online at http://www.youtube.com/watch?v=NNV8-k5UvpY..

244 Available online at http://www.youtube.com/watch?v=NNV8-k5UvpY.

245 Personal notes from William Hamilton from a 1991 interview with Sgt Suggs. See also William Hamilton, *Project Aquarius: The Story of An Aquarian Scientist* (Authorhouse, 2005) pp. 85-86.

246 Personal notes supplied by William Hamilton of a 1991 interview with Commander Suggs' son, Sgt Charles Suggs, Jr. See also, Bill Hamilton, *Project Aquarius: The Story of An Aquarian Scientist,* p. 85.

247 Cited in Timothy Good, *Alien Contact: Top-Secret UFO Files Revealed* (William Morrow and Co., 1993) p. 75

248 Timothy Good, *Need to Know: UFOs, the Military and Intelligence* (Pegasus Books, 2007) p. 208.

249 My summary of a phone interview conducted in February 2004

250 Timothy Good, *Need to Know,* p. 208.

251 Christoforo Barbato, "Vatican and UFO: Secretum Omega," https://tinyurl.com/ybjny9w2 (accessed 12/31/18).

252 Luca Scantamburlo, "Planet X and the "JESUIT FOOTAGE" Classified "SECRETUM OMEGA." First Indirect Confirmation!" https://tinyurl.com/yaf7cwba (accessed 12/31/18).

253 "Testimony of Don Phillips," *Disclosure,* ed., Stephen Greer (Crossing Point, 2001) p. 379

254 William Cooper, *Behold A Pale Horse* (Light Technology Publishing, 1991) 381-96.

255 Salter, *Life With a Cosmos Clearance* (Light Technology Publishing, 2003) p. 8. See also Robert Dean Interview, "Command Sergeant Major Robert Dean, NATO's Secret UFO Assessment & Setting the Record Straight" *Exopolitics Journal* (April 2006) 213-232). Available online at: http://www.exopoliticsjournal.com/Journal-vol-1-3-Dean.pdf (accessed 12/31/18).

[256] Cooper, "Origin, Identity, and Purpose of MJ-12," in *Behold A Pale Horse*, 201-202 . Online version available at: http://www.bibliotecapleyades.net/sociopolitica/esp_sociopol_mj12_1.htm (accessed 12/31/18).

[257] Donald Keyhoe, *Aliens from Space: The Real Story of Unidentified Flying Objects* (Doubleday and Co., 1973) pp. 129-30.

[258] Keyhoe, *Aliens from Space,* p.131.

[259] Cooper, "Origin, Identity, and Purpose of MJ-12," in *Behold A Pale Horse*, 201-202 . Online version available at: http://www.bibliotecapleyades.net/sociopolitica/esp_sociopol_mj12_1.htm (accessed 12/31/18).

[260] For discussion of a discrepancy between William Cooper's testimony that the extraterrestrials met with President Eisenhower at Homestead AFB as opposed to Edwards AFB, as claimed by other whistleblower examined in this chapter, see Michael Salla, *Galactic Diplomacy: Getting to Yes with ET* (Exopolitics Institute, 2013), pp. 62-63

[261] The letter is reprinted in Timothy Good, *Alien Contact, p.* 275.

[262] John Spencer, "Light, Gerald," *The UFO Encyclopedia: Inexplicable Sightings, Alien Abductions, Close Encounters, Brilliant Hoaxes* (Avon Books, 1991) p. 188.

[263] "A Covenant With Death by Bill Cooper," http://www.alienshift.com/id40.html (accessed 12/31/18). Also in William Cooper, *Behold a Pale Horse, p.* 203.

[264] For analysis of the four national security figures that attended the 1954 meeting with Eisenhower, see Michael Salla, *Galactic Diplomacy: Getting to Yes with ET*, pp. 61-62.

## ENDNOTES CHAPTER 10

[265] Cited by Art Campbell, "Eisenhower's Incredible Journey: Did Ike Meet with the ETs?" http://www.ufocrashbook.com/eisenhower.html (accessed 10/25/18).

[266] For discussion of the 1954 Edwards Air Force Base meeting, see Michael Salla, *Galactic Diplomacy: Getting to Yes with ET* (Exopolitics Institute, 2013) pp. 47-63.

[267] See Art Campbell, "Eisenhower's Incredible Journey: Did Ike Meet with the ETs?" http://www.ufocrashbook.com/eisenhower.html (accessed 10/25/18).

[268] Bill Kirklin's paper was published anonymously as "Ike and UFOs," in the *Exopolitics Journal* 2:1 (2007): http://exopoliticsjournal.com/vol-2/vol-2-1-Exp-Ike.htm Kirklin's says he was told about Ike's visit in late February, this is likely a minor mistake on his part since the visit occurred on February 11, 1955. (accessed 10/25/18).

[269] Bill Kirklin, "Ike and UFOs," *Exopolitics Journal* 2:1 (2007): http://exopoliticsjournal.com/vol-2/vol-2-1-Exp-Ike.htm (accessed 10/25/18)

[270] Kirklin,"Ike and UFOs:" http://exopoliticsjournal.com/vol-2/vol-2-1-Exp-Ike.htm (accessed 10/25/18).

[271] Kirklin,"Ike and UFOs:" http://exopoliticsjournal.com/vol-2/vol-2-1-Exp-Ike.htm (accessed 10/25/18).

[272] Transcript of electrician's family letter – recorded by Art Campbell and played on Jerry Pippin Show - 6/23/08. Available online at: https://tinyurl.com/yc832wgo (accessed 12/31/18).

[273] Transcript of Staff Sgt Wykoff interviewed by Art Campbell and played on Jerry Pippin Show - 6/23/08. Available online at: https://tinyurl.com/yc832wgo (accessed 12/31/18).

[274] For Kirklin's recollection of the debriefing of base personnel, see Michael Salla, *Antarctica's Hidden History*, pp. 122-23. Also available online: "Ike and UFOs:" http://exopoliticsjournal.com/vol-2/vol-2-1-Exp-Ike.htm (accessed 10/25/18).

[275] Milton William Cooper, "Origin, Identity, and Purpose of MJ-12," http://www.bibliotecapleyades.net/sociopolitica/esp_sociopol_mj12_1.htm (accessed 10/25/18).

[276] Milton William Cooper, "Origin, Identity, and Purpose of MJ-12," http://www.bibliotecapleyades.net/sociopolitica/esp_sociopol_mj12_1.htm (accessed 10/25/18).

[277] Milton William Cooper, "Origin, Identity, and Purpose of MJ-12," http://www.bibliotecapleyades.net/sociopolitica/esp_sociopol_mj12_1.htm (accessed 10/25/18).

[278] Milton William Cooper, "Origin, Identity, and Purpose of MJ-12," in *Behold a Pale Horse,* 203-04. Also available at: http://www.bibliotecapleyades.net/sociopolitica/esp_sociopol_mj12_1.htm (accessed 10/25/18).

[279] For background information on Schneider and his lectures, see :The Phil Schneider Story," http://www.apfn.org/apfn/phil.htm (accessed 10/25/18).

[280] Phil Schneider, "MUFON Conference Presentation, 1995," available online at: http://web.archive.org/web/20000926021403/http://www.anomalous-images.com//text/schneid.html (accessed 10/25/18).

[281] See Chris Stoner, 'The Revelations of Dr Michael Wolf on the UFO Cover Up and ET Reality," (October 2000) https://tinyurl.com/ycvp58c4 (accessed 12/31/18).

[282] See Paola Harris, http://paolaharris.com/newolfint.htm; and Dr Richard Boylan, http://www.drboylan.com/wolfdoc2.html (accessed 10/25/18).

[283] See Richard Boylan, "Official Within MJ-12 UFO-Secrecy Management Group Reveals Insider Secrets," http://www.drboylan.com/wolfdoc2.html (accessed 12/31/18).

[284] "Executive Order 10483 – Establishing the Operations Coordinating Board" https://www.presidency.ucsb.edu/documents/executive-order-10483-establishing-the-operations-coordinating-board (accessed 10/29/18).

[285] Phillip Corso, *The Day After Roswell* (Pocket Books, 1997) p. 292.

[286] Clark McClelland, *The Stargate Chronicles*, ch. 32, http://www.stargate-chronicles.com/site/ (accessed 10/17/17).

[287] Clark McClelland, *The Stargate Chronicles*, ch. 32, http://www.stargate-chronicles.com/site/ (accessed 10/17/17).

[288] I discuss whistleblower testimony that Gray aliens reached an agreement with the Eisenhower administration at Holloman AFB in *Galactic Diplomacy*, pp. 70-75.

[289] Clark McClelland, *The Stargate Chronicles*, ch. 32, http://www.stargate-chronicles.com/site/ (accessed 10/17/17).

[290] Official estimates of the number of German scientists brought to the US under Project Paperclip is 1500. Bill Tompkins claimed that the true number was many thousands.

[291] Private Interview with William Tompkins, April 30, 2017.

[292] Phillip Corso, *The Day After Roswell* (Pocket Books, 1997) p. 292.

[293] Phillip Corso, *The Day After Roswell*, p. 79.

[294] Interview with Corey Goode, May 19, 2014, "Corporate bases on Mars and Nazi infiltration of US Secret Space Program," http://exopolitics.org/corporate-bases-on-mars-and-nazi-infiltration-of-us-secret-space-program/ (accessed 6/30/15).

## ENDNOTES CHAPTER 11

[295] Bill Tompkins furthermore claimed that the drones belonged to the Nordic space fleet which had overflown Los Angeles and allowed the two drones to be shot down in order to show that the flying saucer phenomenon was very real, and possessed advanced technologies that was extraterrestrial in origin. See Michael Salla, *The US Navy's Secret Space Program and Nordic Extraterrestrial Alliance* (Exopolitics Consultants, 2017) p. 8.

[296] See "Documents Dated 1948-1959", http://www.majesticdocuments.com/documents/1948-1959.php (accessed 10/29/18).

[297] http://www.majesticdocuments.com/pdf/eisenhower_briefing.pdf (accessed 10/29/18).

[298] See "Documents Dated 1948-1959", http://www.majesticdocuments.com/documents/1948-1959.php (accessed 10/29/18).

[299] See Leslie Kean, "Project MoonDust and Operation Blue Fly", https://www.bibliotecapleyades.net/sociopolitica/esp_sociopol_mj12_3k.htm (accessed 12/31/18).

[300] See Leslie Kean, "Project MoonDust and Operation Blue Fly", https://www.bibliotecapleyades.net/sociopolitica/esp_sociopol_mj12_3k.htm (accessed 12/31/18).

[301] There has been great controversy over different versions of the Betz Memo being publicly released. For one perspective of this controversy and Memo's history, see The Rejuvenated "Betz Memo" http://ufos-documenting-the-evidence.blogspot.com/2016/05/the-rejuvenated-betz-memo-in-recent.html (accessed 12/31/18).

[302] Leslie Kean, "Project MoonDust and Operation Blue Fly", https://www.bibliotecapleyades.net/sociopolitica/esp_sociopol_mj12_3k.htm (accessed 12/31/18).

[303] In order to make it difficult to track the activities of the 4602d AISS, the USAF changed its designation several times. For further discussion and the connection to Fort Belvoir, see Francis Ridge, "UFOs, the AISS, & My Brush With the 1127th", http://www.nicap.org/brief1127.htm (accessed 12/31/18).

[304] "Special Operations Manuel," p. 3. Available online at http://www.majesticdocuments.com/pdf/som101_part1.pdf (accessed 10/29/18).

[305] Jeff Hecht, *City of Light: The Story of Fiber Optics*, Revised Edition (Oxford University Press, 2004) pp. 55-70.

[306] Philip Corso, *The Day After Roswell* (Pocket Books, 1997) p.4.

[307] "Special Operations Manuel," p. 3. Available online at http://www.majesticdocuments.com/pdf/som101_part1.pdf (accessed 10/29/18).

[308] Arthur Horn, *Humanity's Extraterrestrial Origins: extraterrestrial Influences on Humankind's Biological and Cultural Evolution* (A & L Horn, 1994) 259. See also Dr Arthur Horn, "The Orion Empire," available online at: https://tinyurl.com/yb2ulmhk (accessed 10/29/18).

[309] David Jacobs, *Alien Encounters: First-hand accounts of UFO abductions* (Virgin Books, 1994).

[310] David Jacobs, *Alien Encounters,* 279.

[311] "Testimony of Captain Bill Uhouse," *Disclosure*, ed., Stephen Greer (Crossing Point, 2001) p. 384.

[312] Ryan S. Wood, *MAJIC EYES ONLY: Earth's Encounters with Extraterrestrial Technology* (Wood Enterprises 2005) p. 111.

[313] Ryan S. Wood, *MAJIC EYES ONLY: Earth's Encounters with Extraterrestrial Technology*, p. 112.

[314] "Testimony of Captain Bill Uhouse," *Disclosure*, ed., Stephen Greer, p. 385.

[315] "Testimony of Captain Bill Uhouse," *Disclosure*, ed., Stephen Greer, p. 386.

[316] "Testimony of Captain Bill Uhouse," *Disclosure*, ed., Stephen Greer, pp. 386-87.

[317] See Gene Huff, "The Lazar Synopsis" (http://www.otherhand.org/home-page/area-51-and-other-strange-places/bluefire-main/bluefire/the-bob-lazar-corner/the-lazar-synopsis/ (accessed 10/30/18).

[318] See Charles Hall, *Millennial Hospitality*, Books 1-5. Website: http://www.millennialhospitality.com/index.php (accessed 10/30/18).

[319] See Michael Salla, "Further Investigations of Charles Hall and Tall Whites at Nellis Air Force Base: The David Coote Interviews," http://exopolitics.org/archived/Exo-Comment-36.htm (accessed 10/30/18).

[320] "Charles Hall and the Tall Whites: Another perception of the extraterrestrial phenomenon and the Area 51," https://tinyurl.com/y7bhdx7a (accessed 12/31/18).

[321] "Interview with Charles Hall- Motivations of Tall White ETs & their Exopolitical Significance", https://www.exopolitics.org/interview-with-charles-hall-motivations-of-tall-white-ets-their-exopolitical-significance/ (accessed 10/25/18).

## ENDNOTES CHAPTER 12

[322] For whistleblower accounts of service on Mars, see Michael Salla, *Insiders Reveal Secret Space Programs and Extraterrestrial Life* (Exopolitics Institute, 2015) pp.309-44.

[323] Anne Jacobsen, *Operation Paperclip: The Secret Intelligence Program that brought Nazi Scientists to America* (Little, Brown and Company, 2014) p. 11.

[324] Mark Erickson, *Into the Unknown Together: The DOD, NASA, and Early Spaceflight*, p. 60. Available online at: https://tinyurl.com/y76onu7q (accessed 10/31/18).

[325] Mark Erickson, *Into the Unknown Together: The DOD, NASA, and Early Spaceflight*, p. 19. Available online at: https://tinyurl.com/y76onu7q (accessed 10/31/18).

[326] Mark Erickson, *Into the Unknown Together: The DOD, NASA, and Early Spaceflight*, p. 21. Available online at: https://tinyurl.com/y76onu7q (accessed 10/31/18).

[327] Cited by Mark Erickson, *Into the Unknown Together: The DOD, NASA, and Early Spaceflight*, p. 35. Available online at: https://tinyurl.com/y76onu7q (accessed 10/31/18) p. 52.

[328] Mark Erickson, *Into the Unknown Together: The DOD, NASA, and Early Spaceflight*, p. 35. Available online at: https://tinyurl.com/y76onu7q (accessed 10/31/18).

[329] Richard Hoagland and Mike Bara, *Dark Mission: The Secret History of NASA* (Feral House, 2007) p. ii.

[330] "National Aeronautics and Space Act of 1958" (Unamended) Sec. 305. (i) Available online at: https://history.nasa.gov/spaceact.html (accessed 11/2/18).

[331] Richard Hoagland and Mike Bara, *Dark Mission: The Secret History of NASA*, p. vi.

[332] See John M. Logsdon, Project Apollo: Americans to the Moon", https://history.nasa.gov/SP-4407vol7Chap2.pdf (accessed 11/2/18).

[333] President Kennedy's May 25, 1961 speech is available online at: https://history.nasa.gov/moondec.html (accessed 11/2/18).

[334] Richard Hoagland and Mike Bara, *Dark Mission*, p. 297.

[335] "MSFC History Office", https://history.msfc.nasa.gov/history_fact_sheet.html (accessed 11/1/18).

[336] Biographical information on Dr. Eberhard Rees available at: https://history.msfc.nasa.gov/management/center_directors/pages/rees.html (accessed 11/1/18).

[337] William Tompkins, *Selected by Extraterrestrials*, p. 329.

[338] William Tompkins, *Selected by Extraterrestrials*, p. 332.

[339] See Michael Salla, *Antarctica's Hidden History*, pp. 357-60.

[340] See Michael Salla, *The U.S. Navy's Secret Space Program and Nordic Extraterrestrial Alliance* (Exopolitics Institute, 2017) pp. 127-42.

[341] See "Apollo Program Budget Appropriations", https://history.nasa.gov/SP-4029/Apollo_18-16_Apollo_Program_Budget_Appropriations.htm (accessed 11/3/18).

[342] See Michel Salla, *Antarctica's Hidden History*, pp. 129-30.

343 For Allen Dulles role as MJ-1 see Michael Salla, Kennedy's Last Stand: UFO's, Eisenhower and the JFK Assassination (Exopolitics Institute, 2013) pp. 127-31.

344 50 U.S.C. 403f(a).

345 For discussion of the deep black budget, see Michael Salla, "The Black Budget Report: An Investigation into the CIA's 'Black Budget' and the Second Manhattan Project," http://exopolitics.org/Report-Black-Budget.htm (accessed 11/3/18).

346 "Pentagon unveils $686 billion military budget for FY19", https://www.defensenews.com/breaking-news/2018/02/12/pentagon-unveils-686-billion-military-budget-for-2019/ (accessed 11/3/18).

347 See "Apollo Program Budget Appropriations", https://history.nasa.gov/SP-4029/Apollo_18-16_Apollo_Program_Budget_Appropriations.htm (accessed 11/3/18).

348 See "U.S. Military Spending, 1946–2009", https://www.infoplease.com/us/military-personnel/us-military-spending-1946-2009 (accessed 11/3/18)

349 See "Apollo Program Budget Appropriations", https://history.nasa.gov/SP-4029/Apollo_18-16_Apollo_Program_Budget_Appropriations.htm (accessed 11/3/18).

350 Mark Erickson, *Into the Unknown Together: The DOD, NASA, and Early Spaceflight*, p. 35. Available online at: https://tinyurl.com/y76onu7q (accessed 10/31/18) p. 60.

351 See "Von Braun Station", http://astronautix.com/v/vonbraunstation.html (accessed 11/5/18)

352 Wernher Von Braun, *The Mars Project* (University of Illinois Press; 1962 [1952]).

353 Clark McClelland, *The Stargate Chronicles*, ch. 15, http://tinyurl.com/ox66j9y (accessed 6/30/15).

354 For discussion of Antarctica Germans infiltrating NASA and US Military Industrial Complex, see Michael Salla, *Antarctica's Hidden History: Corporation Foundations of Secret Space Programs,* pp. 125-29.

## ENDNOTES CHAPTER 13

355 See Adam Gruen, "Manned versus Unmanned Space Systems," *The U.S. Air Force in Space, 1945 to the 21st Century (1995)* p. 69. Available online at https://www.scribd.com/document/50601398/The-U-S-Air-Force-in-Space-1945-to-the-21st-Century (accessed 11/6/18).

356 Cited by Mark Erickson, *Into the Unknown Together: The DOD, NASA, and Early Spaceflight*, p. 340. Available online at: https://tinyurl.com/y76onu7q (accessed 10/31/18).

357 "Memorandum for Director, Manned Orbiting Laboratory (MOL) Program, August 25, 1962. http://www.nro.gov/Portals/65/documents/foia/declass/mol/1.pdf (accessed 11/6/18).

[358] See "MOL Program Perspective"
http://www.nro.gov/Portals/65/documents/foia/declass/mol/736.pdf
[359] "Air Force to Develop Manned Orbiting Laboratory,"
http://www.nro.gov/Portals/65/documents/foia/declass/mol/6.pdf (accessed
11/6/18).
[360] "Air Force to Develop Manned Orbiting Laboratory,"
http://www.nro.gov/Portals/65/documents/foia/declass/mol/6.pdf (accessed
11/6/18).
[361] For discussion of the Transtage, see http://www.designation-
systems.net/dusrm/app3/b-10.html (accessed 11/6/18).
[362] For detailed discussion of the Titan II and its use in the Gemini Space
Program, see Mark Hall, "Titan II" http://www.astronautix.com/t/titanii.html
(accessed 11/6/18).
[363] See President Johnson's Statement on MOL"
http://www.nro.gov/Portals/65/documents/foia/declass/mol/129.pdf (accessed
11/6/18).
[364] See March 10, 1964 letter to Major General Ben L. Funk,
http://www.nro.gov/Portals/65/documents/foia/declass/mol/15.pdf (accessed
11/6/18).
[365] "DORIAN Optical Studies",
http://www.nro.gov/Portals/65/documents/foia/declass/mol/93.pdf (accessed
11/6/18).
[366] Astrospies, http://www.pbs.org/wgbh/nova/military/astrospies.html (accessed
7/28/15).
[367] See "Memorandum for the Record- DOD MOL - A Consideration of
International Political Factors",
http://www.nro.gov/Portals/65/documents/foia/declass/mol/108.pdf (accessed
11/7/18).
[368] See Michael Salla, Antarctica's Hidden History: Corporate Foundations of
Secret Space Programs (Exopolitics Consultants, 2018) p. 86.
[369] "Press Guidance on Location of MOL Launches",
http://www.nro.gov/Portals/65/documents/foia/declass/mol/211.pdf (accessed
11/9/18).
[370] "Memorandum for Chairman Revers from Harold Brown, Subject:
Determination of the Launch Site for MOL"
http://www.nro.gov/Portals/65/documents/foia/declass/mol/219.pdf (accessed
11/9/18).
[371] See Major A. Andronov, "American Geosynchronous SIGINT Satellites"
https://fas.org/spp/military/program/sigint/androart.htm (accessed 11/9/18).
[372] "Application of MOL to Astronomical Observations",
http://www.nro.gov/Portals/65/documents/foia/declass/mol/341.pdf (accessed
11/9/18).
[373] See "Agreement Between NASA and DoD Concerning Gemini Program and
MOL" http://www.nro.gov/Portals/65/documents/foia/declass/mol/33.pdf
(accessed 11/6/18).

374 "Letter to Mr. Webb from Mr. McNamara, Subject: Air Force Effort on MOL During Last Several Months"
http://www.nro.gov/Portals/65/documents/foia/declass/mol/47.pdf (accessed 11/6/18).
375 "Memorandum for the Record, Subject: Proposed MOL Press Release",
http://www.nro.gov/Portals/65/documents/foia/declass/mol/114.pdf (accessed 11/7/18).
376 Astrospies, http://www.pbs.org/wgbh/nova/military/astrospies.html (accessed 7/28/15)
377 Mark Erickson, *Into the Unknown Together: The DOD, NASA, and Early Spaceflight*, p. 343. Available online at: https://tinyurl.com/y76onu7q (accessed 10/31/18).
378 Letter from George E. Mueller,
http://www.nro.gov/Portals/65/documents/foia/declass/mol/3.pdf (accessed 11/6/18).
379 Von Braun Station http://astronautix.com/v/vonbraunstation.html (accessed 11/6/18).
380 When combined with the Gemini 2 capsule, the MOL had a total length of 72 feet (22 meters), see http://www.astronautix.com/m/mol.html (accessed 11/7/18).
381 "MOL Directive No. 67-4, MOL Program Advanced Planning Approved by Gen. Ferguson",
http://www.nro.gov/Portals/65/documents/foia/declass/mol/353.pdf (accessed 11/7/18).
382 Full list of declassified documents available at: "Index, Declassified Manned Orbiting Laboratory (MOL) Records" http://www.nro.gov/Freedom-of-Information-Act-FOIA/Declassified-Records/Special-Collections/MOL/ (accessed 11/6/18).
383 "Draft Memorandum for the President, Subject: MOL",
http://www.nro.gov/Portals/65/documents/foia/declass/mol/709.pdf (accessed 11/9/18).
384 "The Department of Defense has Terminated the Air Force MOL Program",
http://www.nro.gov/Portals/65/documents/foia/declass/mol/742.pdf (accessed 11/9/18).
385 "Terminate MOL Except for the "Automatic" Camera System",
http://www.nro.gov/Portals/65/documents/foia/declass/mol/737.pdf (accessed 11/9/18).
386 See "KeyHole (KH) series satellites"
http://rammb.cira.colostate.edu/dev/hillger/KH.htm (accessed 11/17/18)
387 "Terminate MOL Except for the "Automatic" Camera System",
http://www.nro.gov/Portals/65/documents/foia/declass/mol/737.pdf (accessed 11/9/18).
388 "Lockheed Dorian Resupply Study,
http://www.nro.gov/Portals/65/documents/foia/declass/mol/407.pdf (accessed 11/17/18).

[389] "Advanced MOL Planning: Missions and Systems", p. 17, available at: http://www.nro.gov/Portals/65/documents/foia/declass/mol/794.pdf (accessed 11/17/18).

[390] Ronald Reagan, *The Reagan Diaries* (Harper Perennial, 2009) p. 334.

[391] Available online at: http://www.nro.gov/Portals/65/documents/foia/declass/mol/737.pdf (accessed 11/17/18).

[392] See MSFC History Office, https://history.msfc.nasa.gov/vonbraun/bio.html (accessed 11/17/18).

[393] See "Testimony of Dr. Carol Rosin", http://www.illuminati-news.com/ufos-and-aliens/html/carol_rosin.htm (accessed 11/19/18).

[394] See Michael Salla, *Antarctica's Hidden History*, pp. 56-70.

[395] See "Project Blue Beam", http://educate-yourself.org/cn/projectbluebeam25jul05.shtml (accessed 11/19/18)

[396] See Michael Salla, "Secret NRO Space Stations to be Revealed in Limited Disclosure Plan", https://www.exopolitics.org/secret-nro-space-stations-to-be-revealed-in-limited-disclosure-plan/ (accessed 12/31/18).

[397] "Advanced MOL Planning: Missions and Systems", p. 17, available at: http://www.nro.gov/Portals/65/documents/foia/declass/mol/794.pdf (accessed 11/17/18).

[398] "Advanced MOL Planning: Missions and Systems", p. 17, available at: http://www.nro.gov/Portals/65/documents/foia/declass/mol/794.pdf (accessed 11/17/18).

[399] "Advanced MOL Planning: Missions and Systems", p. 23, available at: http://www.nro.gov/Portals/65/documents/foia/declass/mol/794.pdf (accessed 11/17/18).

[400] See European Space Agency, "International Space Station," http://m.esa.int/Our_Activities/Human_Spaceflight/International_Space_Station/Building_the_International_Space_Station3 and Wikipedia, http://en.wikipedia.org/wiki/Assembly_of_the_International_Space_Station (accessed on 6/10/15).

[401] See Clark McClelland, *The Stargate Chronicles*, ch. 15, http://tinyurl.com/ox66j9y (accessed 6/30/15).

[402] Dr Steven Greer, "Testimony of Mr. Mark McCandlish, US Air Force", *Disclosure: Military and Government Witnesses Reveal the Greatest Secrets in Modern History* (Crossing Point, Inc, 2001) pp. 500-01.

[403] Mark McCandlish comments about what Sorensen told him about the accuracy of the depiction at: https://youtu.be/9QNvZN7X7v8?t=3502 (accessed 11/24/18).

[404] For discussion of the Biefeld Brown Effect see Michael Salla, *Insiders Reveal Secret Space Programs and Extraterrestrial Alliances* (Exopolitics Institute, 2015) pp. 19-28.

[405] Mark McCandlish's ARV art is available for purchase as a high resolution print. For more details, please visit http://www.markmccandlish.com/ (accessed 4/23/19).

[406] Paul La Violette, *Secrets of Antigravity Propulsion: Tesla, UFOs and Classified Aerospace Technology* (Bear and Co., 2008) 9.

[407] "A Method of and an Apparatus or Machine for Producing Force or Motion." http://www.checktheevidence.com/Disclosure/Web%20Pages/www.soteria.com/brown/docs/egravity/gravsap1.htm (accessed on 6/10/15).

[408] Dr Steven Greer, "Testimony of Mr. Mark McCandlish, US Air Force", *Disclosure: Military and Government Witnesses Reveal the Greatest Secrets in Modern History*, p. 502.

[409] Dr Steven Greer, "Testimony of Mr. Mark McCandlish, US Air Force", *Disclosure: Military and Government Witnesses Reveal the Greatest Secrets in Modern History*, p. 504.

[410] Dr Steven Greer, "Testimony of Mr. Mark McCandlish, US Air Force", *Disclosure: Military and Government Witnesses Reveal the Greatest Secrets in Modern History*, pp. 505-506.

[411] Dr Steven Greer, "Testimony of Mr. Mark McCandlish, US Air Force", *Disclosure: Military and Government Witnesses Reveal the Greatest Secrets in Modern History*, pp. 500-01, 504.

[412] See Earthfiles, http://www.earthfiles.com/news.php?ID=1501&category=Real+X-Files (accessed on 11/14/17).

[413] See Michael Salla, *Antarctica's Hidden History: Corporate Foundations of Secret Space Programs* (Exopolitics Consultants, 2018) pp. 56-71. A brief biography of Vladimir Terziski is available online at: http://www.whale.to/b/terziski_h.html (accessed on 11/14/17).

[414] Vladimir Terziski presented his findings in a 1992 workshop available online at: https://youtu.be/MPBvHjuJtB8

[415] Rob Arndt, http://www.bibliotecapleyades.net/ufo_aleman/esp_ufoaleman_6.htm (accessed 11/13/17).

[416] For more details see Rob Arndt, "RFZ (Rundflugzeug) of the Thule-Vril type Series 1-7 (1937-1942), http://discaircraft.greyfalcon.us/RFZ.htm (accessed 1/1/19).

[417] Rob Arndt, "Haunebu – H-Great, Hanueburg Device 1939-1945)" http://discaircraft.greyfalcon.us/HAUNEBU.htm (accessed 7/3/17).

[418] "Rob Arndt, "Haunebu – H-Great, Hanueburg Device 1939-1945)" http://discaircraft.greyfalcon.us/HAUNEBU.htm (accessed 11/13/17).

[419] "Rob Arndt, "Haunebu – H-Great, Hanueburg Device 1939-1945)" http://discaircraft.greyfalcon.us/HAUNEBU.htm (accessed 11/13/17).

[420] "Rob Arndt, "Haunebu – H-Great, Hanueburg Device 1939-1945)" http://discaircraft.greyfalcon.us/HAUNEBU.htm (accessed 11/13/17).

[421] Dr Steven Greer, "Testimony of Mr. Mark McCandlish, US Air Force", *Disclosure: Military and Government Witnesses Reveal the Greatest Secrets in Modern History*, p. 504.

[422] Peter W. Merlin, "Taking E.T. Home: Birth of a Modern Myth," *Sunlight: Shedding some light on UFOlogy and UFOs*, Vol 5, No. 6 (2013) http://home.comcast.net/~tprinty/UFO/SUNlite5_6.pdf (accessed on 7/14/15).

[423] Peter W. Merlin, "Taking E.T. Home: Birth of a Modern Myth," *Sunlight: Shedding some light on UFOlogy and UFOs*,Vol 5, No. 6 (2013) http://home.comcast.net/~tprinty/UFO/SUNlite5_6.pdf (accessed on 7/14/15).

[424] Rich gave a speech where it is claimed he made this comment. See http://www.unexplained-mysteries.com/forum/index.php?showtopic=63914 (accessed on 7/14/15).

[425] The two documents were uploaded to the internet on December 17, 2017, by Corey Goode who obtained them from a confidential source. See "Defense Intelligence Reference Documents", https://spherebeingalliance.com/blog/defense-intelligence-reference-documents.html (accessed 11/22/18).

[426] See Michael Salla, "Advanced Technology Reports Leaked by Corey Goode Confirmed by Leading Scientist", https://www.exopolitics.org/advanced-technology-reports-leaked-by-corey-goode-confirmed-by-leading-scientist/ (accessed 11/22/18).

[427] Ronald Reagan, *The Reagan Diaries* (Harper Perennial, 2099) p. 334.

## ENDNOTES CHAPTER 15

[428] Edgar Fouche's documents are available online at: http://www.checktheevidence.com/pdf/Edgar%20Fouches%20Military%20Documents.pdf (1/2/19).

[429] Edgar Fouche's book was published in 1998 as *Alien Rapture: The Chosen* (Galde Press). and was co-written with Brad Steiger.

[430] Edgar Rothschild Fouche, "Secret Government Technology," http://www.bibliotecapleyades.net/ciencia/ciencia_extraterrestrialtech08.htm (accessed on 6/18/15).

[431] "Aurora Timeline," http://tinyurl.com/oaf9s62 (accessed on 6/18/15).

[432] See Nick Cook, *The Hunt for Zero Point,* 14.

[433] Sweetman, Bill. "Secret Warplanes of Area 51." http://www.popsci.com/military-aviation-space/article/2006-10/top-secret-warplanes-area-51 (accessed on 6/18/15).

[434] Edgar Rothschild Fouche, "Secret Government Technology," http://www.bibliotecapleyades.net/ciencia/ciencia_extraterrestrialtech08.htm (accessed on 6/18/15).

[435] For discussion of the use of stealth technology as a cover for antigravity technology, see Michael Salla, *Insiders Reveal Secret Space Programs and Extraterrestrial Alliances* (Exopolitics Institute, 2015) pp. 28-40.

[436] Edgar Rothschild Fouche, "Secret Government Technology," https://tinyurl.com/y4avzh3u (accessed on 6/12/15).

[437] Edgar Rothschild Fouche, "Secret Government Technology," https://tinyurl.com/y4avzh3u (accessed on 6/18/15).

[438] Edgar Rothschild Fouche, "Secret Government Technology," https://tinyurl.com/y4avzh3u (accessed on 6/18/15).

[439] Edgar Rothschild Fouche, "Secret Government Technology," https://tinyurl.com/y4avzh3u (accessed on 6/18/15).

[440] Edgar Rothschild Fouche, "Secret Government Technology," https://tinyurl.com/y4avzh3u (accessed on 6/18/15).

[441] Edgar Rothschild Fouche, "Secret Government Technology," https://tinyurl.com/y4avzh3u (accessed on 6/18/15).

[442] Edgar Fouche presentation on Area 51, https://www.scribd.com/document/54982740/Edgar-Fouche-Lecture-in-Print (accessed on 11/30/18).

[443] See "TR-3B Questions & Answers – Edgar Fouche" https://tinyurl.com/yxu28j6g (accessed on 11/30/18).

[444] See "TR-3B Questions & Answers – Edgar Fouche" https://tinyurl.com/yxu28j6g (accessed on 11/30/18).

[445] See Sunday Express Newspaper Aritcle, September 17, 1995. http://www.ufoevidence.org/documents/doc418.htm (accessed on 6/20/15).

[446] See Edgar Rothschild Fouche, "Secret Government Technology," https://tinyurl.com/y4avzh3u (accessed on 6/18/15)

[447] See Sunday Express Newspaper Aritcle, September 17, 1995. http://www.ufoevidence.org/documents/doc418.htm (accessed on 6/20/15).

[448] "Fast or Slow," http://explorerplanet.blogg.no/1418483151_fast_or_slow_part_3.html (accessed on 6/21/15).

[449] Wikipedia, "G-Force," https://en.wikipedia.org/wiki/G-force#Human_tolerance_of_g-force (accessed on 6/21/15).

[450] "Fast or Slow," http://explorerplanet.blogg.no/1418483151_fast_or_slow_part_3.html (accessed on 6/21/15).

[451] "Questions for Corey/GoodETxSG – 4/4/2015," http://exopolitics.org/secret-space-programs-more-complex-than-previously-revealed/ (accessed 7/29/15).

[452] "Questions for Corey/GoodETxSG – 4/4/2015," http://exopolitics.org/secret-space-programs-more-complex-than-previously-revealed/ (accessed 7/29/15).

## ENDNOTES CHAPTER 16

[453] Information about the USAF's sponsorship of the Stargate SG-1 series is drawn from Wikipedia, https://en.wikipedia.org/wiki/Stargate_SG-1 (accessed 12/7/18).

[454] See Corey Goode interview on Gaia TV's Cosmic Disclosure "Portals: The Cosmic Web", http://www.gaia.com/video/portals-cosmic-web#play/107251 (accessed 12/8/18).

[455] See Corey Goode interview on Gaia TV's Cosmic Disclosure "Portals: The Cosmic Web", http://www.gaia.com/video/portals-cosmic-web#play/107251 (accessed 12/8/18).

[456] See Corey Goode interview on Gaia TV's Cosmic Disclosure "Portals: The Cosmic Web", http://www.gaia.com/video/portals-cosmic-web#play/107251 (accessed 12/8/18).

[457] See "Portals: Connecting the Cosmic Highway" https://www.gaia.com/video/portals-connecting-cosmic-highway (accessed 1/2/19).

[458] See http://www.conscioushugs.com/ (accessed 12/9/18).

[459] "Mystery Of Dr. Dan Burisch - Beginning To End", https://rense.com/general42/mssy.htm (accessed 12/9/18).

[460] William Hamilton III, *Project Aquarius: The Story of an Aquarian Scientist* (Authorhouse, 2005).

[461] "Dan Burisch: Interview transcript – Part 1" http://projectcamelot.org/lang/en/dan_burisch_stargate_secrets_interview_transc ript_1_en.html (accessed 12/10/18).

[462] "Dan Burisch: Interview transcript – Part 2" http://projectcamelot.org/lang/en/dan_burisch_interview_transcript_2_en.html (accessed 12/8/18).

[463] See "An Interview with 'Henry Deacon', a Livermore Physicist" http://projectcamelot.org/livermore_physicist.html (accessed 12/8/18).

[464] "A further update from 'Henry Deacon' http://projectcamelot.org/livermore_physicist_3.html (accessed 12/8/18).

[465] "An Exopolitical Perspective on the Preemptive War against Iraq", http://exopolitics.org/archived/Study-Paper2.htm (accessed 12/8/18).

[466] "Electric Sun", https://www.gaia.com/video/electric-sun (accessed 12/8/18).

[467] Corey Goode interview on Gaia TV's Cosmic Disclosure "Portals: The Cosmic Web", http://www.gaia.com/video/portals-cosmic-web#play/107251 (accessed 12/8/18).

[468] "Cosmic Disclosure: Remnants of Ancient Mars", November 10, 2015 https://www.gaia.com/video/remnants-ancient-mars (accessed 12/9/18).

[469] See Michael Salla, "Alleged Time Traveler Runs for U.S. President: Real Deal or CIA PsyOp?" https://www.exopolitics.org/alleged-time-traveler-runs-for-u-s-president-real-deal-or-cia-psyop/ (accessed 1/2/19).

[470] See Michael Salla, "Jump Room to Mars: Did CIA Groom Obama & Basiago as future Presidents?"https://www.exopolitics.org/jump-room-to-mars-did-cia-groom-obama-basiago-as-future-presidents/ (accessed 1/2/19).

[471] Gautham Shenoy, "The interstellar contributions of Kip Thorne, the man Carl Sagan contacted for physics advice" https://factordaily.com/interstellar-kip-thorne-nobel-prize/ (accessed 12/11/2018).

[472] Gautham Shenoy, "The interstellar contributions of Kip Thorne, the man Carl Sagan contacted for physics advice" https://factordaily.com/interstellar-kip-thorne-nobel-prize/ (accessed 12/11/2018).

[473] M.S. Morris and K.S. Thorne, "Wormholes in spacetime and their use for interstellar travel: A tool for teaching general relativity", *American Journal of Physics*, Vol 56 (1988): pp. 395-412.

[474] Gautham Shenoy, "The interstellar contributions of Kip Thorne, the man Carl Sagan contacted for physics advice" https://factordaily.com/interstellar-kip-thorne-nobel-prize/ (accessed 12/11/2018).

[475] Paul Davies, *How to Build a Time Machine* (Penguin Books, 2003) p. 69.

476 Eisenhower Briefing Document is available online at:
http://www.majesticdocuments.com/pdf/eisenhower_briefing.pdf (accessed
12/14/18).
477 See Stanton T. Friedman, *Top Secret/Majic: Operation Majestic-12 and the
United States Government's UFO Cover-up* (De Capo Press, 2005) pp. 26-40.
478 The two documents are available at
https://spherebeingalliance.com/blog/defense-intelligence-reference-
documents.html (accessed 12/10/18).
479 "To The Stars Academy of Arts & Science, Tom Delonge, and the Secret
DoD UFO Research Program" https://www.theblackvault.com/casefiles/to-the-
stars-academy-of-arts-science-tom-delonge-and-the-secret-dod-ufo-research-
program/ (accessed 12/10/18).
480 "Those Defense Intelligence Reference Documents", http://ufos-
scientificresearch.blogspot.com/2018/06/those-defense-intelligence-
reference.html?m=1 (accessed 12/10/18).
481 The full list is available online at: https://tinyurl.com/ybs6du4v (accessed
12/10/18).
482 "AATIP or AAWSA", https://ufos-
scientificresearch.blogspot.com/2018/05/aatip-or-aawsa.html (accessed
12/10/18).
483 "Those Defense Intelligence Reference Documents", http://ufos-
scientificresearch.blogspot.com/2018/06/those-defense-intelligence-
reference.html?m=1 (accessed 12/10/18).
484 A detailed analysis of the public emergence of the DIA documents,
Greenewald's discovery of Goode's role, what Knapp, Koi and other researchers
have concluded about the documents appears in a September 7, 2018 blog post
by Mike Waskosky titled: "Corey Goode's DIA Documents: The Unreported
AATIP Revealing". https://www.disclosurecolorado.org/corey-goode-dia-
documents/ (accessed 12/10/18).
485 Eric W. Davis, "Traversable Wormholes, Stargates, and Negative Energy",
Defense Intelligence Reference Document (April 6, 2010) p. v. Available at:
https://spherebeingalliance.com/blog/defense-intelligence-reference-
documents.html_(accessed 12/14/18).
486 Eric W. Davis, "Traversable Wormholes, Stargates, and Negative Energy",
Defense Intelligence Reference Document (April 6, 2010) p. viii. Available at:
https://spherebeingalliance.com/blog/defense-intelligence-reference-
documents.html_(accessed 12/14/18).
487 Eric W. Davis, "Traversable Wormholes, Stargates, and Negative Energy",
Defense Intelligence Reference Document (April 6, 2010) p. viii. Available at:
https://spherebeingalliance.com/blog/defense-intelligence-reference-
documents.html_(accessed 12/14/18).
488 Eric W. Davis, "Traversable Wormholes, Stargates, and Negative Energy",
Defense Intelligence Reference Document (April 6, 2010) p. 4. Available at:
https://spherebeingalliance.com/blog/defense-intelligence-reference-
documents.html_(accessed 12/14/18).

489 Eric W. Davis, "Traversable Wormholes, Stargates, and Negative Energy", Defense Intelligence Reference Document (April 6, 2010) p. 9. Available at: https://spherebeingalliance.com/blog/defense-intelligence-reference-documents.html (accessed 12/14/18).
490 The first "chronovisor" was allegedly constructed by the Benedictine monk Father Pellegrino Ernetti in the 1950s. See Peter Krassa, *Father Ernetti's Chronovisor : The Creation and Disappearance of the World's First Time Machine* (New Paradigm Books, 2000). Dan Burisch discussed Looking Glass technology in "Dan Burisch: Interview transcript – Part 1", http://projectcamelot.org/lang/en/dan_burisch_stargate_secrets_interview_transc ript_1_en.html (accessed 12/10/18).
491 Andrew Basiago has claimed that he was part of an intelligence gathering program called Project Pegasus which used chronovisors and other stargate related technologies for intelligence gathering purposes. See Michael Salla, "Alleged Time Traveler Runs for U.S. President: Real Deal or CIA PsyOp?" https://www.exopolitics.org/alleged-time-traveler-runs-for-u-s-president-real-deal-or-cia-psyop/ (accessed 12/14/18).

## ENDNOTES CHAPTER 17

492 An analysis of Michael Relfe and Randy Cramer's testimonies is available in Michael Salla, Insiders Reveal Secret Space Programs and Extraterrestrial Alliances (Exopolitics Institute, 2015) pp. 309-44.
493 For a chronology of articles and interviews of Corey Goode, visit: https://www.exopolitics.org/secret-space-programs-sphere-being-alliance-corey-goode-testimony/ (accessed 12/15/18).
494 William Tompkins, *Selected by Extraterrestrials: My life in the top secret world of UFOs., think-tanks and Nordic secretaries* (Createspace, 2015).
495 See Michael Salla, *Insiders Reveal Secret Space Programs and Extraterrestrial Alliances* (Exopolitics Institute, 2015).
496 See Michael Salla, Military Abduction & Extraterrestrial Contact Treaty – Corey Goode Briefing Pt 2, https://www.exopolitics.org/military-abduction-extraterrestrial-contact-treaty-corey-goode-briefing-pt-2/ (accessed 12/15/18).
497 Corey Goode, "Latest Intel and Update", https://spherebeingalliance.com/blog/latest-intel-and-update.html (accessed 12/15/18).
498 Corey Goode and David Wilcock, "Endgame Part II: The Antarctic Atlantis & Ancient Alien Ruins", https://spherebeingalliance.com/blog/endgame-part-ii-the-antarctic-atlantis-and-ancient-alien-ruins.html (accessed 12/15/18).
499 See Michael Salla, "Partial Disclosure and Competing Secret Space Programs," https://www.exopolitics.org/partial-disclosure-and-competing-secret-space-programs/ (accessed 12/15/18).
500 Corey Goode and David Wilcock, "Endgame Part II: The Antarctic Atlantis & Ancient Alien Ruins", https://spherebeingalliance.com/blog/endgame-part-ii-the-antarctic-atlantis-and-ancient-alien-ruins.html (accessed 12/15/18).

[501] See Michael Salla, "US Air Force Officials Investigate Claims of Secret Navy Space Program", https://www.exopolitics.org/us-air-force-officials-investigate-claims-of-secret-navy-space-program/ (accessed 12/17/18).

[502] Corey Goode and David Wilcock, "Endgame Part II: The Antarctic Atlantis & Ancient Alien Ruins", https://spherebeingalliance.com/blog/endgame-part-ii-the-antarctic-atlantis-and-ancient-alien-ruins.html (accessed 12/15/18).

[503] William Tompkins, *Selected by Extraterrestrials: My life in the top secret world of UFOs., think-tanks and Nordic secretaries.*

[504] For more details on the Sphere Being Alliance see Michael Salla, *Insiders Reveal Secret Space Programs & Extraterrestrial Alliances*, pp. 278-81.

[505] For more details see Michael Salla, "Tom DeLonge & UFO Disclosure: Rocking the Secret Space Programs Boat", https://www.exopolitics.org/sekret-machines-the-secret-space-programs-onion/ accessed 12/17/18).

[506] Tom DeLonge and AJ Hartley, *Sekret Machines* [Kindle Locations 95-100].

[507] Tom DeLonge and A.J. Hartley, *Sekret Machines Book 1: Chasing Shadows* (To The Stars, 2016).

[508] See Michael Salla, "Wikileaks Reveals USAF General involved in UFO & Secret Space Program Disclosures", https://www.exopolitics.org/wikileaks-reveals-usaf-general-involved-in-ufo-secret-space-program-disclosures/ (accessed 12/17/18).

[509] Wikileaks, "General McCasland," https://wikileaks.org/podesta-emails/emailid/3099 (accessed 12/17/18)

[510] "Major General William N. McCasland", https://www.af.mil/About-Us/Biographies/Display/Article/104776/major-general-william-n-mccasland/ (accessed 12/18/18).

[511] See https://dpo.tothestarsacademy.com/ (accessed 12/18/18).

[512] "Corey Goode Mega-Update – Part II" https://spherebeingalliance.com/blog/ancient-builder-race-recovering-humanitys-billion-year-legacy-part-2.html (accessed 12/18/18).

[513] See Michael Salla, "Is Tom DeLonge's To The Stars Academy a Deep State Operation?" https://www.exopolitics.org/is-tom-delonges-to-the-stars-academy-a-deep-state-operation/ (accessed 12/17/18).

[514] See Special Aerospace Services website: https://specialaerospaceservices.com/ (accessed 12/18/18).

[515] See Michael Salla, *U.S. Navy's Secret Space Program and Nordic Extraterrestrial Alliance.*

## ENDNOTES CHAPTER 18

[516] See "Forty-Seven Pages From the DIA: Why Should We Care?" https://exonews.org/47-pages-from-the-dia-why-should-we-care/ (accessed 1/3/19).

[517] The DIA document is available online at: https://tinyurl.com/ychkyupo (accessed 12/18/18).

[518] See, "The Majestic Documents" http://www.majesticdocuments.com/ (accessed 12/18/18).

[519] Harry Truman's Executive Order establishing Operation Majestic 12 is available online at: http://www.majesticdocuments.com/pdf/truman_forrestal.pdf (accessed 12/18/18).

[520] See Michael Salla, *Kennedy's Last Stand: Eisenhower, UFOs, MJ-12 & JFK's Assassination* (Exopolitics Institute, 2013) 128-30.

[521] "Homeland Security Advisory Council Members", https://www.dhs.gov/homeland-security-advisory-council-members (accessed 12/18/18).

[522] "Assessment of the Situation/Statement of Position on Unidentified Flying Objects", https://tinyurl.com/ychkyupo (accessed 12/18/18).

[523] "Assessment of the Situation/Statement of Position on Unidentified Flying Objects", https://tinyurl.com/ychkyupo (accessed 12/18/18).

[524] "Assessment of the Situation/Statement of Position on Unidentified Flying Objects", https://tinyurl.com/ychkyupo (accessed 12/18/18).

[525] "Assessment of the Situation/Statement of Position on Unidentified Flying Objects", https://tinyurl.com/ychkyupo (accessed 12/18/18).

[526] "Assessment of the Situation/Statement of Position on Unidentified Flying Objects", https://tinyurl.com/ychkyupo (accessed 12/18/18).

[527] "Assessment of the Situation/Statement of Position on Unidentified Flying Objects", https://tinyurl.com/ychkyupo (accessed 12/18/18).

[528] "Assessment of the Situation/Statement of Position on Unidentified Flying Objects", https://tinyurl.com/ychkyupo (accessed 12/18/18).

[529] "Assessment of the Situation/Statement of Position on Unidentified Flying Objects", https://tinyurl.com/ychkyupo (accessed 12/18/18).

[530] See Nick Redfern, "Taking a Look at the Interplanetary Phenomenon Unit", https://mysteriousuniverse.org/2018/08/taking-a-look-at-the-interplanetary-phenomenon-unit/ (1/2/19).

[531] "Assessment of the Situation/Statement of Position on Unidentified Flying Objects", https://tinyurl.com/ychkyupo (accessed 12/18/18).

[532] "Assessment of the Situation/Statement of Position on Unidentified Flying Objects", https://tinyurl.com/ychkyupo (accessed 12/18/18).

[533] "Assessment of the Situation/Statement of Position on Unidentified Flying Objects", https://tinyurl.com/ychkyupo (accessed 12/18/18).

[534] For detailed discussion of the 1954 meeting at Edwards AFB, see Michael Salla, *Galactic Diplomacy: Getting to Yes with ET* (Exopolitics Institute, 2013) pp. 47-80.

[535] For discussion of the failed diplomatic discussions at Edwards AFB, see Michael Salla, *Galactic Diplomacy: Getting to Yes with ET*, pp. 47-80.

[536] See William Tompkins, *Selected by Extraterrestrials* (Createspace, 2015).

[537] For description of the different groups of extraterrestrials Smith met, see Michael Salla, "Insider Reveals More about Extraterrestrials working in Classified Programs", https://www.exopolitics.org/insider-reveals-more-about-extraterrestrials-working-in-classified-programs/ (accessed 12/18/18).

[538] "Aliens at Home on Earth, *Cosmic Disclosure*, August 7, 2018, https://www.gaia.com/video/aliens-home-earth (accessed 12/18/18).

[539] For analysis of multiple witness reports of extraterrestrials living among us, see Michael Salla, *Galactic Diplomacy: Getting to Yes with ET*, pp. 199-232.

[540] For example, see John Greenewald, "New Majestic-12 (MJ-12) Briefing Documents Released June 2017" http://www.theblackvault.com/casefiles/new-majestic-12-mj-12-briefing-documents-released-june-2017/# (accessed 12/18/18).

[541] See "Forty-Seven Pages From the DIA: Why Should We Care?" https://exonews.org/47-pages-from-the-dia-why-should-we-care/ (accessed 1/3/19).

[542] See "Forty-Seven Pages From the DIA: Why Should We Care?" https://exonews.org/47-pages-from-the-dia-why-should-we-care/ (accessed 1/3/19).

[543] "TV REVIEWS; 'The Ring of Truth,' On Ways of Perception", https://www.nytimes.com/1987/10/20/arts/tv-reviews-the-ring-of-truth-on-ways-of-perception.html?pagewanted=1 (accessed 12/18/18).

[544] See "Forty-Seven Pages From the DIA: Why Should We Care?" https://exonews.org/47-pages-from-the-dia-why-should-we-care/ (accessed 1/3/19).

[545] "Eisenhower Briefing Document", http://www.majesticdocuments.com/pdf/eisenhower_briefing.pdf (accessed 12/18/18).

[546] See Stanton T. Friedman, *Top Secret/Majic: Operation Majestic-12 and the United States Government's UFO Cover-up* (De Capo Press, 2005) pp. 26-40.

[547] Donald Howard Menzel, *A field guide to the stars and planets: Including the moon, satellites, comets, and other features of the universe* (Houghton Mifflin, 1964).

[548] "Dr. Carl Sagan", https://starchild.gsfc.nasa.gov/docs/StarChild/whos_who_level2/sagan.html (accessed 12/18/18).

[549] For example, see John Greenewald, "New Majestic-12 (MJ-12) Briefing Documents Released June 2017" http://www.theblackvault.com/casefiles/new-majestic-12-mj-12-briefing-documents-released-june-2017/# (accessed 12/18/18).

[550] Cosmic Disclosure, "E.T. Detention & Interrogation", https://www.gaia.com/video/et-detention-interrogation (accessed 12/19/18).

[551] For discussion of another Cosmic Disclosure episode where Corey Goode and Emery Smith talk about human looking extraterrestrials living among us, see Michael Salla, "Military Insiders Confirm Thousands of Extraterrestrials Live Among Us", https://www.exopolitics.org/military-insiders-confirm-thousands-of-extraterrestrials-live-among-us/ (accessed 12/19/18).

[552] Cosmic Disclosure, "E.T. Detention & Interrogation", https://www.gaia.com/video/et-detention-interrogation (accessed 12/19/18).

# ENDNOTES CHAPTER 19

[553] A page has been created on my Exopolitics.org website where articles and videos featuring JP's photos, along with originals, can be found for independent analysis. https://www.exopolitics.org/jp-articles-photos-videos/ (accessed 12/20/18).

[554] Helmut Lammer and Marion Lammer, *MILABS: Military Mind Control and Alien Abduction* (Illuminet Press, 2000)

[555] My original article announcing JP is available online, "Photos of Antigravity UFO Near MacDill AFB support claims of USAF SSP", https://www.exopolitics.org/photos-antigravity-ufo-fort-macdill-usaf-ssp/ (accessed 12/20/18).

[556] Howard Altman, "MacDill chuckling as UFO website reports 'flying triangles' at base", http://www.tampabay.com/news/military/macdill/howard-altman-base-chuckling-as-ufo-website-reports-flying-triangles-at/2338274 (accessed 12/20/18).

[557] See "Memorandum for MacDill AFB Personnel" https://tinyurl.com/yalmyt4a (accessed 12/21/18).

[558] See "Memorandum for MacDill AFB Personnel" https://tinyurl.com/yalmyt4a (accessed 12/21/18).

[559] "Irma approaching Tampa Bay area with hurricane-force winds; 5 dead, 3.5M without power" https://abcnews.go.com/US/hurricane-irma-close-landfall-florida-keys-dead-state/story?id=49738525 (accessed 12/20/18).

[560] See Dane Wigginton, "Massive US Senate Document On National And Global Weather Modification", https://www.geoengineeringwatch.org/massive-us-senate-document-on-national-and-global-weather-modification/ (accessed 12/21/18).

[561] Nick Begich and Jeane Manning, *Angels Don't Play This Haarp: Advances in Tesla Technology* (Earthpulse, 1995).

[562] Dane Wigginton, "Geoengineering Microwave Transmissions And Their Connection To Hurricane Florence", https://www.geoengineeringwatch.org/category/haarp-2/ (accessed 12/21/18).

[563] Dane Wigginton, "Hurricane Irma Manipulation: Objectives And Agendas" https://www.geoengineeringwatch.org/hurricane-irma-manipulation-objectives-and-agendas/ (accessed 12/21/18).

[564] Nick Collins, "Crop circles 'created using GPS, lasers and microwaves'," *The Telegraph*, https://www.telegraph.co.uk/news/science/science-news/8671207/Crop-circles-created-using-GPS-lasers-and-microwaves.html (accessed 12/21/18).

[565] "Maser Crop Circles & Weapons," http://www.orwelltoday.com/cropcircleweapon.shtml (accessed 12/21/18).

[566] "Maser Crop Circles & Weapons," http://www.orwelltoday.com/cropcircleweapon.shtml (accessed 12/21/18).

[567] "Joint Polar Satellite System Common Ground System", https://www.raytheon.com/capabilities/products/jpss (accessed 12/21/18).

[568] "Connecting the Raytheon, AMS, Lockhead, HAARP, NOAA, General Dynamics and DARPA dots...."
https://www.geoengineeringwatch.org/html/weatherreportedbyraytheon.html (accessed 12/21/18).
[569] Corey Goode's responses to my questions were originally published in "Was Hurricane Irma Steered by Maser Satellites in Weather War against USA?"
https://www.exopolitics.org/was-hurricane-irma-steered-by-maser-satellites-in-weather-war-against-usa/ (accessed 12/21/18).
[570] To read all of Corey Goode's responses to my questions see "Was Hurricane Irma Steered by Maser Satellites in Weather War against USA?"
https://www.exopolitics.org/was-hurricane-irma-steered-by-maser-satellites-in-weather-war-against-usa/ (accessed 12/21/18).
[571] See "Memorandum for MacDill AFB Personnel"
https://tinyurl.com/yalmyt4a (accessed 12/21/18).
[572] Howard Altman, "MacDill Air Force Base reopens with little damage"
http://www.tampabay.com/news/weather/hurricanes/macdill-air-force-base-reopens-with-little-damage/2337114 (accessed 12/21/18).
[573] Benjamin Fulford, "High level and high stakes Mexican stand off continues despite 911 weather warfare attack on US",
https://benjaminfulford.net/2017/09/11/high-level-high-stakes-mexican-stand-off-continues-despite-911-weather-warfare-attack-us/ (accessed 12/21/18).
[574] Victor Ferreira, "Researchers find secret, warm oasis beneath Antarctica's ice that could be home to undiscovered species"
https://nationalpost.com/news/world/researchers-find-secret-warm-oasis-beneath-antarcticas-ice-that-could-be-home-to-undiscovered-species (accessed 12/21/18).
[575] See Michael Salla, *Antarctica's Hidden History: Corporate Foundations of Secret Space Programs*, pp, 83-114.
[576] See Michael Salla, *Antarctica's Hidden History: Corporate Foundations of Secret Space Programs*, pp, 129-32.
[577] I first discussed the leaked DIA document in "New Majestic Document Reveals US Diplomatic Relations with Extraterrestrials",
https://www.exopolitics.org/majestic-document-reveals-us-diplomatic-relations-with-extraterrestrials/ (accessed 12/21/18).
[578] I first wrote about JP's photos and skype messages about the October 19, 2017 photos in "Covert Disclosure of Antigravity Craft near MacDill AFB"
http://www.exopolitics.org/disclosure-of-antigravity-craft-near-macdill-afb/ (accessed 12/21/18).
[579] For a video showing close ups of the October 19, 2017 photos, go to:
https://www.youtube.com/watch?v=PPGvBfW1-CY (accessed 12/21/18).
[580] I first wrote about JP's October 23, 2017 photos and encounter in "Covert Disclosure of Antigravity Rectangle Weapons Platforms by USAF Special Operations", https://www.exopolitics.org/covert-disclosure-rectangle-craft-usaf-spec-ops/ (accessed 12/21/18).
[581] Photos of the March 16, 2018 sighting available online here:
https://www.exopolitics.org/photos-of-triangle-shaped-antigravity-craft-over-

orlando-florida/ . Photos of the March 23 sighting available here:
https://www.exopolitics.org/more-flying-triangles-photographed-near-orlando-
florida/ (accessed 12/20/18).

[582] See Michael Salla, *Galactic Diplomacy: Getting to Yes with ET* (Exopolitics
Institute, 2013) pp. 199-232.

[583] See Michael Salla, "Russian PM not joking – extraterrestrials live among us
according to MIB documentary", https://www.exopolitics.org/russian-pm-not-
joking-extraterrestrials-live-among-us-according-to-mib-documentary/
(accessed 12/21/18).

[584] For an online video showing all the photos he sent, along with zooms of two
of the photos go to: https://www.youtube.com/watch?v=2Q1QryFjofg (accessed
12/21/18).

[585] I discuss this in detail in *Antarctica's Hidden History: Corporate
Foundations of Secret Space Programs, pp.* 115-36.

# ENDNOTES CHAPTER 20

[586] The KGMB videoclip was uploaded in a tweet by political analyst Nick
Short, https://twitter.com/PoliticalShort/status/952288600271478784 (accessed
12/26/18).

[587] *Star Advertiser*, "'Wrong button' sends out false missile alert,"
https://www.staradvertiser.com/2018/01/13/breaking-news/emergency-officials-
mistakenly-send-out-missile-threat-alert/ (accessed 1/2/19).

[588] *Star Advertiser*, "HI-EMA's Miyagi resigns, 'button pusher' fired in
aftermath of false missile alert",
https://www.staradvertiser.com/2018/01/30/breaking-news/ige-to-announce-
results-of-false-missile-alert-investigation/ (accessed 1/2/19).

[589] "False Ballistic Missile Alert Investigation for January 13, 2018",
https://dod.hawaii.gov/wp-content/uploads/2018/01/report2018-01-29-
181149.pdf (accessed 12/24/18) pp. 1-2.

[590] "False Ballistic Missile Alert Investigation for January 13, 2018",
https://dod.hawaii.gov/wp-content/uploads/2018/01/report2018-01-29-
181149.pdf (accessed 12/24/18) p. 10.

[591] "False Ballistic Missile Alert Investigation for January 13, 2018",
https://dod.hawaii.gov/wp-content/uploads/2018/01/report2018-01-29-
181149.pdf (accessed 12/24/18) p. 10.

[592] See: https://twitter.com/awarenessadvent/status/952295221106110464 For
CNN story confirming sirens going off in different areas in Honolulu, see:
"Missile threat alert for Hawaii a false alarm"
https://www.cnn.com/2018/01/13/politics/hawaii-missile-threat-false-
alarm/index.html (accessed 12/26/18).

[593] "Emails detail how senior U.S. military officers grappled with false Hawaii
missile alert",
https://www.washingtonpost.com/news/checkpoint/wp/2018/02/17/emails-
detail-how-senior-u-s-military-officers-grappled-with-false-hawaii-missile-alert/
(accessed 2/20/19)

594 https://archive.4plebs.org/pol/thread/156845778/ (accessed 12/24/18).
595 "The False Flag Formula – 15 Ways to Detect a False Flag Operation", http://freedom-articles.toolsforfreedom.com/false-flag-formula-15-ways-to-detect/ (accessed 1/2/19).
596 "Dark Cabal Launches Nuclear Missile At Hawaii - Alliance Stops Attack - January 13, 2018", http://www.ascensionwithearth.com/2018/01/dark-cabal-launches-nuclear-missile-at.html (accessed 1/2/19).
597 Robert David Steele's website lists DefDog as the alias of a recurring author and describes him/her as a "Serving Officer and Closet Revolutionary" https://phibetaiota.net/authors/ (accessed 12/23/18).
598 "DefDog: There Was a Missile, Fired by Israel, China, or the US, Government is Lying", https://phibetaiota.net/2018/01/defdog-there-was-a-missile-government-is-lying/ (accessed 1/2/19).
599 Biographical information available at: https://davejanda.com/about/ (accessed 1/2/19).
600 See "Arrests & Prosecutions Coming for Elite", https://www.youtube.com/watch?v=SIO4vWYtk6s&feature=youtu.be&t=24m (accessed 1/2/19).
601 "HAWaii tourists say that they saw something blow up in the sky", https://www.youtube.com/watch?v=Bhto4f-NiQs&feature=youtu.be (accessed 1/2/19).
602 "HAWaii tourists say that they saw something blow up in the sky", https://www.youtube.com/watch?v=Bhto4f-NiQs&feature=youtu.be (accessed 1/2/19).
603 "HAWaii tourists say that they saw something blow up in the sky", https://www.youtube.com/watch?v=Bhto4f-NiQs&feature=youtu.be (accessed 1/2/19).
604 See Michael Salla, "QAnon is US Military Intelligence that recruited Trump for President to prevent Coup D'etat", https://www.exopolitics.org/qanon-is-us-military-intelligence-that-recruited-trump-for-president-to-prevent-coup-detat/ (accessed 12/23/18).
605 "Edward Snowden: Whistleblower or CIA Spy?" http://www.ahijackedlife.com/qanon-moves-from-4chan-to-8chan-confirmation/ (accessed 12/23/18)
606 See posts # 498, 500, 502, 506, 510 & 511, at: https://qanon.pub/?q=defcon (accessed 12/22/18).
607 Both posts available online at: https://qanon.pub/#500 & https://qanon.pub/#498 (accessed 12/22/18).
608 Available online at: https://qanon.pub/#520 (accessed 12/22/18).
609 https://justinformednews.com/2018/08/30/edward-snowden-whistleblower-or-cia-spy/ (accessed 12/23/18).
610 "Inside President Trump's Trip to Asia", https://www.whitehouse.gov/articles/president-trumps-trip-asia/ (accessed 12/22/18).

[611] For description of Bulk Data Transfer see, "What is Bulk Data Transfer", http://www.learn.geekinterview.com/data-warehouse/data-extraction/what-is-bulk-data-transfer.html (accessed 12/23/18).

[612] "Was Hawaii False Alarm Necessary to Free Captive NSA Documents?" https://onehope2016.wordpress.com/2018/01/18/was-hawaii-false-alarm-necessary-to-free-captive-nsa-documents/ (accessed 1/2/19).

[613] Source: https://qanon.pub/#538 (accessed 12/24/18).

[614] Kimiko de Freytas-Tamura, "Days After Hawaii's False Missile Alarm, a New One in Japan" https://www.nytimes.com/2018/01/16/world/asia/japan-hawaii-alert.html (accessed 1/2/19).

[615] Benjamin Fulford, "Future of world being negotiated in next two weeks as Super Blue Blood Moon approaches", https://kauilapele.wordpress.com/2018/01/25/full-article-benjamin-fulford-1-22-18-future-of-world-being-negotiated-in-next-two-weeks-as-super-blue-blood-moon-approaches/ (accessed 1/2/19).

[616] Source: https://qanon.pub/#725 (accessed 12/24/18).

[617] See "Executive Order Blocking the Property of Persons Involved in Serious Human Rights Abuse or Corruption", https://www.whitehouse.gov/presidential-actions/executive-order-blocking-property-persons-involved-serious-human-rights-abuse-corruption/ (accessed 12/24/18).

[618] https://archive.4plebs.org/pol/thread/156845778/ (accessed 12/24/18).

[619] Michael Salla, "The Secret Navy behind the Ballistic Missile Attack on Hawaii", https://www.exopolitics.org/the-secret-navy-behind-the-ballistic-missile-attack-on-hawaii/ (accessed 12/24/18)

[620] "US Navy shoots down ballistic missile in test", -(accessed 12/22/18).

[621] David B. Larter, "Another US Navy ballistic missile intercept reportedly fails in Hawaii", https://www.defensenews.com/breaking-news/2018/01/31/second-navy-sm-3-block-iia-ballistic-missile-intercept-hawaii-report/ (accessed 1/2/19).

[622] "Was a Nuclear Missile Attack on Hawaii Thwarted by a Secret Space Program?" https://www.exopolitics.org/was-a-nuclear-missile-attack-on-hawaii-thwarted-by-a-secret-space-program/ (accessed 1/2/19).

[623] See "Covert Disclosure of Antigravity Rectangle Weapons Platforms by USAF Special Operations," https://www.exopolitics.org/covert-disclosure-rectangle-craft-usaf-spec-ops/ (accessed 1/2/19).

[624] "Maui Space Surveillance Site (MSSS)", https://www.globalsecurity.org/space/systems/msss.htm (accessed 1/2/19).

[625] "Solaris Modalis Commentary: Photographs Taken During Hawaii Missile Incident", https://solarismodalis.com/solaris-modalis-commentary-photographs-taken-during-hawaii-missile-incident/ (accessed 1/2/19).

## ENDNOTES CHAPTER 21

[626] For estimate of 80% of military space assets belonging to USAF see

"Proposed Space Force to protect, expand U.S. cosmic capabilities" http://apgnews.com/community-news/proposed-space-force-to-protect-expand-u-s-cosmic-capabilities/ (accessed 4/21/19)

[627] *NBC News*, https://www.youtube.com/watch?v=5lEaLcumd08 (accessed 12/29/18).

[628] Oriana Pawlyk, "It's Official: Trump Announces Space Force as 6th Military Branch", https://www.military.com/daily-news/2018/06/18/its-official-trump-announces-space-force-6th-military-branch.html (accessed 12/28/18).

[629] *Roll Call*, "Air Force Opposes Creation of Space Corps", https://www.rollcall.com/policy/air-force-opposes-creation-space-corps (accessed 12/28/18).

[630] Letter to Committee on Armed Services, http://static.politico.com/11/99/eab4f5be445fa2d0749ae8b798f1/mattis-ndaa-heartburn-letter.pdf (accessed 12/28/18).

[631] Phillip Swarts, "Space Corps proposal has murkier path forward in the Senate", https://spacenews.com/space-corps-proposal-has-murkier-path-forward-in-the-senate/ (accessed 12/28/18).

[632] "Trump floats idea for a military 'space force'" https://www.youtube.com/watch?v=6lrJhatw3K4 (accessed 12/28/18).

[633] See Valerie Insinna, "Mattis supportive of new combatant command for space operations", https://www.defensenews.com/space/2018/08/07/pentagon-setting-up-new-combatant-command-for-space-operations-mattis-confirms/ (accessed 12/27/18).

[634] See Loren Grush, "Space Command is coming back, but Space Force still needs approval from Congress", https://www.theverge.com/2018/12/18/18146433/trump-space-force-combatant-command-department-of-defense-usstratcom (accessed 12/27/18).

[635] "Final Report on Organizational and Management Structure for the National Security Space Components of the Department of Defense", p. 6. https://partner-mco-archive.s3.amazonaws.com/client_files/1533834803.pdf (accessed 12/27/18).

[636] "Final Report on Organizational and Management Structure for the National Security Space Components of the Department of Defense", p. 8. https://partner-mco-archive.s3.amazonaws.com/client_files/1533834803.pdf (accessed 12/27/18).

[637] "Final Report on Organizational and Management Structure for the National Security Space Components of the Department of Defense", p. 10. https://partner-mco-archive.s3.amazonaws.com/client_files/1533834803.pdf (accessed 12/27/18).

[638] For discussion of Goode and Relfe's experiences, see Michael Salla, *Insiders Reveals Secret Space Programs and Extraterrestrial Alliances* (Exopolitics Institute, 2015) pp. 309-44. For Michael Gerloff's experiences, see Michael Salla, "Covert Recruitment into Space Marines '20 and Back' Program", http://www.exopolitics.org/pt-2-covert-recruitment-space-marines/ (accessed 12/28/18).

[639] "Final Report on Organizational and Management Structure for the National Security Space Components of the Department of Defense", p. 12. https://partner-mco-archive.s3.amazonaws.com/client_files/1533834803.pdf (accessed 12/27/18).

[640] "Text of a Memorandum from the President to the Secretary of Defense Regarding the Establishment of the United States Space Command", https://www.whitehouse.gov/briefings-statements/text-memorandum-president-secretary-defense-regarding-establishment-united-states-space-command/ (accessed 12/27/18)

[641] "President Donald J. Trump is Establishing America's Space Force", https://www.whitehouse.gov/briefings-statements/president-trump-establishing-americas-space-force/(accessed 2/19/19).

[642] https://qanon.pub/#2222 (accessed 12/28/18).

[643] "Project Magnet", https://www.bibliotecapleyades.net/sociopolitica/esp_sociopol_mj12_3g7.htm (accessed 12/28/18).

[644] https://qanon.pub/#2225 (accessed 12/28/18).

[645] "Final Report on Organizational and Management Structure for the National Security Space Components of the Department of Defense", p. 6. https://partner-mco-archive.s3.amazonaws.com/client_files/1533834803.pdf (accessed 12/27/18).

[646] For detailed description of the Navy's cooperation with Nordics, see Michael Salla, *The US Navy's Secret Space Program and Nordic Extraterrestrial Alliance.*

[647] See two volume book series, "The Mars Records", http://www.themarsrecords.com (accessed 12/29/18)

[648] For discussion of Q and Deep State satellites being taken offline, see David Wilcock, "Stunning New Briefings: Spy Satellites Down, Deep State Arrests Finally Imminent?" https://divinecosmos.com/davids-blog/22005-stunning-new-briefings-spy-satellites-down-deep-state-arrests-finally-imminent/3/ (accessed 12/29/18).

[649] For estimates of the CIA's black budget, see Michael Salla, "The Black Budget Report", https://exopolitics.org/archived/Report-Black-Budget.htm (accessed 1/2/19).

[650] See Robert Wood, "47 pages from the DIA: Why Should We Care?" https://exonews.org/47-pages-from-the-dia-why-should-we-care/ (accessed 1/3/19).

[651] Nick Redfern, E.T. and the Brookings Institution, https://mysteriousuniverse.org/2015/06/e-t-and-the-brookings-institution/ (accessed 12/29/18).

[652] See "Earth Catastrophe Cycle | Solar Micronova," https://www.youtube.com/watch?v=jTUJ7GtEx0Y (accessed 1/2/19).

[653] For discussion of a polar shift, see Michael Salla, *Antarctica's Hidden History: Corporate Foundations of Secret Space Programs*, pp. 289-292.

# INDEX

## A

abductee, 194

Abella, Alex, 48, 51, 431, 432, 434

Advanced Design, 36, 83

age-regressed, xv, 408

Air Accident Report, 65, 67, 69, 71

Air Force One, 167, 170, 172, 173, 174, 440

Air Force Special Ops, 338, 367, 368, 398

Aldebaran, 134, 138

Allen, Franklin, 162

Antarctic Germans, 146, 164, 168, 181, 184, 206-208, 211, 212, 217, 219, 220-23, 225, 229, 230, 242, 244, 249, 275, 299, 312, 322, 324, 331, 405, 408, 409, 414, 415, 418

Antarctica, ii, xiv, xv, 44, 87, 88, 91, 92, 97-99, 103-106, 110, 112, 117, 126, 133, 134, 136-41, 168, 175, 180, 181-183, 212, 219, 220, 223, 229, 230, 231, 232, 258, 267, 286, 299, 303, 309, 354, 357, 358, 414, 436, 438,439, 442, 445-47, 449, 450, 460, 461, 465

antigravity, 27, 42, 44, 45, 83

Area 51, 5, 45, 46, 249, 258, 261, 267, 268, 269, 270, 284, 309, 427, 433, 444, 451, 452

Argentina, 103

Arianni, 134, 135, 136, 138

Army Air Corps, 21

Army Air Force, US, 3, 14, 16, 17, 19, 20, 21, 27-31, 33, 38, 35, 37, 44, 45, 47, 50-53, 55-58, 64, 65, 70, 71, 73, 84, 93, 94, 107, 109, 112, 114, 116, 117, 146, 187, 411, 432, 435

Army G2, 21, 24

Arnold, General Hap, 2, 36, 55, 411

Arnold, Kenneth, 94, 104

ARVs, 256, 257, 258, 259, 261, 262, 265, 304, 419

Aryans, 135

Ashtar, 130, 131

Ashtar Command, 128, 138

astrospies, 229, 447, 448

Atoms for Peace, 143, 145, 146, 147, 163, 168, 439

Aurora, Project, 268, 269, 270, 451

Australia, xiii, 425

## B

Bara, Mike, 210, 211, 213, 445

Barbato, Cristoforo, 157, 440

Basiago, Andrew, 287, 288, 453, 455

Begich, Nick, 348, 459

Ben Rich, 95, 245, 263, 274, 435

Berkner, Dr. Lloyd, 76

Berlitz, Charles, 58, 432

Betelgeuse, 175

Biefeld, Dr. Alfred, 252

Biefeld-Brown Effect, 252

Blanchard, William, 59, 60, 61, 62

Boeing, 40

Book of Enoch, 225, 245

Botta,Rear Admiral Rico, 27, 36, 37

Brandenberg, Dr. John, 133

Bravo test. See Hydrogen bomb

Brazel, Mack, 59, 60

Britain, 6, 29, 30

Bronk, Dr Detlev, 75

Brown, Thomas Townsend, 253

Bureau of Aeronautics, 27

Burisch, Daniel, 284, 286, 288, 453, 455

Bush, President George, 327

Bush, Vannevar, 20, 31, 75

Byrd, Rear Admiral Richard, 91, 92, 93, 94, 97, 104, 110, 112, 134, 135, 136, 137, 436, 438, 439

# C

Caltech, 289

Campbell, Art, 169, 441, 442

Campbell, Joseph, 13

Cantwheel, Thomas, 8, 10, 26, 27, 427, 429

Cape Girardeau, Missouri, 6, 7, 8, 10, 11, 33, 57, 71

Carey, General Michael, 307, 432

Carey, Thomas, 5, 59, 432, 433

Castle Bravo, nuclear test, 147, 150, 151, 154, 161, 323

Chile, 91

China Lake, 52

Churchill, Winston, 6, 427

CIA, xiv, xvi, 8, 33, 39, 45, 64, 98, 100-105, 117, 219, 220, 221, 229, 269, 273, 287, 288, 314, 327, 330, 360, 361, 385, 391, 392, 393, 396, 402, 417, 419, 430, 431, 433, 436, 437, 446, 453, 455, 462, 465

Clancarty, Lord, 154

Clinton administration, 220

Colbohm,Franklin, 37, 38

Cook, Nick, 268

Cooper, William, 158, 159, 160, 161, 175, 176, 440, 441, 442

Cornine, Lance, 23

Corso, Philip, 62, 63, 64, 178, 179, 180,

183, 184, 191, 192, 433, 442, 443, 444

Cosmic Disclosure, 280, 282, 300, 302, 325, 331, 368, 452, 453, 458

Counter Intelligence Corps, US Army, 8, 14, 26, 57, 69, 71

Cutler, Robert, 77

Cylindrical UFO, 358

# D

DARPA, 267, 287, 288, 460

Davis, Dr. Eric, 265, 291, 293, 294, 296

Dayton, Ohio, 2, 3, 4, 5, 45, 64

Deacon, Henry, 284, 286, 288, 453

Dean, Robert, 158, 440, 441

Debus, Dr. Kurt, 181, 182, 217, 218, 219, 223, 232

Deep State, xiii, xiv, xv, xvi, 146, 147, 164, 309, 311, 333-36, 374, 377, 385, 386, 389, 391-94, 396, 398, 400-402, 405, 408, 409, 412, 415-19, 421, 456, 465

Defense Intelligence Agency, 291, 301, 312, 313, 314, 357, 360, 416, 420

Defense Intelligence Reference Documents, 264, 292, 293, 451, 454

DeLonge, Tom, 306, 307, 308, 309, 310, 418, 456

Department of Defense, 56, 71, 210, 227, 231, 233, 277, 338, 389, 403, 406, 428, 448, 464, 465

Department of the Air Force, xvi, 56, 87, 278, 403, 411, 421

Department of the Navy, 55, 56, 411

Department of the Space Force, xvi, 411, 421

Department of War, 55, 56

Directed Energy Weapon, 318, 398, 399, 401

Donald Douglas, 29, 30, 35, 39, 48, 430

Donovan, William, 20

Double Top Secret, 32, 34
Douglas Aircraft Company, 29, 30, 34,
	35, 36, 37, 38, 39, 40, 41, 42, 79, 81,
	83, 84, 86, 87, 89, 228, 236, 237,
	238, 324
Draco Federation, 135
Draconians, 88, 137, 416
Dryden, Hugh, 209, 210
Dulce, New Mexico, 177, 178

## E

Earth, 130, 135, 137, 154, 159, 160,
	161, 175, 176
*EBE*, 191, 192
Edwards, 147, 148, 150, 151, 153, 154,
	155, 161, 162, 172
Edwards Air Force base, 148
Einstein-Rosen bridge, 288
Eisenhower Briefing Document, 74, 75,
	76, 133, 188, 290, 328, 454, 458
Eisenhower, Dwight, 74, 75, 77
Eisenhower, President Dwight, 55, 65,
	68, 71, 74, 76, 77, 128, 133, 139-58,
	161-65, 167-78, 180-84, 188, 192,
	193, 199, 207, 208, 210, 211, 213,
	214, 219, 220, 249, 290, 313, 323,
	324, 328, 330, 433, 438, 439, 441,
	443, 446, 454, 457, 458
Eisenhower,President Dwight, ii
electrogravitics, 269, 271
Erickson, Mark, 50, 209, 431, 432, 445,
	. 446, 448
Estimate of the Situation, 429
Etherians, 162
Exopolitics, iv
Exopolitics Institute, iv
Extraterrestrial Biological Entities, 191
extraterrestrial civilizations, 153
extraterrestrials, 139

## F

Fairchild Industries, 241, 242
Fechino, Frank, 120, 121, 437
First Contact, 148, 153
Flying Disc, 65, 66
flying rectangle, 364, 365, 366, 367
flying saucer, 5, 6, 10, 11, 15, 20, 27,
	29, 30, 33, 35, 37, 41, 43, 45, 46, 47,
	48, 50, 51, 52, 53, 56, 57, 58, 59, 60,
	62, 66, 70, 92, 96-102, 104-106, 109
	114, 116, 117, 120-23, 125-27, 137,
	138, 154, 155, 164, 167, 168, 170-
	73, 176, 183, 187, 195-99, 208, 242,
	249, 250, 251, 255-59, 263-65, 274,
	304, 309, 311, 319, 324, 331, 413,
	431, 443
Flying Saucers, 91, 98, 101, 257, 260,
	263
flying triangle, 269, 274, 275, 341, 342,
	344, 345, 346, 348, 352, 355, 364
Flying Triangle, 342
FOIA,Documents, 23
Forrestal, James, 56, 73, 74, 75
Fouche, Edgar, 267-76, 339, 346, 451,
	452
Fourth Reich, xiv, xv, 207, 212, 216,
	219, 220, 221, 231
Friedman, Stanton, 58, 75, 77, 434
Fulford, Benjamin, 357, 394, 460, 463

## G

Gabriel Green. *See* Green, Gabriel
Galactic Federation, 128, 129, 131,
	132, 133, 136, 138, 141, 143, 144,
	146, 147, 148, 161
Gemini capsule, 228, 229, 233
General Electric, 228, 237, 238, 239,
	240, 241, 243, 245, 246, 247, 248
Georgia, 148, 167, 170
German Antarctic colony, xv, 88, 136,
	137, 138, 142, 184

German Navy, 1

Giant Rock, 127

Good, Timothy, 110, 155, 440, 441

Goode, Corey, xv, 134-39, 184, 243, 245, 247, 275, 280-82, 284-86, 288, 291-93, 295, 299, 300-311, 331, 332, 335, 338, 352, 356, 357, 361, 368, 370, 373, 408, 415, 418, 438, 439, 443, 451, 452, 453, 454, 455, 456, 458, 460, 464

Gray, Gordon, 75

Grays, 172, 193

Greada Treaty, 177

Guild, William, 23

# H

HAARP, 348, 349, 352, 354, 355, 356, 357, 460

Hagerty, James, 167

Hall, Charles, 199, 200, 250, 444

Hanuebu, 261

Haunebu, 103

Haut, Walter, 58, 59, 60, 61, 432, 433

Hawaii missile, alert, 386, 389, 390, 391, 392, 393, 397, 461

HI-EMA, 379, 380, 381, 382, 384, 461

Hiroshima, 132, 135, 150

Hitler, Adolf, 141

Hoagland, Richard, 210, 211, 213, 445

Holden, Bill, 150, 440

Holloman, 165, 167, 169, 170, 171, 173, 174, 175

Hollywood, 439

Horn, Tom, 192, 444

Horten brothers, 95

Horten Ho 229, 95

Howe, Linda Moulton, 45, 103

Huffman, William, 2, 3, 6, 7, 8

human-looking extraterrestrials. See Nordics

Hunsaker, Dr. Jerome, 75, 76

Hurricane Irma, 336, 342, 344, 346, 347, 348, 352, 355, 356, 357, 358, 361, 398, 405, 417, 459, 460

hydrogen bomb, 128, 150, 154, 161, 438

# I

Intercontinental Ballistic Missiles, 141, 239

International Space Station, 247, 449

Interplanetary Corporate Conglomerate, xv, 353, 405, 418

Interplanetary Phenomenon Unit, 8, 13, 21, 23, 24, 26, 27, 30, 33, 37, 44, 45, 46, 57, 62, 68, 71, 179, 188, 321, 457

# J

Jacobs, Dr David, 193, 444

Japan, 15, 32, 34, 38, 64, 135, 247, 393, 394, 395, 396, 401, 402, 417, 463

Jesuits, 157, 158

Johnson, President Lyndon, 228

JP, Anonymous Photographer, 335, 336, 337, 338, 339, 340, 341, 342, 343, 344, 345, 346, 347, 352, 357, 358, 359, 360, 361, 362, 363, 364, 365, 366, 367, 368, 369, 370, 371, 372, 373, 374, 398, 416, 459, 460

Jumper, General John, 278

# K

Kammler, Hans, 223, 249

Kennedy Space Center, 182, 215, 218, 221, 223, 229, 249

Kennedy, President John F., ii, 218

Kewper, 45, 46, 64, 103, 105, 117, 229

Keyhoe, Donald, 102, 103, 120, 123, 159, 160, 429, 436, 437, 441

Kirklin, Bill, 148, 169, 170, 171, 174, 176, 439, 441, 442

Kirtland AFB, 324, 325, 330, 331

Kissner, Andrew, 110, 111, 112, 114, 115, 116, 117, 437

Klemperer, Dr. Wolfgang, 83, 84, 85, 86, 87, 435

Korean War, 145, 195

## L

Lee Van Atta, 91, 93

LeMay, General Curtis, 45, 47, 48, 50, 51, 56, 57, 69, 70, 71, 98, 431

Leslie, Desmond, 154, 155

Light, Gordon, 161, 162, 163, 440, 441

Lockheed, 40, 95, 140, 203, 217, 228, 238, 241, 245, 263, 265, 271, 273, 274, 275, 310, 352, 353, 357, 418, 435, 449

Los Alamos National Laboratory, 65, 199, 325

Los Angeles Air Raid, 14, 15, 17, 27, 31, 33, 36, 37, 45, 51, 114, 184, 187, 321, 428

Lubbock, UFO, 104, 105, 106

## M

MacDill, AFB, 334-38, 341, 342, 344-48, 352, 355-358, 360, 361, 362, 364, 367, 368, 371, 372, 375, 398, 405, 416, 417, 459, 460

MacIntyre, Francis, 157, 162

*Majestic*, 74, 75, 77, 79, 434

Majestic 12 Group, 77-79, 81, 87, 88, 89, 133, 179, 220, 290, 326, 329

Majestic documents, 8, 43, 65, 74, 76, 314, 416, 428, 429, 430, 431, 433, 434, 435, 457

MAJIC, 71, 74, 175, 188, 196, 225, 328, 444

Maldek, 132, 133, 164

Manhattan Project, 11, 20, 72, 327, 428, 446

Mann, Charlette, 6, 7

Manned Orbiting Laboratory, 225, 226, 228, 233, 245, 299, 447, 448

Mantel, Thomas, 118

Marcel, Jesse, 58, 59, 60, 432

Marconi, Guglielmo, 6, 260

Mars, 438, 439

Marshall, General George, 15, 17, 20, 21, 24, 38

Marshall Space Flight Center, 213, 214, 217, 219, 221, 241

Marx,Wesley, 39

Maser satellites, 350, 351, 352, 405, 417

Mattis, General James, 404, 405, 406, 464

McCandlish, Mark, 251, 253, 254, 255, 256, 257, 258, 259, 449, 450

McCasland, General William, 308, 309, 456

McClelland, Clark, 139, 140, 181, 182, 223, 439, 443, 446, 449

McElroy, Henry. *See* New Hampshire

McNamara, Robert, 231

Medvedev, Dimitri, 370

Meier,Billy, 137

Menzel, Dr. Donald, 75, 290, 327, 328, 329, 458

micronova, xvii, 421

Military-Industrial Complex, xiv, 182, 184, 185, 216, 301, 353, 357, 358, 361, 373, 405, 418

mind-wiped, xv

MJ-12, 441, 442

MJ-12 Committee, 220

MJ-12 SSP, 77

MOL program, 226, 227, 231, 235, 236, 237, 238, 239, 240, 241, 242, 243, 247

Montague, General Robert, 75

Montauk Project, 288
Moore, William, 58, 148, 432, 439
Morrison, Dr. Philip, 327, 328, 329
Muroc. *See* Edwards
Mussolini, Bennito, 6, 427

**N**

NACA, 42, 209, 210
Nagasaki, 135
NASA, xiv, 139, 181, 182, 207-219, 221-
    23, 226, 227, 230-33, 240-42, 244,
    247, 248-50, 265, 285, 299, 306,
    328, 413, 431, 432, 445, 446, 448
National Advisory Committee for
    Aeronautics, 209
National Archives, 77, 169, 434
National Reconnaissance Office, 229,
    235, 237, 245, 269, 301, 357
national security, 150, 151, 152, 163
National Security Act, 56, 411
NATO, 441
Naval Air Station San Diego, 71, 311
Naval Intelligence, 21
Navy,US, 13, 15, 16, 21, 27, 33, 35, 37,
    38, 42, 50, 51, 52, 53, 431
Nazi Germany, 6, 13, 15, 27, 29, 95,
    106, 137, 168, 183, 197, 206, 242,
    249, 258, 262
Nazi SS, 137, 213, 214, 215, 217, 219,
    221, 223, 258, 259, 260, 261
Nellis Air Force Base, 199, 267, 444
New Hampshire, 151
Nordic extraterrestrials, 128, 132, 134,
    138, 143, 147, 148, 153, 154, 161,
    162, 163, 164, 165, 167, 175, 177,
    180, 305, 312, 323, 324, 330, 334,
    336, 337, 360, 372, 373, 400, 401,
    414-17, 419, 420, 428, 465
Northrup, 40
Nourse, Edwin, 162
NRO, 229, 237, 238, 240-48, 250, 265,

269, 273, 276, 301, 312, 334, 349,
    390, 420, 449
NSA, 227, 269, 273, 301, 312, 334, 390,
    391, 392, 420, 463
nuclear weapons, 132, 160

**O**

Odell, Col W.C., 160
Office of Naval Intelligence, 14, 71,
    158, 187, 428
Office of Scientific Intelligence, CIA, 98
Office of Strategic Services, 20, 33
Operation Blue Fly, 26, 188, 189, 443,
    444
Operation Highjump, 87, 91, 92, 93, 99,
    104, 110, 134, 137, 197, 435, 436
Operation Majestic Twelve, 74, 77
Operation Paperclip, 45, 47, 49, 92,
    180, 181, 182, 184, 212, 232, 249,
    258, 259, 260, 445
Oppenheimer, Robert, 132, 438
Orion, 175, 444
Orsic, Maria, 47, 134, 136-38, 141

**P**

Pacific Command, US, 377, 378, 380,
    381, 383, 395
PACOM. *See* US Pacific Command
Palm Springs, California, 147, 148, 149,
    151, 162, 163
Palmdale, California, 148
Parker, Dana, 29, 35
Peenemunde, 103
Pentagon, 59, 61, 62, 112, 123, 128,
    129, 182, 202, 211, 212, 221, 226,
    228, 229, 392, 394, 403, 406, 407,
    414, 419, 446
Phillips, Don, 158, 440
Pleiadians, 137
pole shift, 421
Pope Pius XII, 157

Project Blue Beam, 242, 244, 449
Project Bluebook, 24, 97, 98, 118, 121
Project Grudge, 24
Project Moon Dust, 26, 188, 189
Project RAND, 3, 30, 31, 38-42, 44, 45,
   48, 50, 53, 57, 73, 79, 81- 84, 89
Project Sigma, 159
Project Sign, 24, 429, 435, 436

## Q

QAnon, 389, 394, 395, 396, 412, 417,
   420, 462

## R

Ramey, Roger, 59, 60, 61
RAND Corporation, 50, 81, 82, 432, 435
Raymond,Arthur, 37, 38
Raytheon Corporation, 351
Reagan, President Ronald, 225, 240,
   243, 246, 265, 276, 349, 350, 449,
   451
Relfe, Michael, 300, 408, 415, 455, 464
Reptilian extraterrestrials, 106, 139,
   175, 206, 286, 311, 312, 324, 333,
   334, 374, 414, 416
Rods of God, 244
Rome, 157
Roosevelt, President Franklin, 10, 11,
   15-22, 24, 27, 30-34, 36, 37, 87,
   428-30
Roswell Army Air Field, 57
Roswell, UFO, 46, 53, 55, 57-60, 62-65,
   68-73, 81, 84, 87-89, 98, 114, 116,
   117, 168, 172, 175, 179, 183, 184,
   187, 191, 193, 308, 313, 319, 373,
   412, 413, 432, 433, 442, 443, 444
Ruppelt, Edward, 97-99, 103, 118, 119,
   121, 122, 126, 127, 129, 435, 436,
   437
Russia, 229, 247, 372, 438
Ryan Wood, 7, 8, 24, 42, 67, 88, 188,

196, 314, 427, 428, 429, 433, 434,
   435
Ryan, General Michael, 278

## S

Sagan, Dr. Carl, 288, 289, 290, 291,
   294, 327, 328, 329, 453, 454, 458
Salla, Dr Michael, iv, 444
Salter, Daniel, 158, 440
Samford, Major General John, 126
San Diego, 36, 37, 149
Santa Monica, 36, 41, 82, 432, 435
Sather, Jordan, 391, 392
Scantamburlo, Luca, 157, 440
Schmitt, Donald, 5, 59, 62, 432,

Schneider, Phil, 177, 442
Schumann, Winfried Otto, 47, 49, 249,
   259, 431
Scientific American, 1, 2, 191
secret space program, xiv, 134, 137,
   165, 181, 223, 243, 247, 278, 281,
   282, 297, 300, 303, 305, 308, 310,
   311, 336, 348, 357, 368, 414, 416,
   419, 420
Selected by Extraterrestrials, 428, 430,
   431, 432, 434
Sellen, Kent, 255, 256, 257
Servizio Informazione del Vaticano,
   157
Short Grays, extraterrestrials, 191-93,
   195, 199
Sigmund, 299, 301, 302, 303, 305, 306,
   309, 310, 311, 312, 331, 333, 335
Signal Corps, US Army, xiv, 2, 35
Skunk Works, 95, 245, 274, 435
Skylab, 213, 215, 218, 221, 249, 299
Smith, Emery, 281, 324, 325, 370
Smith, Walter, 43, 431
Smith, Wilbert, 413
Snowden, Edward, 391, 393, 462

Solar Warden, xv, xvi, 134, 301, 309, 310, 311, 331

SOM1-01, 191

Sorensen, Brad, 250, 251, 255, 256, 257, 258, 259, 261, 262, 265, 449

South America, 103

South Pole, 140

Soviet Union, 19, 29, 30, 101, 118, 132, 143, 146, 150, 209, 229, 230, 322, 439

Spaatz, General Carl, 55, 56, 65, 79, 81

Space Command, 264, 265, 278, 307, 357, 399, 407, 408, 409, 410, 420, 464, 465

Space Force, xiii, xv, xvi, xvii, 403, 404, 405, 406, 407, 408, 410, 411, 412, 413, 414, 417, 419, 420, 464, 465

Sparrowhawk, 85

Special Operations Manual, 188, 189, 190, 191

Sputnik, 209, 210, 211

SR-74, 268

SR-75, 268

Stalin, Joseph, 146

Stargate SG-1, 277, 278, 279, 280, 281, 282, 286, 295, 297, 415, 452

State Federal Relations and Veterans Affairs committee. See New Hampshire

Stein. See Kewper

Steinman, William, 23

Stevens, Wendelle, 149

Stringfield, Leonard, 7, 63, 120, 427, 433

Suggs, Charles, 153, 440

Sweetman, Bill, 268, 451

## T

Tall Gray, 168, 176, 177, 180, 195

Tall Grays, extraterrestrials, 177, 192, 193, 199

Tall Whites, 199, 200, 201, 202, 203, 204, 205, 250, 414, 444

Tassel, George Van, 127, 128, 129, 133, 138, 438,

Teller, Dr. Edward, 132

Tesla, Nikola, 254, 267, 313, 317, 318, 319, 330, 450, 459

think tank, 42, 83

Thorne, Professor Kip, 289, 290, 291, 294, 296, 453, 454

Tillman, Barrett, 47, 48, 431

Tompkins William,

Tompkins, William, xv, 14-16, 19, 27, 36, 37, 40-42, 44, 82-84, 87, 88, 137-39, 182-84, 187, 215-18, 300, 305, 311, 324, 331, 357, 370, 373, 428, 430-32, 434, 439, 443, 445, 455-57

TR-3B, 267, 269, 270, 271, 272, 273, 274, 275, 276, 305, 309, 338, 339, 343, 344, 345, 361, 362, 364, 369, 400, 415, 452

traversable wormholes, 277, 280, 285, 288, 289, 290, 291, 293, 294, 296

Truman administration, 88, 112, 114, 141, 164, 321

Truman, President Harry, 56, 65, 68, 71, 73, 75, 87, 88, 109, 112, 114, 126, 128, 140, 141, 146, 147, 162, 164, 182, 314, 321, 433, 457

Trump, Donald, xv, xvi, 326, 331, 355, 356, 384, 389, 391-95, 403-406, 410-12, 414, 417, 418, 462-65

Twining, General Nathan, 65-75, 77, 79, 84, 88, 89, 98, 133, 141, 145, 163, 183, 431, 433

## U

U.S. Air Force, 128, 154, 158, 159

U.S. Navy, 153, 158

UFO, ii, 13, 44, 153, 158, 191, 428, 437,
439, 440, 441, 442, 444
Uhouse, Bill, 195, 196, 197, 198, 250,
444
UN General Assembly, 144
US Air Force, *see* USAF
US Marine Corps, 10, 55, 153, 195, 331
US Navy, xiv, xv, 10, 14, 27, 28, 33, 42,
56, 81, 83, 85, 87, 89, 106, 110, 137,
300, 305, 324, 396, 402, 414, 415,
431, 435, 436, 439, 443, 463, 465
US Space Command, 277, 279, 420
USAF, xiii, xv, 21, 38, 46, 51, 56, 70, 71,
79, 81, 82, 84, 97, 98, 100, 107, 117,
120, 128, 129, 133, 138, 140, 141,
147, 155, 159, 163, 164, 168, 169,
176, 177, 179, 184, 187, 188, 191,
195, 196, 199-209, 212, 217, 222,
223, 226, 228, 231-33, 237, 240,
241-45, 247-50, 259, 263-65, 267,
269, 271, 275-79, 282, 284, 288,
297, 299, 300-10, 312, 314, 321,
323, 324, 325, 330, 331, 333-36,
348, 356, 357, 358, 359, 361, 367-
74, 377, 385, 396, 398, 399, 400-
403, 405, 409, 410, 412, 414-20,
444, 452, 456, 459, 460, 463, 464,
USS Akron, 85, 86, 435
USS Macon, 86
USSR. Soviet Union

## V

V2 rockets, 109
Van Flandern, Dr. Thomas, 132, 133,
438
Vandenberg, General Hoyt, 75, 79,
133, 141
Vatican, 157, 158, 440
Von Braun Station, 234, 235, 236, 240,
241, 243, 446, 448
von Braun, Dr. Werner, 51, 140, 181,

183, 213-17, 219, 222, 223, 232,
234, 241, 242, 245, 247
von Kárman, Dr. Theodore, 84
Vril, 46, 47, 134, 136, 137, 141, 258,
259, 261, 262, 265, 450

## W

Warp Drive, 264, 291
Washington DC Flyover, 123, 128, 133,
134, 138, 140, 144
Washington, DC, 129, 139, 140, 141,
167, 425, 437
Webster, William Hedgcock, 314, 326,
327, 330
White Hot Report, 11, 41, 42, 43, 71,
73, 74, 79, 84, 88, 89, 164, 428
White Sands Proving Ground, 65, 68,
99, 109, 110, 113, 114, 180
Wilcock, David, 280, 281, 282, 283,
285, 331, 455, 439, 456, 465
Wolf, Michael, 178, 442
Wood, Robert and Ryan, 24, 429
Wood,Dr Robert, 42, 326, 431
World War II, 3, 5, 8, 11, 14, 21, 29, 30,
33, 35, 37, 52, 55, 77, 84, 87, 95, 97,
99, 133, 136, 137, 168, 195, 213,
214, 242, 257, 258, 261, 311, 357,
373, 395, 411, 428, 429, 430, 431
Wormholes, 264, 277, 288, 290, 291,
293, 453, 454, 455
Wright brothers, 2, 3, 35, 36
Wright Field, 3-5, 8, 9, 24-28, 33, 37,
40, 45-48, 57, 62-64, 66, 69-71, 249,
261, 427
Wright Patterson AFB, 5, 24, 63, 97, 98,
100, 128, 168, 177, 189, 195, 249,
258, 308
Wykoff, Staff Sergeant, 174, 442

## Z

Zeppelin, 84

Printed in Great Britain
by Amazon

62538162R00281